S0-AKY-163

Cyclopolymerization and Polymers with Chain–Ring Structures

Cyclopolymerization and Polymers with Chain-Ring Structures

George B. Butler, EDITOR
University of Florida

Jiri E. Kresta, EDITOR
University of Detroit

Based on a symposium cosponsored by the Divisions of Polymer Chemistry and Organic Coatings and Plastics Chemistry at the 181st ACS National Meeting, Atlanta, Georgia, March 30–April 1, 1981.

ACS SYMPOSIUM SERIES 195

AMERICAN CHEMICAL SOCIETY
WASHINGTON, D. C. 1982

Library of Congress Cataloging in Publication Data

Cyclopolymerization and polymers with chain–ring structures.
(ACS symposium series, ISSN 0097–6156; 195)

Includes bibliographies and index.

1. Polymers and polymerization—Congresses. 2. Cyclic compounds—Congresses.
I. Butler, George B., 1916– . II. Kresta, Jiri E., 1934– . III. American Chemical Society. Division of Polymer Chemistry. IV. American Chemical Society. Division of Organic Coatings and Plastics Chemistry. V. Series.

QD380.C9 1982 547.7 82–11331
ISBN 0–8412–0731–3 ACSMC8 195 1–459 1982

Copyright © 1982

American Chemical Society

All Rights Reserved. The appearance of the code at the bottom of the first page of each article in this volume indicates the copyright owner's consent that reprographic copies of the article may be made for personal or internal use or for the personal or internal use of specific clients. This consent is given on the condition, however, that the copier pay the stated per copy fee through the Copyright Clearance Center, Inc. for copying beyond that permitted by Sections 107 or 108 of the U.S. Copyright Law. This consent does not extend to copying or transmission by any means—graphic or electronic—for any other purpose, such as for general distribution, for advertising or promotional purposes, for creating new collective work, for resale, or for information storage and retrieval systems. The copying fee for each chapter is indicated in the code at the bottom of the first page of the chapter.

The citation of trade names and/or names of manufacturers in this publication is not to be construed as an endorsement or as approval by ACS of the commercial products or services referenced herein; nor should the mere reference herein to any drawing, specification, chemical process, or other data be regarded as a license or as a conveyance of any right or permission, to the holder, reader, or any other person or corporation, to manufacture, reproduce, use, or sell any patented invention or copyrighted work that may in any way be related thereto.

PRINTED IN THE UNITED STATES OF AMERICA

QD380
C9
1982
CHEM

ACS Symposium Series

M. Joan Comstock, *Series Editor*

Advisory Board

David L. Allara
Robert Baker
Donald D. Dollberg
Robert E. Feeney
Brian M. Harney
W. Jeffrey Howe
James D. Idol, Jr.
Herbert D. Kaesz
Marvin Margoshes
Robert Ory
Leon Petrakis
Theodore Provder
Charles N. Satterfield
Dennis Schuetzle
Davis L. Temple, Jr.
Gunter Zweig

FOREWORD

The ACS SYMPOSIUM SERIES was founded in 1974 to provide a medium for publishing symposia quickly in book form. The format of the Series parallels that of the continuing ADVANCES IN CHEMISTRY SERIES except that in order to save time the papers are not typeset but are reproduced as they are submitted by the authors in camera-ready form. Papers are reviewed under the supervision of the Editors with the assistance of the Series Advisory Board and are selected to maintain the integrity of the symposia; however, verbatim reproductions of previously published papers are not accepted. Both reviews and reports of research are acceptable since symposia may embrace both types of presentation.

This book is dedicated to George Bergen Butler,
a pioneer of cyclopolymerization chemistry,
on the anniversary of his 65th birthday.

GEORGE BERGEN BUTLER was born in Liberty, Mississippi, in April 1916. He studied at Mississippi College, Mississippi (B.A. 1938), and at the University of North Carolina at Chapel Hill (Ph.D. 1942). In 1946, after working briefly at the Rohm and Haas Co., he joined the faculty of the Chemistry Department at the University of Florida as an instructor. He was promoted to assistant professor in 1947, to associate professor in 1951 and to full professor in 1957. During his academic career, over 80 students have obtained advanced degrees under his guidance, and he has published hundreds of scientific papers.

His research contributions to polymer chemistry span many areas, from studies of alternating copolymers of maleic anhydride and vinyl ether ("pyran copolymer") having interferon inducing capability and antitumor activity, to polymerization of triazoline diones and synthesis of the anti-aromatic 3,3-dimethoxycyclopropene.

His best known contribution to polymer science is the discovery of the cyclopolymerization reaction in 1956, during the study of free radical polymerization of divinyl monomers. For this pioneering work he was awarded the 1980 American Chemical Society Witco Award for Polymer Chemistry. Among other awards, he also received the Florida Section ACS Award in 1963, and the Herty Award from the Georgia Section of the ACS in 1978.

Besides his research work, he is an organizer and co-editor of *Reviews in Macromolecular Chemistry* and the *Journal of Macromolecular Science—Reviews.* He also serves on the editorial boards of the *Journal of Macromolecular Science—Chemistry, Macromolecules,* the *Journal of Polymer Science, and Macromolecular Synthesis.*

On the occasion of his 65th birthday, the contributors to this book wish Professor Butler many productive years in his future research activities in polymer chemistry.

CONTENTS

PREFACE

CYCLOPOLYMERIZATION HAS BEEN DEFINED as a chain-growth polymerization reaction of 1,X-bis-(multiple bonded) monomers which introduces cyclic structures into the main chain of the resulting polymers. However, polymers containing cyclic units in the main chain can be synthesized by a wide variety of methods. On this basis, the organizers of the symposium upon which this book is based decided to broaden its scope to include both polymers formed by cyclopolymerization and polymers containing chain–ring structures.

Although cyclopolymerization has been extensively investigated, because of its complexity and the variety of monomers capable of participating in this mode of chain propagation, considerable research emphasis continues in this area. The microstructure of cyclopolymers studied by modern spectroscopic techniques, particularly ^{13}C NMR, has yielded significant results. Aspects not totally understood and which continue to be investigated include ring size, theoretical interpretations of the driving force for control of ring size, the relative importance of kinetic versus thermodynamic control, electronic interactions, steric effects, macrocyclic polymers, and charge–transfer complexes in cyclocopolymerization. This volume includes 16 papers dealing with the various aspects of cyclopolymers and cyclopolymerization.

Sixteen papers dealing with polymers containing chain–ring structures are also included. During an extended period in which considerable emphasis was placed on development of thermally stable polymers, a wide variety of synthetic methods have been developed, and numerous polymers having interesting and occasionally superior properties have been reported. Among these reactions are the novel cross-linking or chain extension trimerization reactions of various end groups, and a variety of condensation reactions leading to heterocyclic structures in the main polymer chain.

The editors believe that the scientific community will find this book helpful as a source of advances and developments in the area of cyclopolymerization, and the preparation and properties of polymers with chain–ring structures.

GEORGE B. BUTLER
University of Florida
Gainesville, FL

JIRI E. KRESTA
University of Detroit
Detroit, MI

1

Biologically Active Synthetic Anionic Polymers

DAVID S. BRESLOW

Hercules Research Center, Wilmington, DE 19899

Maleic anhydride and vinyl ether undergo a cyclic alternating copolymerization in a 2:1 ratio. ^{13}C NMR has shown the polymer to contain a mixture of tetrahydrofuran and tetrahydropyran rings in an approximately 0.8:1 ratio. The copolymer shows a remarkable variety of biological activities. It is an antitumor agent; it induces the formation of interferon; it has antiviral, antibacterial, and antifungal activity; it is an anticoagulant and an anti-inflammatory agent; and it aids in the removal of plutonium from the liver. The copolymer is an immunostimulant, and appears to act by stimulating macrophages. A study of the effect of molecular weight and molecular weight distribution on toxicity and biological activity has led to the synthesis of an active copolymer with low toxicity, which is being investigated clinically as an antitumor agent.

Maleic anhydride and vinyl ether undergo a radical-catalyzed cyclic alternating copolymerization in a 2:1 ratio. This copolymer, frequently designated in the literature as DIVEMA or as pyran copolymer, has shown unusual and exciting biological activity, and has therefore undergone intensive investigation.

Structure

The 2:1 copolymer, first reported in 1958 (1), was assumed to have the tetrahydropyran structure, I in Scheme 1 (2). Although this structure had been widely accepted (3), there was no convincing evidence for it. In fact, a number of reports in the literature suggested that the kinetically controlled route leading to a tetrahydrofuran ring, II in Scheme 1, would be more likely. Thus, it was shown by ESR that the 5-hexenyl radical, formed by photolysis of 6-heptenoyl peroxide, cyclizes

0097-6156/82/0195-0001$06.00/0
© 1982 American Chemical Society

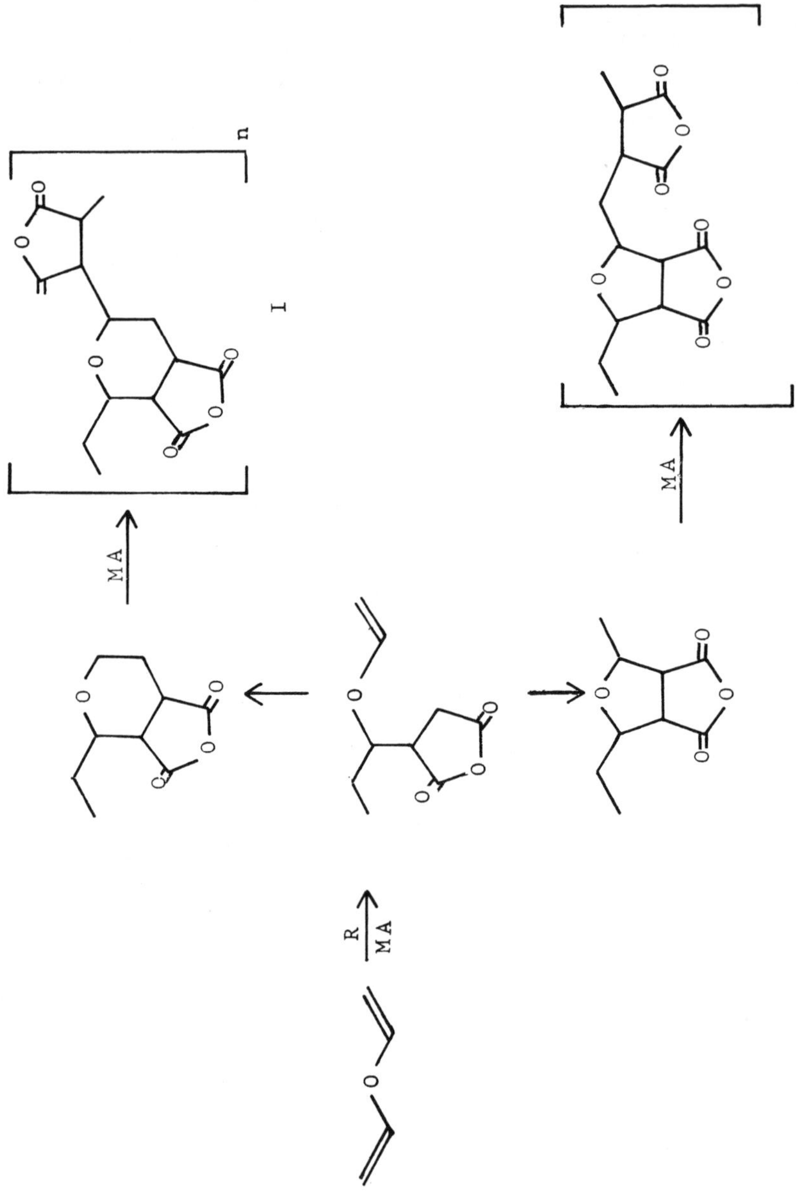

SCHEME 1

exclusively to the cyclopentylmethylene radical (4), whereas hexenyl radicals stabilized by carbonyl or nitrile groups cyclize to cyclopentane and/or cyclohexane derivatives, depending on the substitution pattern (5). Similarly, methyldiallylamine undergoes cyclopolymerization exclusively to form a polymeric pyrrolidine, whereas methyl-substituted derivatives give polymers containing both five- and six-membered rings (6).

Since it was quite apparent from the literature that the structure of the maleic anhydride-vinyl ether copolymer could not be predicted, we set out to determine the structure by ^{13}C NMR, using model compounds to aid in assigning the resonances (7). Figure 1 shows the ^{13}C NMR spectrum of the polymer, prepared in benzene with carbon tetrachloride as a chain-transfer agent, after hydrolysis in D_2O. It consists of seven sets of peaks, labeled A-H. By a single frequency off-resonance decoupling experiment, it was possible to show that peaks A and B are triplets due to methylene carbons, peaks C-F are doublets due to methine carbons, and peaks G and H are due to carbonyls. Definitive assignments were made by comparison with the spectra of the model compounds. 2,6-Dimethyltetrahydropyran-3,4-dicarboxylic acid was prepared, as shown in Scheme 2, as the six-membered ring model; only one isomer, identified as all-*trans* by proton NMR, was isolated.

$CH_3CH = CHCHO$ $\xrightarrow{H_3O^+}$ [2,6-dimethyl-dihydropyran, CHO] $\xrightarrow{Ag_2O}$

[CO_2H derivative] $\xrightarrow{EtOH}$ [CO_2Et derivative] $\xrightarrow{1.\ KCN;\ 2.\ KOH}$

[all-trans 2,6-dimethyltetrahydropyran-3,4-dicarboxylic acid: CH_3, H, O, CO_2H, CO_2H]

Scheme 2

2,5-Dimethyltetrahydrofuran-3,4-dicarboxylic acid was prepared, as shown in Scheme 3, as the five-membered ring model; three isomers were isolated, but the stereochemistry could not be assigned unequivocally. Inspection of the ^{13}C NMR spectra of the model compounds, plus that of commercially available

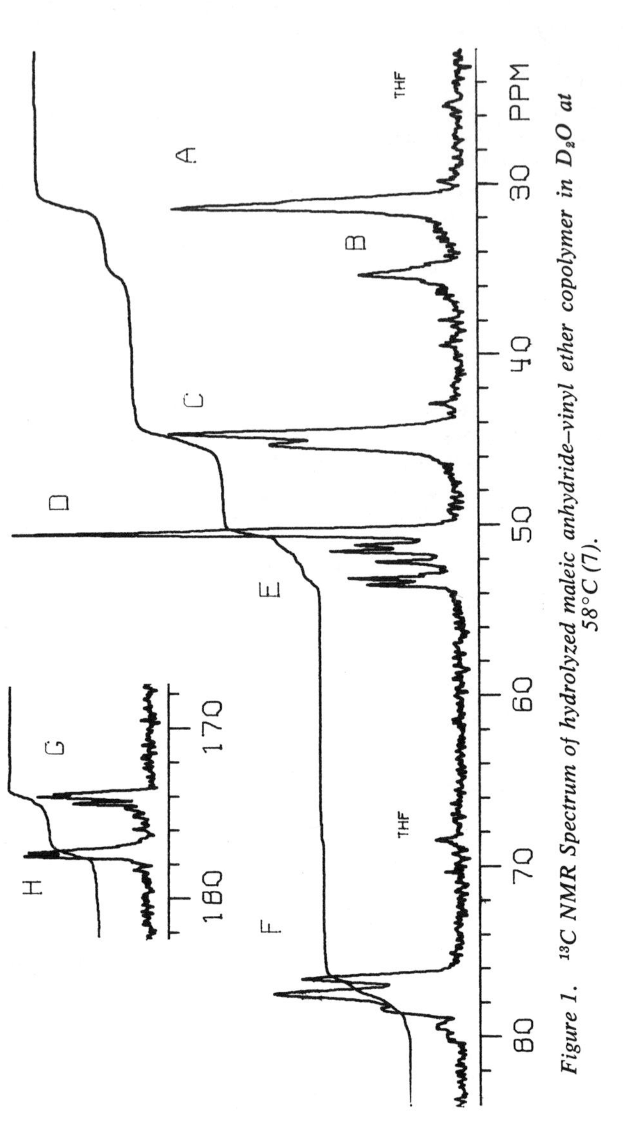

Figure 1. ^{13}C *NMR Spectrum of hydrolyzed maleic anhydride–vinyl ether copolymer in* D_2O *at 58°C (7).*

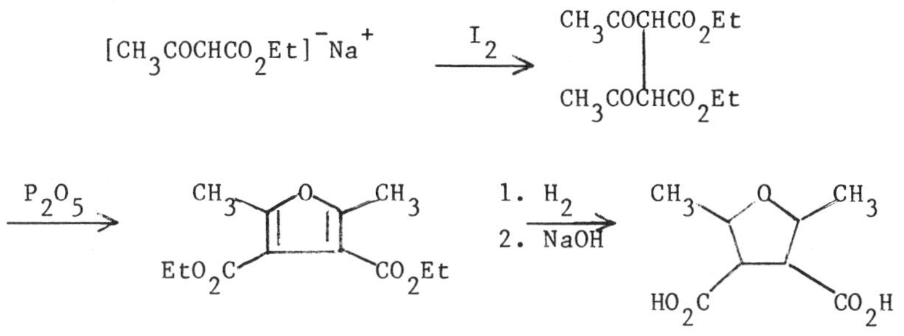

$[CH_3COCHCO_2Et]^-Na^+ \xrightarrow{I_2}$ $CH_3COCHCO_2Et$ | $CH_3COCHCO_2Et$

$\xrightarrow{P_2O_5}$ (furan: CH_3, CH_3, EtO_2C, CO_2Et) $\xrightarrow{1.\ H_2;\ 2.\ NaOH}$ (tetrahydrofuran: CH_3, CH_3, HO_2C, CO_2H)

Scheme 3

dimethylsuccinic acid (meso plus racemic), showed only one unique carbon atom, at 34.6 ppm (using dioxane at 67.0 ppm as an internal reference), assignable to the methylene carbon of the tetrahydropyran ring. This corresponds to peak B in the polymer spectrum. However, peak areas (from suppressed NOE experiments) showed that the polymer could not consist solely of tetrahydropyran rings; calculations show the ratio of five- to six-membered rings to be approximately 0.8 to 1. We feel there is insufficient evidence at the present time to make definite assignments of stereochemistry to the polymer (7).

Kunitake and Tsukino (8) also used ^{13}C NMR to determine ring size in the copolymer, but they calculated chemical shifts by extrapolating published data on simpler cyclic compounds. By this method they concluded that a copolymer prepared in chloroform contained only tetrahydrofuran rings. However, their published spectrum of the copolymer contained the same peak B as ours, and therefore the copolymer must consist of both five- and six-membered rings. The spectrum of a copolymer prepared in acetone-carbon disulfide was too poorly resolved to show peak B, but Kunitake and Tsukino estimated the polymer to contain about 90% tetrahydropyran rings.

Biological Activity

The 2:1 maleic anhydride-vinyl ether copolymer shows a bewildering variety of biological activities (9). It was first shown to have antitumor activity against a number of solid tumors (10). It induces the formation of interferon (11), and therefore shows antiviral activity. It has been shown to be active against over 20 viruses, including a number of cancer-inducing viruses and herpes simplex (9,12). Quite surprisingly, it has also been shown to have antibacterial activity against both gram-positive and gram-negative organisms (13). It also shows antifungal activity (14) and acts as an anticoagulant (15). It inhibits adjuvant disease in rats (16), a syndrome believed to be related to rheumatoid arthritis in humans. In conjunction with certain chelating agents, it has been shown to be active in removing polymeric plutonium from the liver (17).

This broad spectrum of activity quite obviously led to considerable activity by biologists and physicians to explain the polymer's mode of action. One of the first discoveries showed that, under certain conditions, it greatly accelerated the rate of phagocytosis, i.e., the rate of removal of foreign bodies from the blood stream (18). Quite surprisingly, it showed the same activity in immunosuppressed mice as in normal mice (19). It was shown to inhibit reverse transcriptase in birds, reptiles, and mammals (20). A considerable amount of work in a number of laboratories finally reached the conclusion that the major mode of action involves the activation of macrophages (21, 22), large white blood cells which have the function of removing foreign bodies from the blood stream. This hypothesis is generally accepted, but by no means proven (9).

Narrow Molecular Weight Distribution Polymer

The initial polymer investigated was prepared using benzoyl peroxide in a nonsolvent for the polymer, and the polymerization was carried to a very high conversion. As a result, the polymer had a very broad molecular weight distribution. This polymer underwent Phase I clinical evaluation as a chemotherapeutic agent against cancer. Unfortunately, it showed a number of undesirable side effects (23), and it was decided in this laboratory to see if we could decrease the toxicity without losing the antitumor activity. Degradation of the polymer showed that decreasing the molecular weight did indeed decrease the toxicity of the polymer (24). However, before going further we had to develop a means for characterizing the polymer. This was done by converting it to its methyl ester and running size exclusion chromatography (SEC) on a preparative scale. These fractions, which had reasonably narrow molecular weight distributions on an analytical instrument, were then compared with polystyrene standards and shown to fit the universal calibration curve in the usual fashion. In order to obtain sufficiently large quantities of polymer for both characterization and biological evaluation, a solution polymerization with limited conversion was devised; a comparison of the two polymers by SEC is shown in Figure 2; the original polymer is NSC 46015 and the new polymer MVE-2. Biological testing showed that this new polymer did indeed have low toxicity and maintained its antitumor activity (24). Meanwhile, however, research done at the National Cancer Institute, as well as at other laboratories, demonstrated that the most desirable way of using this material was as an immunoadjuvant in conjunction with other treatments (25, 26). The goal is to reduce the tumor burden by some means - surgery, radiation, chemotherapy - and

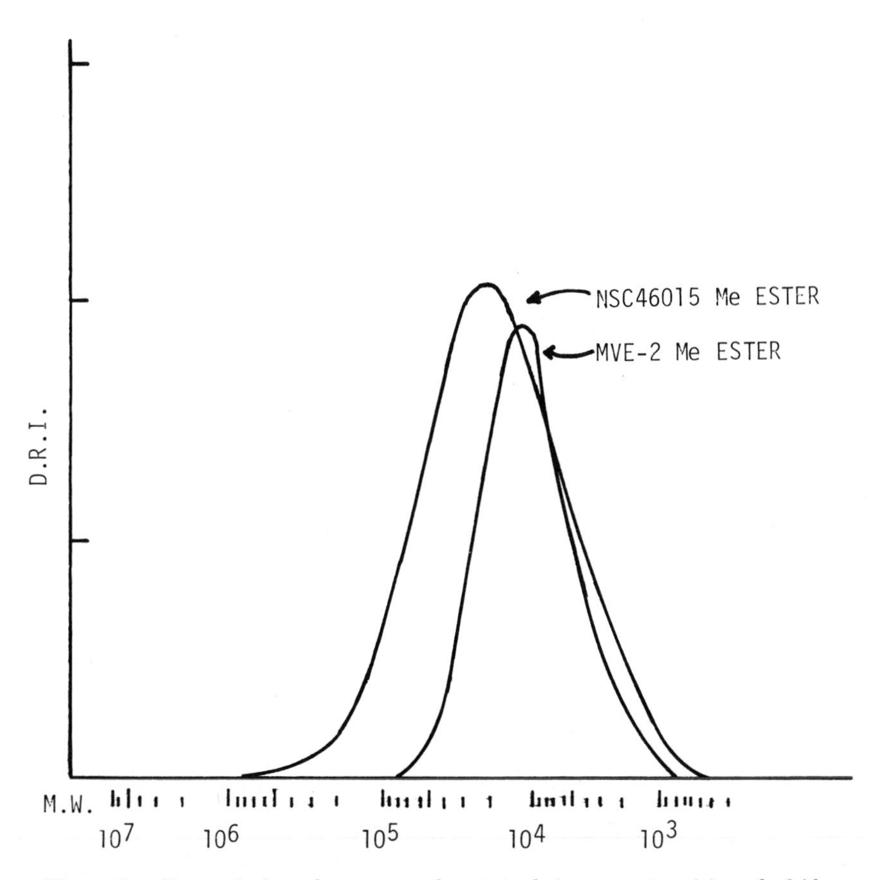

Figure 2. Size exclusion chromatography of methyl esters of maleic anhydride–vinyl ether copolymers. Key: NSC46015, broad; and MVE-2, narrow MW distribution.

then to use the polymer to stimulate the body to get rid of the last few remaining cancer cells, which is necessary to prevent subsequent growth of the cancer. An IND (Investigative New Drug) clearance has been obtained from the FDA and the polymer is now being investigated clinically in a Phase I study in several hospitals.

Literature Cited

1. Butler, G. B., Abstracts 133rd Meeting American Chemical Society, San Francisco, CA, April 13-18, 1958, p. 6R.
2. Butler, G. B. J. Polym. Sci. 1960, 48, 279.
3. Butler, G. B.; Corfield, G. C.; Aso, C. "Progress in Polymer Science"; Vol. 4 (ed. A. D. Jenkins); Pergamon Press, Oxford, 1975; p. 129 ff.
4. Kochi, J. K.; Krusic, P. J. J. Am. Chem. Soc. 1969, 91, 3940.
5. Julia, M. Accts. Chem. Res. 1971, 4, 386.
6. Solomon, D. H. J. Polym. Sci., Polym. Symp. 1975, 49, 175.
7. Freeman, W. J.; Breslow, D. S., "Biological Activities of Polymers", ACS Symposium Series, in press.
8. Kunitake, T.; Tsukino, M. J. Polym. Sci. Polym. Chem. Ed. 1979, 17, 877.
9. Breslow, D. S., Pure & Appl. Chem. 1976, 46, 103.
10. National Cancer Institute, private communication.
11. Merigan, T. C. Nature 1967, 214, 416.
12. Morahan, P. S.; Cline, P. F.; Breinig, M. C.; Murray, B. K. Antimicrob. Agents Chemother. 1979, 15, 547.
13. Giron, D. J.; Schmidt, J. P.; Ball, R. J.; Pindak, F. F. Antimicrob. Agents Chemother. 1972, 1, 80.
14. Regelson, W.; Munson, A. E. Ann. N.Y. Acad. Sci. 1970, 173, 831.
15. Shamash, Y.; Alexander, B. Biochim. Biophys. Acta 1969, 194, 449.
16. Kapusta, M. A.; Mendelson, J. Arthritis Rheum. 1969, 12, 463.
17. Guilmette, R. A.; Lindenbaum, A. "Health Effects of Plutonium and Radium", (ed. W. S. S. Jee); J. W. Press, Salt Lake City, 1976; p. 223.
18. Munson, A. E.; Regelson, W.; Lawrence, W.; Wooles, W. R. J. Reticuloendothel. Soc. 1970, 7, 375.
19. Hirsch, M. S.; Black, P. M.; Wood, M. L.; Monaco, A. P. Proc. Soc. Exp. Biol. Med. 1970, 134, 309.
20. Papas, T. S.; Pry, T. W.; Chirigos, M. A. Proc. Nat. Acad. Sci. 1974, 71, 367.
21. Braun, W.; Regelson, W.; Yajima, Y.; Ishizuka, M. Proc. Soc. Exp. Biol. Med. 1970, 133, 171.

22. Schultz, R. M.; Papamatheakis, J. D.; Stylos, W. A.; Chirigos, M. A. Cell. Immunol. 1976, 25, 309.
23. Regelson, W.; Shnider, B. I.; Colsky, J.; Olson, K. B.; Holland, J. F.; Johnson, C. L., Jr.; Dennis, L. H. "Immune Modulation and Control of Neoplasia by Adjuvant Therapy", (ed. M. A. Chirigos); Raven Press, New York, 1978; p. 469.
24. Breslow, D. S.; Edwards, E. I.; Newburg, N. R. Nature 1973, 246, 160.
25. Mohr, S. J.; Chirigos, M. A.; Fuhrman, F. S.; Pryor, J. W. Cancer Res. 1975, 35, 3750.
26. Mohr, S. J.; Chirigos, M. A.; Smith, G. T.; Fuhrman, F. S. Cancer Res. 1976, 36, 2035.

RECEIVED March 1, 1982.

2

Temperature Dependence of Cyclization Ratios in Cyclopolymerization

MARINO GUAITA

Istituto di Chimica Macromolecolare dell'Università di Torino,
Via G. Bidone 36, 10125 Torino, Italy

The hypotheses which have been suggested to explain the ease of formation of ring structures in the free radical cyclopolymerization of unconjugated dienes are reviewed and discussed. The agreement between calculated entropy changes occurring in intramolecular and intermolecular chain propagation reactions and experimental differences between the activation entropies of the same reactions gives support to the interpretation that the former reactions are favoured over the latter ones by a smaller entropy decrease.

It is well known that, in the cyclopolymerization of unconjugated dienes which can form 5- and 6-membered ring structural units, these units often largely prevail over the linear ones, that is the units deriving from monomer molecules which have contributed to the polymer chain growth with only one of their unsaturations. To account for this fact, several hypotheses have been suggested, which will be reviewed and discussed in the present paper.

Interaction across the space between reacting groups.

The simplest, very tempting idea, which has been suggested by Butler (1) to explain why the cyclic units can so easily be formed, is that the monomer molecules such as 1,5- and 1,6-dienes can already be in a pseudo-

0097-6156/82/0195-0011$06.00/0
© 1982 American Chemical Society

cyclic conformation because of an interaction across the space between the two unsaturations, which can be schematically represented as follows:

(1)

If such an equilibrium were strongly shifted on the right, and/or the attack of an active species on the ring form were energetically more favoured than the attack on the open one, the propagation reaction might be represented by the scheme:

R· + ⟶ R (2)

and the resulting polymer would be predominantly formed by cyclic structural units.

Although several studies (2,3) have given support to the possibility of interactions between unconjugated unsaturations, a strong objection to generalizing such an interpretation of the cyclopolymerization process comes from the obsereved dependence of the polymer structure on the monomer concentration. In fact, if equilibrium (1) would be operative in controlling the polymer structure, no change would be expected in the structures of polymers obtained from monomer solutions of low and of high concentration, because no influence can be exerted by the monomer concentration on equilibrium (1). On the other hand, almost without exceptions it has been found that the content of cyclic structural units in the polymer increases, when the monomer concentration in the solution from which the polymer is obtained decreases. This fact leads to the concept of a competition between intramolecular and intermolecular chain propagation reactions.

Instead of considering the addition of a monomer molecule to the growing chain as represented in the scheme (2), this addition is represented stepwise in the following scheme (3). In the first step only one of the monomer double bonds is involved, to form radi-

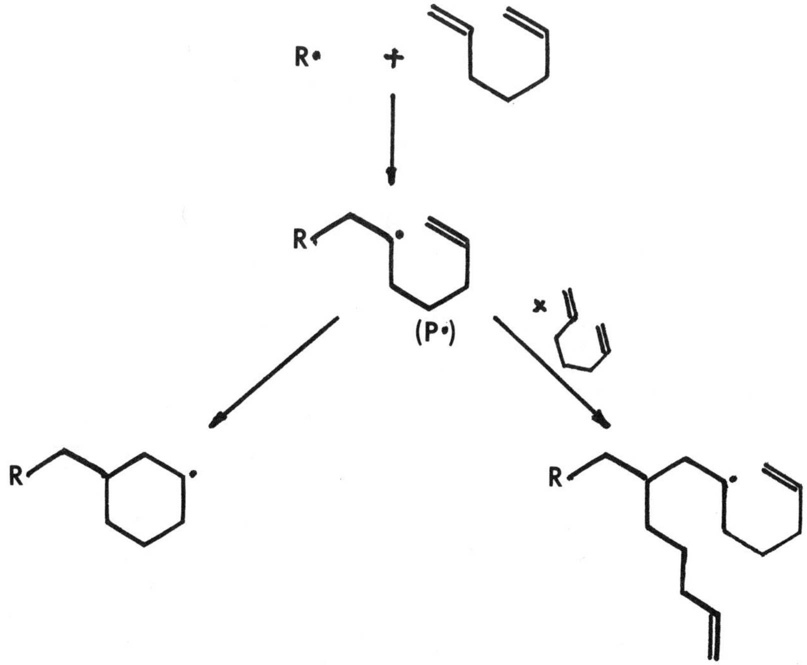

R• +

(P•)

cal P• . Radical P• can either yield a cyclic structure, by reacting intramolecularly with the residual double bond, or attack one of the double bonds of another molecule; in this second case, instead of a cyclic structural unit, an unsaturated linear structural unit is left behind in the growing chain.

It is clear that the polymer composition, in terms of the ratio between saturated cyclic structural units and unsaturated linear structural units, is controlled by the competition between intra- and intermolecular propagation reactions. It is clear as well that the change of the monomer concentration does not affect the monomolecular cyclization, whereas it has an influence on the bimolecular addition of P• radical onto a monomer molecule: specifically, by decreasing the monomer concentration, the bimolecular addition becomes less favoured in agreement with the experimental findings.

This can be immediately put in a quantitative form

in the case of symmetrical unconjugated dienes, that is of dienes with two identical double bonds. In this case, the ratio between the mole fraction f_c of cyclic structural units and the mole fraction f_u of unsaturated structural units is obviously given by the ratio between the rate of the reactions yielding the two kinds of structural units:

$$\frac{f_c}{f_u} = \frac{k_c [P\bullet]}{2\,k_p [P\bullet][M]} \qquad (4)$$

where k_c and k_p are the rate constants of the intramolecular and the intermolecular reactions, respectively, and $[M]$ is the monomer molar concentration. The factor 2 depends on the presence of two points of attack, the two unsaturations, of $P\bullet$ radicals onto the monomer molecule.

Assuming that the polymer is comprised only of cyclic and unsaturated structural units, $f_u + f_c = 1$, and letting $r_c = k_c/k_p$ one obtains from eq. (4):

$$\frac{1}{f_c} = 1 + \frac{2}{r_c}[M] \qquad (5)$$

This simple relation shows that, by increasing the monomer concentration, the mole fraction f_c must decrease. Eq. (5) has been found to fit satisfactorily the experimental data, in the cases in which the composition of cyclopolymers obtained from symmetrical dienes has been investigated as a function of the monomer concentration (4-7).

Focussing attention on the parameter r_c, which will be called "cyclization ratio", one can see that it is the double bond concentration (equal to twice the monomer concentration) at which $f_c = f_u = 0.5$. In other words, in the polymer obtained from a solution in which twice the monomer concentration is equal to r_c , the structural units are 50 % cyclic and 50 % unsaturated: in the competition between intra- and intermolecular chain propagation there is no winner. It might be of interest, therefore, to take into consideration some experimental values of r_c.

In Table I some of these values have been collec-

Table I - Cyclization ratios in free radical cyclopolymerizations.

monomer	polymerization temperature (°C)	r_c (mol/l)	Ref.
acrylic anhydride	70	9.79	(4)
methacrylic anhydride	60	97.32	(8)
diphenylhexadiene	70	8.70	(9)

ted. One can see that they are very high, indicating that even in the polymers obtained from undiluted monomers the cyclic structural units prevail over the linear ones. On the other hand, Butler and Raymond (10) calculated the probabilities of being in the neighbourhood of the reaction site for the double bond pendant from the reactive chain end, and for a double bond of an unreacted molecule, and found that, from a purely statistical point of view, only at a monomer concentration as low as 1 or 2 mol/l should the double bond pendant from the active chain end have a greater opportunity to meet the active site than any other double bond in the solution. Therefore, the high values of the cyclization ratios mean that, in concentrated solutions, the active chain end can distinguish that one of the many double bonds around it is unique and, of course, that the reaction of the active end with this unique double bond is easier than with any other double bond.

Of the ideas which have been put forward to interpret such a situation, it seems reasonable not to give much importance to the possibility that an inductive effect would be operative at a position three or four atoms removed from the reaction site: if such an effect would exist, it would be very weak, and would not justify such a dramatic increase of the reactivity of the pendant double bond, as is particularly evident in the case of methacrylic anhydride, but also in the cases of the other monomers.

Another, more realistic, possibility would be the analog of what it has been already said for the monomer molecules: an interaction across the space between the reaction site and the pendant double bond (1), as is schematically represented here:

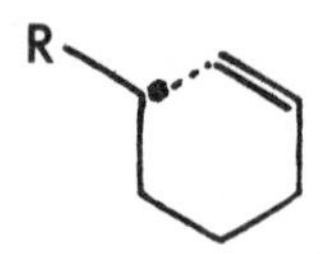

Without entering into details of the nature of such an interaction, it seems possible to argue that, if the interaction can exist, it can exist as well between the reaction site and one double bond of an unreacted molecule, as is shown in the following scheme:

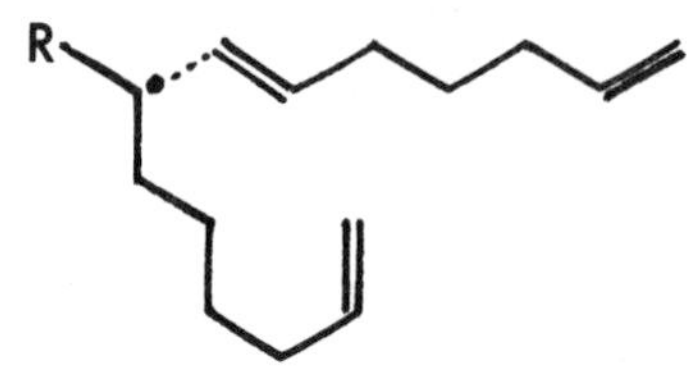

Thus, an interaction across the space might represent a possible reaction path, but it would not explain, alone, the preferential addition of the active chain end on the pendant double bond.

Effect of a large substituent at the active chain end.

A quite different idea has been suggested by Gibbs and Barton (11) to justify the predominance of intramolecular cyclizations: on the basis of literature data (12,13) regarding the rate constants for chain propagation and termination in the free radical polymerization of methyl acrylate and butyl acrylate, and particularly giving much importance to the large difference in the values of k_p and k_t as reported by Matheson (12) for methyl acrylate and by Melville and Bickel (13) for butyl acrylate, Gibbs and Barton (11) postulated that the presence of a large substituent at the active chain end in the cyclopolymerization of unconjugated dienes might hinder the approach of unreacted molecule.

The main drawback of this point of view lies in the uncertainty of the values of the rate constants relative to butyl acrylate. As one can see in Table II, the constants calculated from the data of Bagdasaryan

Table II - Rate constants in free radical polymerizations at 30°C.

monomer	k_p	k_t	Ref.
methyl acrylate	720	$2.2 \cdot 10^6$	(12)
methyl acrylate	590	$3.6 \cdot 10^6$	(14)
butyl acrylate	14	$1.8 \cdot 10^4$	(13)
butyl acrylate	570	$8.3 \cdot 10^6$	(15)

(14) for methyl acrylate do not greatly differ from those of Matheson (12); on the contrary, the constants calculated for butyl acrylate according to Bengough and Melville (15) are several times higher than those calculated according to Melville and Bickel (13). It is quite obvious that, if the Melville-Bengough constants are correct, the large difference between the values of the constants relative to methyl and butyl acrylates disappears, and the hypothesis of Gibbs and Barton loses its experimental support. It is of interest to note that the Bengough-Melville k_p value agrees, whereas that of Melville-Bickel does not, with the calculated pre-exponential factor of the constant k_p, and this might be taken as an evidence, though indirect of the greater reliability of the Bengough-Melville result.

Difference between intra- and intermolecular chain propagation reactions.

The hypotheses considered so far share the attempt of explaining the ease of formation of the ring structures in cyclopolymerization taking into account only the relative ability of the double bond pendant from the active chain end, and of the double bonds of an unreacted monomer molecule, to reach the reaction site. No attention is paid to the fact that the intramolecular and intermolecular chain propagation reactions are basically different, in the sense that the former involves an entropy decrease due to the loss of some rotational degrees of freedom at the end of the active chain, whereas the latter involves an entropy decrease depending on the fact that two independent molecular entities become "fused" together. The question to be answered is: which of the two entropy decreases is the largest?

To be able to answer this question is very interesting because for the reactions considered, which can be schematically represented as follows:

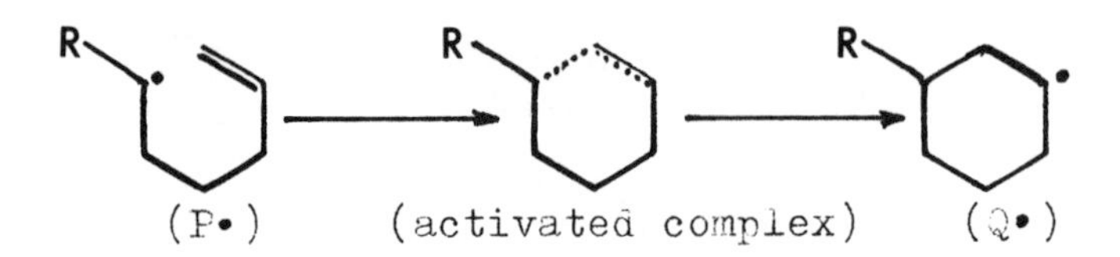

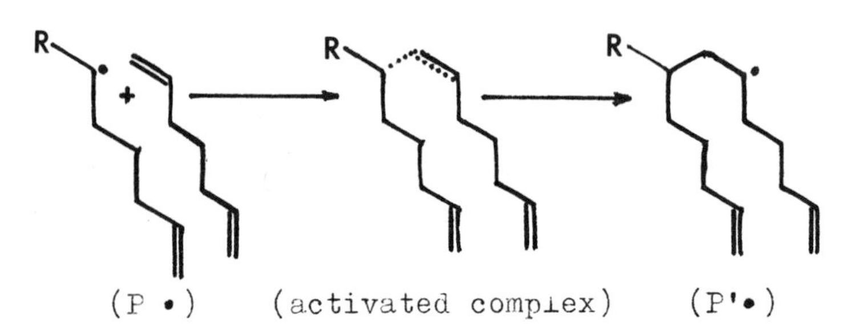

one might tentatively assume that the entropies of the reaction products are close to the entropies of the activated complexes. In other words, the entropy change ΔS_c occurring in the cyclization should be approximately equal to the activation entropy $\Delta S_c^{\ddagger}$ of the reaction:

$$\Delta S_c = S_{Q\bullet} - S_{P\bullet} \simeq S_{ac,c} - S_{P\bullet} = \Delta S_c^{\ddagger}$$

where $S_{P\bullet}$, $S_{Q\bullet}$ and $S_{ac,c}$ are the molar entropies of P• and Q• radicals and of the activated complex, respectively. Similarly, for the intermolecular propagation reaction:

$$\Delta S_p = S_{P'\bullet} - S_{P\bullet} \simeq S_{ac,p} - S_{P\bullet} = \Delta S_p^{\ddagger}$$

where $S_{P'\bullet}$ is the molar entropy of the growing radical, increased by one monomeric unit with respect to P•, and $S_{ac,p}$ is the molar entropy of the activated complex of this reaction. Therefore, if one is able to evaluate ΔS_c and ΔS_p, an estimate of the corresponding activation entropies becomes possible.

An experimental evaluation of the difference between the activation entropies is obtained, in principle, from the investigation of the temperature dependence of the cyclization ratio r_c, which can be expressed in the form:

$$r_c = \exp\left(\frac{\Delta S_c^{\ddagger} - \Delta S_p^{\ddagger}}{R}\right) \exp\left(-\frac{E_c - E_p}{RT}\right)$$

By plotting $\ln r_c$ as a function of $1/T$ a straight line should be obtained, and the difference between the activation entropies should be given by the intercept. The problem is that, as it is well known, especially when one considers reaction taking place in condensed phases, it is difficult to decide if one is handling true activation entropies and energies, or apparent activation quantities. In other words, quite frequently one does not know if the constants being treated are pure, single, rate constants, rather that a combination of several rate constants. Of course, the agreement between the experimental difference between the activation entropies for intra- and intermolecular chain propagation, and the calculated difference between the entropy decreases occurring in the same reactions, can be taken as an indication of the reliability of both the experimental and the calculated values.

The entropy changes occurring in intra- and intermolecular chain propagation reactions can be easily calculated if some more or less drastic assumptions are made. First of all, it is well known that, from an experimental point of view, free radical polymerizations seem scarcely influenced by the environmental conditions, and practically occur with the same rate in solvating and non-solvating, or in polar and nonpolar media. This suggests that the free energy change associated with a given step of the growth of the polymer chain does not significantly depend on the actual value of the free energy of the different species involved. It might then be accepted that the rate of the propagation reactions, either intra- or intermolecular, would be similar in solution, as well as in a hypothetical ideal gas state. The reason for such an assumption is that the eavaluation of the entropy of the reactants and of the reaction products can be carried out more easily if they are considered in the ideal gas state.

It is well known that entropy, and in general the thermodynamic properties, of a given molecule can be built up as the sum of definite contributions which can be assigned to the various groups present in the molecule. As an example, consider the molecule of butane, made of two methyl and two methylene groups: each methyl group contributes to the molar entropy in the ideal gas state for 118.8 J/mol K (16) and each methylene group contributes for 39.2 J/ mol K (16). Summing up, and subtracting the term Rln2 (being 2 the symmetry number of this molecule), one obtains for the molar entropy of butane the value 310.4 J/mol K, to be compared with the experimental value 310.2 J/ mol K (17).

It follows that, if in a chemical reaction a molecule is only partly modified, the part of the molecule which remains unchanged contributes in the same way to the total entropy of both the reactants and of the products. As a consequence, the entropy change accompanying a given reaction is simply the difference between the contribution to the total entropy of those groups which are modified in the reaction.

Calculation of ΔS_p. The situation is particularly simple (18) in the case of the intermolecular chain propagation reactions, generally represented in the following scheme:

$$R{-}CH_2{-}\underset{X}{\overset{Y}{C}}\cdot \; + \; CH_2{=}\underset{X}{\overset{Y}{C}} \longrightarrow R{-}CH_2{-}\underset{X}{\overset{Y}{C}}{-}CH_2{-}\underset{X}{\overset{Y}{C}}\cdot \qquad (6)$$

Taking S_R as the molar entropy of the R part of the growing polymer chain which remains unchanged after the reaction, S_{mu} as the molar entropy of the monomeric unit formed in the reaction, and S_{ru} as the molar entropy of the radical unit at the end of the chain, one obtains:

$$\Delta S_p = S_R + S_{mu} + S_{ru} - (S_R + S_{ru} + S_M) = S_{mu} - S_M$$

In other words, as far as the entropy change is concerned, the propagation reaction is equivalent to the insertion of a monomer molecule in the inner part of a growing chain.

It is possible to go further considering that the entropy contribution of the lateral substituents X and Y in the monomer molecule and in the monomeric unit can be regarded as the same, except for the effects, which can however be assessed, which depend upon the possible existence of conjugation between the monomer double bond and one of the substituents, and the changes occurring in the symmetry of the molecules and in the number of possible optical isomers, before and after the reaction. Therefore it can be shown (18) that:

$$\Delta S_p = S(-CH_2-C\lessgtr) - S(CH_2=C<) + S_{conj} + R \ln \sigma \nu \qquad (7)$$

where $S(-CH_2-C\lessgtr)$ and $S(CH_2=C<)$ are the entropy contributions of the groups which are modified in the reaction, S_{conj} is the lowering of the monomer entropy due to conjugation, σ is the symmetry number of the monomer molecule (the symmetry number of the monomeric unit formed in reaction (6) is usually 1), and ν is the number of optical isomers formed in the reaction. Of course, eq. (7) is not limited to the intermolecular additions taking place in the free radical polymerization of unconjugated dienes, but should be valid for the chain propagation taking place in the free radical polymerization of any monomer.

In Table III the experimental activation entropy of the chain propagation in the polymerization of some typical monomers is compared with the calculated entropy change accompanying the reaction. Considering the large experimental errors affecting the evaluation of the absolute rate constants (for example, by the rotating sector method, or by non-stationary kinetics) the agreement is quite good, supporting the assumed similarity between the reaction in the gas phase and in solution, and showing that the activation entropies satisfactorily compare with the entropy changes resulting from the complete reaction.

Table III - Experimental activation entropy $\Delta S_p^{\ddagger}$ and calculated entropy change ΔS_p for the chain propagation in the free radical polymerization of some monomers.

monomer	$-\Delta S_p^{\ddagger}$ (J/mol K)	$-\Delta S_p$ (J/mol K)
styrene	119.3	124.0
vinyl acetate	139.1	129.9
methyl acrylate	135.3	129.9
methyl methacrylate	143.7	141.3

Turning to the comparison between the rate constants for the chain propagation in the free radical polymerization of methyl and butyl acrylayes, it can be observed that both these reactions should occur with the same entropy decrease, because identical double bonds are involved. From the experimental data by Melville and Bickel (13) and by Bengough and Melville (14) relative to butyl acrylate, 4 pairs of activation energy and entropy can be calculated, which are collected in Table IV. It is evident that the experimental activation entropy which is closest to the calculated ΔS_p for alkyl acrylates (i.e. the ΔS_p value reported for methyl acrylate in Table III) is -124.4 J/mol K, whereas all the other activation entropies seem to be too high. The rate constant calculated at 30°C from

Table IV - Experimental activation energy and entropy and rate constant at 30°C for the chain propagation in the free radical polymerization of butyl acrylate.

Ref.	E_p (kJ/mol)	$-\Delta S_p^{\ddagger}$ (J/mol K)	k_p
(13)	23.07	149.9	27
(13)	9.63	199.7	14
(15)	23.07	124.4	579
(15)	9.60	178.9	173

this value of the activation entropy and the corresponding value E_p = 23.07 kJ/mol is k_p = 579, close to the value k_p = 590 calculated at 30°C from the data of Bagdasaryan (14) relative to methyl acrylate: in other words, it seems not likely that the chain propagations in the polymerization of methyl and butyl acrylates are strongly different.

Calculation of ΔS_c. The evaluation of the entropy change occurring in the cyclization reactions is a little more complicated than the evaluation of ΔS_p, in as much as in the cyclization reactions the product radical is different from the radical undergoing cyclization.

It seems reasonable to assume even in the present case that the part of the growing chain not directly involved in the cyclization reaction does not contribute to ΔS_c. Therefore, considering the following schemes:

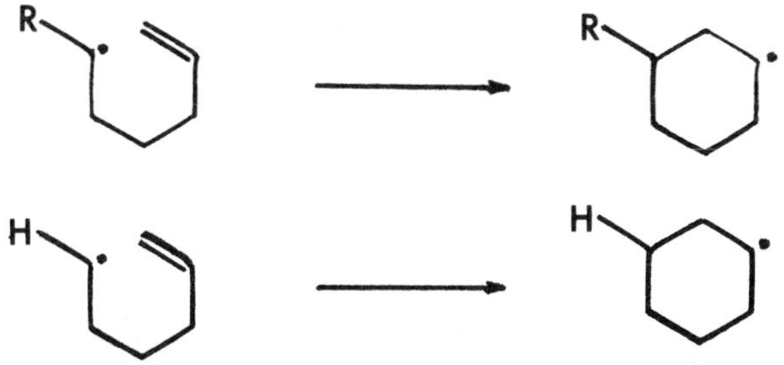

(P•) (Q•)

where, as usual, R stands for the polymer chain, one might expect that the two cyclization reactions are endowed with the same entropy decrease. As a consequence:

$$\Delta S_c = S_{Q\bullet} - S_{P\bullet}$$

The difference between the molar entropies of Q• and P• radicals has been related (16) to the difference between the molar entropies of QH and PH molecules, which are obtained from Q• and P• radicals by saturating the lone electron with a H atom. The following relation has been suggested(16):

$$\Delta S_c = S_{QH} - S_{PH} + R \ln \frac{\sigma_{QH} \ \sigma_{P\bullet}}{\sigma_{PH} \ \sigma_{Q\bullet}} \qquad (8)$$

where $\sigma_{Q\bullet}$, $\sigma_{P\bullet}$, σ_{QH} and σ_{PH} are the symmetry numbers of the species indicated.

Eq. (8) has been used to evaluate the entropy changes occurring in the cyclization reactions (carried out in the ideal gas state) represented in Figure 1. The most evident feature of the values of the entropy decreases calculated for the cyclization reactions is that they are decidedly lower than the entropy decreases calculated according to eq. (7), associated with the intermolecular additions represented as well in Figure 1. This fact obviously means that, if the cyclization reactions are in competition with the intermolecular additions, they are strongly favoured as far as the entropy changes are involved. As a consequence, when the ring structure which can be formed is not endowed with high tensions on the valence bonds and angles (i.e. when the ring structure has not a strongly higher energy than the open structure formed in the intermolecular addition), the cyclization reaction is faster. This is certainly not the case in the polymerization of butadiene, where the formation of the 3-membered rings would involve a very small entropy decrease, but is totally suppressed by the high energy of the strained valence angles: as it is well known, the chain propagation reactions in the polymerization of butadiene exclusively occur by intermolecular additions.

On the contrary, when 5- and 6-membered rings can be formed, i.e. rings practically free of tensions, the decrease of entropy occurring in the cyclization is still strongly lower than that occurring in the intermolecular addition, and the cyclization are favoured on this account.

Comparison between calculated differences $\Delta S_c - \Delta S_p$ and experimental differences $\Delta S_c^{\ddagger} - \Delta S_p^{\ddagger}$.

The final point to be discussed is the possibility of transferring the above conclusion to actual cyclopolymerization processes, that is to see if, as it has been already verified for the intermolecular chain propaga-

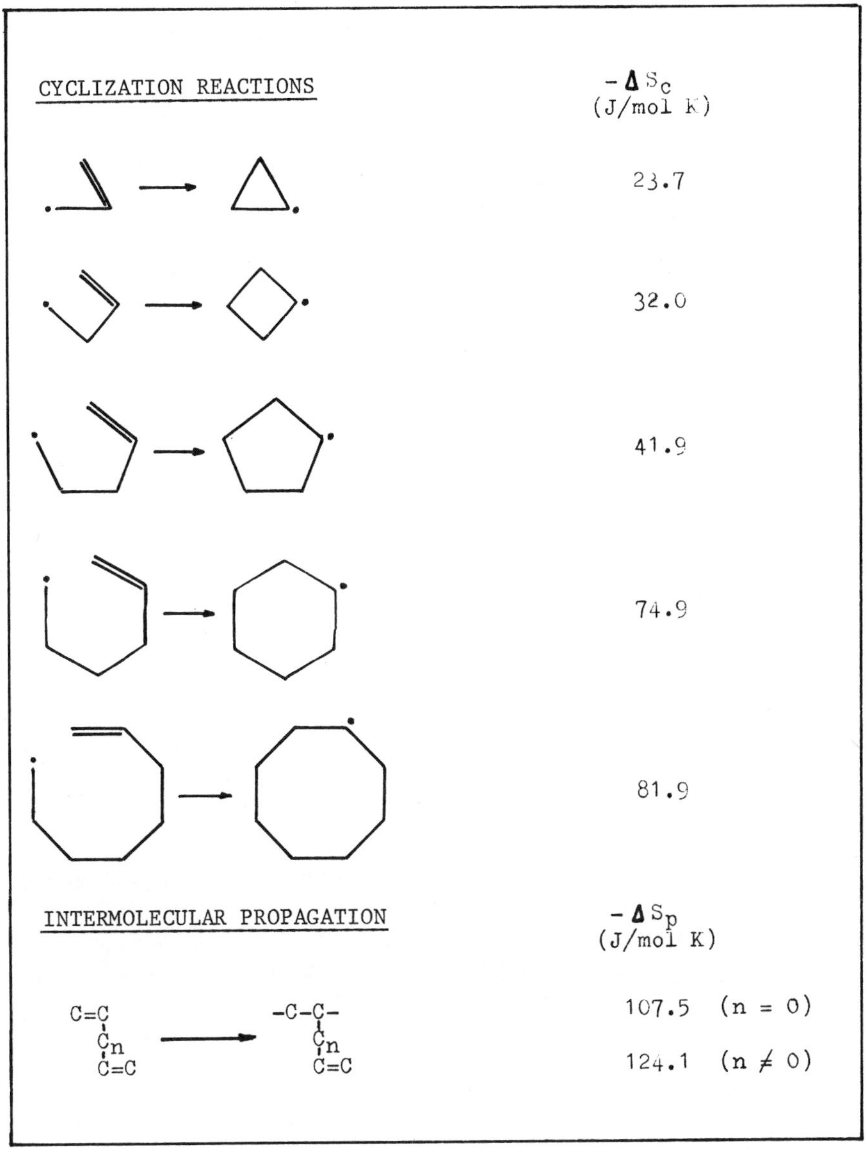

CYCLIZATION REACTIONS	$-\Delta S_c$ (J/mol K)
	23.7
	32.0
	41.9
	74.9
	81.9

INTERMOLECULAR PROPAGATION	$-\Delta S_p$ (J/mol K)
C=C–C_n–C=C → –C–C(–C_n–C=C)–	107.5 (n = 0)
	124.1 (n ≠ 0)

Figure 1. Calculated entropy changes occurring in intramolecular cyclizations and intermolecular propagations.

tion reactions, a close connection exists between the activation entropy and the entropy change occurring in cyclization reactions.

It can be observed that, as far as entropy decrease of the cyclization reactions is concerned, it should mainly depend on the number of bonds which change in the reaction their internal rotation modes with vibrational modes (this is, of course, the reason why the absolute values of ΔS_c in Figure 1 increase by increasing the ring size), whereas ΔS_c should be scarcely affected by the substitution of one or more of the methylene groups between the carbon radical and the double bond with different divalent atoms or atomic groups, such as -O- , >C=O , >NH , etc.. This is equivalent to say that the contribution to the entropy of the internal rotation around a given single bond is practically the same irrespective of the height of the potential barriers hindering rotation. Actually, for potential barriers fro 4 to 16 kJ/mol, that is from one to the other of the limits of the heights of the potential barriers found for the rotation around the most common single bonds of organic molecules, the entropy contribution of the internal rotation around a single bond varies by 4 J/mol K, which means a change of a few per cent in the value of ΔS_c. Therefore, one can use the values of ΔS_c calculated for cyclization reactions involving hydrocarbons such as those represented in Figure 1, together with the values of ΔS_p calculated for the corresponding intermolecular additions, to evaluate the difference $\Delta S_c - \Delta S_p$, and to compare it with the experimental difference between the activation entropies, as obtained from the temperature dependence of the cyclization ratios of real cyclopolymerizations.

In Table V are shown, for some monomers undergoing cyclopolymerization, the differences $\Delta\Delta S = \Delta S_c - \Delta S_p$ calculated (8,9,16) according to the procedures previously described, and the experimental differences $\Delta\Delta S^{\ddagger}$ between the corresponding activation entropies. In the case of divinyl ether, a monomer which undergoes cyclopolymerization yielding 5- and 6-membered ring units (19) in a ratio increasing with increasing temperature (20), the calculated (21) diffe-

Table V - Calculated $\Delta\Delta S$ and experimental $\Delta\Delta S^{\ddagger}$ for the chain propagation reactions in the free radical cyclopolymerization of some monomers.

monomer	ring size	$\Delta\Delta S$ (J/mol K)	$\Delta\Delta S^{\ddagger}$ (J/ mol K)
diphenylhexadiene	5	88.8	86.0 ± 10.8
acrylic anhydride	6	49.2	46.2 ± 4.4
methacrylic anhydride	6	60.6	66.5 ± 15.6
o-divinylbenzene	7	53.5	46.6 ± 14.8
divinyl ether	5,6	30.2	31.0 ± 2.0

rence $\Delta\Delta S = \Delta S_{c5} - \Delta S_{c6}$ between the entropy changes occurring in the 5- and 6-membered ring closure is compared with the experimental difference $\Delta\Delta S = \Delta S_{c5} - \Delta S_{c6}$ between the activation entropies of these two reactions.

The agreement between calculated and experimental differences is evidently very good. This means that it is correct to assume that the entropy decreases occurring in both cyclizations and intermolecular additions are close to the corresponding activation entropies. But it means also that the cyclization reactions taking place in the usual cyclopolymerization processes are effectively favoured by a smaller activation entropy.

Literature cited

1. Butler, G.B. J. Polymer Sci. 1960, 48, 279.
2. Butler, G.B.; Brooks, T.W. J. Org. Chem. 1963, 28, 2699.
3. Butler, G.B.; Raymond, M.A. J. Org. Chem. 1965, 30, 2410.
4. Mercier, J; Smets, G. J. Polymer Sci. 1962, 57, 763.
5. Gibbs, W.E. J. Polymer Sci. 1964, A2, 4815.
6. Minoura, Y; Mitoh, M. J. Polymer Sci. 1965, A3, 2149.
7. Chiantore, O; Camino, G.; Chiorino, A; Guaita, M. Makromol. Chem. 1977, 178, 119.
8. Chiantore, O; Camino, G; Chiorino, A.; Guaita, M. Makromol. Chem. 1977, 178, 125.

9. Costa, L.; Guaita, M. "Cyclopolymerization and Polymers with Chain-Ring Structures"; American Chemical Society: Washington, DC, 1982.
10. Butler, G.B.; Raymond, M.A. J. Polymer Sci. 1965, A3, 3413.
11. Gibbs, W.E.; Barton, J.M. "The Mechanism of Cyclopolymerization of Unconjugated Diolefins", in "Kinetics and Mechanisms of Polymerizations", edited by Ham, G.E.; Edward Arnold (Publisher) Ltd, London, 1967; vol. I; p. 133.
12. Matheson, M.; Auer, E.; Bevilacqua, E.; Hart, E. J. Am. Chem. Soc. 1951, 73, 5395.
13. Melville, H.W.; Bickel, A.F. Trans. Faraday Soc. 1949, 45, 1049.
14. Sinitsyna, Z.A.; Bagdasaryan, Kh.S. Zh. F. Ch. 1958, 32, 1319.
15. Bengough, W.J.; Melville, H.W. Proc. Roy. Soc. (London) 1954, 225, 330.
16. Guaita, M. Makromol. Chem. 1972, 157, 111.
17. Stull, D.R.; Westrum, E.F.; Sinke, G.C. "The Chemical Thermodynamics of Organic Compounds"; John Wiley & Sons, London, 1969; p.245.
18. Guaita, M. Makromol. Chem. 1972, 154, 191.
19. Aso, C.; Ushio, S.; Sogabe, M. Makromol. Chem. 1967, 100, 100.
20. Guaita, M.; Camino, G.; Trossarelli, L. Makromol. Chem. 1971, 149, 75.
21. Costa, L.; Chiantore, O.; Guaita, M. Polymer 1978, 19, 202.

RECEIVED March 8, 1982.

3

Intramolecular Addition Modes in Radical Cyclopolymerization of Some Unconjugated Dienes

AKIRA MATSUMOTO, KUNIO IWANAMI, TAKAO KITAMURA, and MASAYOSHI OIWA

Kansai University, Department of Applied Chemistry, Faculty of Engineering, Suita, Osaka 564, Japan

GEORGE B. BUTLER

University of Florida, Center for Macromolecular Science, Gainesville, FL 32611

The factors influencing intramolecular addition modes were investigated in detail in the radical cyclopolymerization of some unconjugated dienes. Thus, in diallyl phthalate, only the cyclic polymer of an 11-membered ring was obtained, whereas 10-membered ring formation was observed in diallyl aliphatic dicarboxylates. In dimethallyl dicarboxylates the intramolecular head-to-head(hh) addition was highly enhanced compared with corresponding diallyl dicarboxylates. In acrylic anhydride, the 5-membered ring formation was favored with the increase of the polarity of solvent, the elevation of temperature, and the decrease of monomer concentration; thus, the ring size of cyclic structure could be freely controlled. In methacrylic anhydride, the analogous results were obtained at elevated temperatures. In allyl methacrylate and allyl atropate, the enhanced tendency to cyclize and the highly occurring intramolecular hh addition were observed above the ceiling temperatures of the corresponding alkyl α-substituted acrylates.

These results are discussed from the thermodynamical standpoint.

It is well known that head-to-tail(ht) addition for the attack of a growing chain radical on a monomer is predominant compared with other types of addition modes such as head-to-head(hh), tail-to-head, and tail-to-tail in the radical polymerization of vinyl monomers, thus forming the polymer exclusively of ht structure. This fact has been explained by some qualitative considerations of resonance stability and steric factors leading to the intermediate formation of the more stable free radical and, furthermore, by theoretical treatment based on molecular orbital theory(1). In this connection, most of the early literature on the structure of

0097-6156/82/0195-0029$06.00/0
© 1982 American Chemical Society

cyclic polymers obtained in the cyclopolymerization of 1,6-dienes have assumed that a six-membered ring is formed in conformity with expected thermodynamic stability of the secondary or tertiary radical formed via intramolecular ht addition compared with the required primary radical for hh addition.

However, numerous reports of predominant five-membered ring formation via intramolecular hh addition of the uncyclized radical to the internal double bond have been recently published; in particular, the cyclopolymerizations of N-substituted dimethacrylamides(2-5), diallyl amines and their salts(6-11), divinyl acetals (12), and divinyl phosphonates(13) could be presented as typical examples. These findings are closely relevant to the recent development of analytical procedures especially including ^{13}C-NMR spectroscopy although chemical analyses and IR spectroscopy were employed earlier for the determination of cyclic structures. Here it should be noted that completely opposing results can be seen in the cyclopolymerization of divinyl formal and diallyl ammonium salts: Thus, Tsukino and Kunitake(12) have reported recently the exclusive five-membered ring formation from a detailed investigation of the ^{13}C-NMR spectrum of cyclic poly(divinyl formal) as opposed to the earlier results of Matsoyan(14) who had determined by chemical means that the cyclic structure was a six-membered ring, although Minoura and Mitoh(15) have claimed some five-membered ring formation from a higher content of 1,2-glycol structure in poly(vinyl alcohol) obtained by hydrolysis of poly(vinyl formal) than that in commercial poly(vinyl alcohol). Chemical evidence has been also presented to support the cyclopolymerization hypothesis of diallyl ammonium salts by Butler, Crawshaw, and Miller(16) contrary to recent spectroscopic evidences(6-11).

Another approach to the study of the cyclic units formed in these cyclopolymerizations is through the study of the radical cyclization reaction of selected model compounds. Thus, extensive studies have shown that the five-membered ring is predominant, while new evidence indicates that radical stability exerts a marked influence on the ring size. These conflicting aspects of cyclopolymerization have been discussed in detail by Butler(17); he has pointed out that considerably more investigations will be necessary before definite conclusions can be drawn with respect to the ratio of five- to six-membered rings in the many cyclopolymers already synthesized, and a satisfactory explanation for these extensive variations from one system to another is available.

Thus, we investigated in detail the factors influencing intramolecular addition modes in the radical cyclopolymerization of some unconjugated dienes, including diallyl and dimethallyl dicarboxylates, acrylic and methacrylic anhydrides, and allyl α-substituted acrylates and discussed the selectivity of reaction mode of intramolecular hh or ht addition in terms of the thermodynamical standpoint. The present article gives a summary of our recent work including published papers(18-23).

Cyclopolymerization of Diallyl and Dimethallyl Dicarboxylates

In our continuing studies of the cyclopolymerization of diallyl dicarboxylates(24), we attempted to determine the ring size of the cyclic unit of the resulting polymer. During the course of this investigation, it has been clearly demonstrated that an anomalous hh addition occurs remarkably in the radical polymerization of monoallyl and monomethallyl carboxylates(18,21): Figure 1 shows a typical example of the highly occurring hh addition in the radical polymerization of allyl acetate(AAc) and methallyl acetate. This has been interpreted from the following standpoints(21): In the radical polymerization of allylic compounds, which are typical unconjugated monomers, the resonance stabilization of the growing secondary or tertiary radical by the substituent is not so significant. In other words, the difference in the resonance stability between two types of intermediate radicals, the secondary or tertiary head and primary tail radicals, is small, implying the reduced reaction selectivity governed by the thermodynamic stability of the resulting radicals in comparison with the cases of common vinyl monomers. This discussion suggests the significance of the steric factor for the selectivity of addition modes in the radical polymerization of allylic compounds because steric effect tends to favor the hh or tt structure, that is, the alternate hh or tt polymer should be comparatively free from steric repulsion between substituents made evident in the inspection of the molecular model (25,26). Thus, the effect of the side group steric hindrance arising from the acid moiety of monoallyl carboxylates, including AAc(18), allyl benzoate(21), allyl propyl succinate(APSu)(20), allyl propyl phthalate(APP)(20), and allyl propyl cis-1,2-cyclohexanedicarboxylate(APCH)(20) was examined and the contents of hh linkage of the polymers obtained in bulk at 80°C were estimated to be 8.5, 11, 14, 15, and 19%, respectively. Consequently, the occurrence of hh addition was enhanced with an increase in the bulkiness of the acid moiety of monoallyl carboxylates as the side group of the resulting polymers in conformity with our expectation.

Such a striking feature of allylic compounds that the selectivity of addition modes is governed significantly by any other factors than the thermodynamic stability of the resulting radicals has been quite reflected in the intramolecular addition modes in the cyclopolymerization of diallyl and dimethallyl dicarboxylates as follows: The polymerizations of three kinds of diallyl dicarboxylates, including diallyl phthalate(DAP)(19), diallyl succinate(DASu)(20), and diallyl cis-1,2-cyclohexanedicarboxylate(DACH)(20), all of which can form the same ring size, were carried out in bulk or in benzene solutions at 80°C, using benzoyl peroxide as the initiator. Figure 2 shows the relationships between the percentage of hh linkage and the content of uncyclized unit, i.e., 100 - f_c; the results have been discussed from the following three points: First, the contents of hh linkage were quite high for all uncyclized polymers(f_c = 0) which were prepared from the model polymeri-

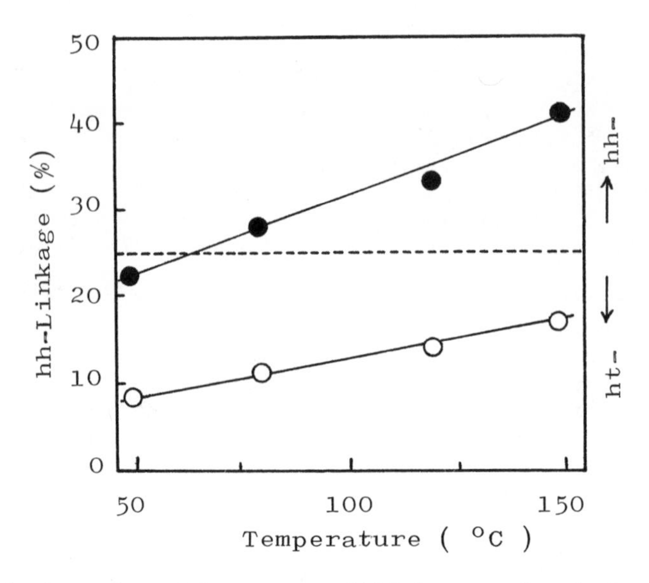

Figure 1. Dependence of the percentage of hh linkage on polymerization temperature in bulk polymerizations. Key: ○, *allyl acetate; and* ●, *methallyl acetate. (Reproduced, with permission, from Ref.* 21. *Copyright 1981, J. Polym. Sci. Polym. Lett. Ed.)*

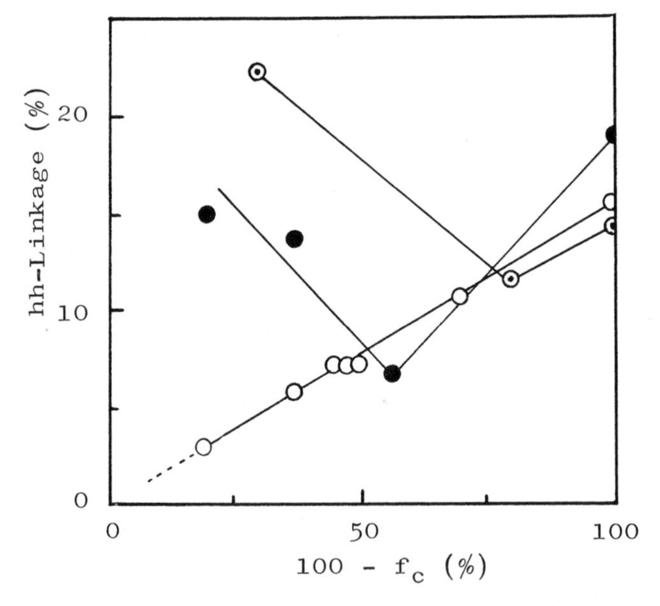

Figure 2. Relationship between the percentage of hh linkage and the content of uncyclized unit, 100 — f_c, obtained for cyclopolymerizations in benzene solutions at 80°C. Key: ○, *diallyl phthalate;* ⊙, *diallyl succinate; and* ●, *diallyl* cis-*1,2-cyclohexanedicarboxylate. (Reproduced, with permission, from Ref.* 21. *Copyright 1981, J. Polym. Sci. Polym. Lett. Ed.)*

zations of APP, APSu, and APCH corresponding to monoallyl counterparts of DAP, DASu, and DACH, respectively, as discussed above in detail. Second, the contents of hh linkage tended to decrease with an increase in the cyclized unit; this is ascribed to the reduced steric crowding by the incorporation of cyclized structure into the resulting polymer because the carbon numbers of backbone polymer chain increase by four with the formation of cyclized unit. Third, the minimum hh linkage was observed in the cyclopolymerization of DASu and DACH, whereas in the cyclopolymerization of DAP the content of hh linkage decreased only linearly with increased cyclization. That is, the cyclized DASu and DACH polymers contained high hh linkages, whereas on the contrary, cyclic poly-DAP had no hh linkage. Here it should be remembered that the content of hh linkage of the cyclized polymer was determined from the ^{13}C-NMR spectrum of poly-AAc derived by saponification and subsequent esterification of the cyclized diallyl dicarboxylate polymer. Consequently, two possible reaction paths, i.e., an intramolecular hh addition of uncyclized head radical and an intermolecular hh addition of cyclized head radical should be considered for the incorporating process of hh linkage into the derived poly-AAc chain. The discrimination of these two reactions is not easy, but the possibility of the occurrence of an intermolecular hh addition of cyclized head radical may be omitted by considering the steric crowding in the resulting cyclized polymer. Thus, the following three types of typical cyclopolymers will be given schematically:

```
          R R             R R
          | |             | |
~CH2-C-C-CH2-CH2-C-C-CH2~                      (A)
          | |             | |
          |_|             |_|

          R        R        R        R
          |        |        |        |
~CH2-C-CH2-C-CH2-C-CH2-C~                      (B)
          |        |        |        |
          |________|        |________|

R                R R                R
|                | |                |
~C-CH2-CH2-C-C-CH2-CH2-C~                      (C)
|                | |                |
|________________| |________________|
```

where R is H for diallyl dicarboxylates and CH_3 for dimethallyl dicarboxylates. The polymers of A, B, and C types are obtainable via intramolecular hh and intermolecular tt additions, intramolecular ht and intermolecular ht additions, and intramolecular tt and intermolecular hh additions, respectively. Evidently, the A-type polymer is most favorable in terms of steric repulsion between large cyclic units.

On the other hand, no hh linkage of the cyclic poly-DAP means that the intramolecular cyclization of uncyclized radical and the intermolecular propagation of cyclized radical proceed only via ht addition as completely opposed to our expectation. The nonoccurrence of intramolecular hh addition is attributed to the fact that the occurrence of hh addition, i.e., the formation of ten-membered ring, requires the sacrifice of resonance stabilization between carbonyl double bond and aromatic benzene ring because of the loss of coplanarity as is expected from the molecular model(19,20). Moreover, the selective intermolecular ht addition of cyclized head radical in the cyclopolymerization of DAP would be as a result of nonoccurrence of sterically less favorable intermolecular hh addition leading to the C-type polymer.

For the cyclopolymerization of dimethallyl dicarboxylates the formation of the B-type polymer should be considerably sterically hindered compared with diallyl dicarboxylates from the similar standpoints for the monomethallyl polymerization in which a predominant occurrence of an anomalous hh addition has been observed and discussed in detail from a thermodynamical standpoint(21); thus, the increase in the occurrence of intramolecular hh addition was expected. The polymerizations of dimethallyl phthalate(DMAP) and dimethallyl cis-1,2-cyclohexanedicarboxylate(DMACH) were conducted in bulk at 80°C, using benzoyl peroxide as the initiator. The contents of hh linkage of the resulting polymers were estimated to be 19% for DMAP and 23% for DMACH in a markedly high percentage compared with 6.5% for both DAP and DACH of the corresponding diallyl dicarboxylates. From these results, it has been suggested that an intramolecular hh addition in the cyclopolymerization of unconjugated dienes be promoted above the ceiling temperature for ht addition in the polymerization of monoene counterparts; this was developed for the cyclopolymerization of allyl α-substituted acrylates and methacrylic anhydride as will be discussed later.

Cyclopolymerization of Allyl α-Substituted Acrylates

In connection with the preceding discussion,it should be recalled that the ceiling temperatures in the polymerization of methacrylates as typical α,α-disubstituted ethylenes are quite low compared with those in common monosubstituted ethylenes; for example, the ceiling temperature in methyl methacrylate is 164°c in bulk(26) and it decreases with dilution. Thus, it is expected that in the radical polymerization of allyl methacrylate(AMA) the cyclization and the five-membered ring formation are quite favored above the ceiling temperature of a ht propagation of methacrylyl group because the methacrylyl group in two polymerizable double bonds of AMA molecule is predominantly attacked first by the propagating radical and successively, the uncyclized methacrylyl radical produced undergoes intramolecular cyclization or intermolecular propagation(27): First, the intermolecular propagation of the uncyclized methacrylyl radical will be quite equilibrated under the polymerization conditions($\Delta G_{ht} \geq 0$) and thus, the intramolecular cyclization as a competitive reaction be relatively favored. Figure 3 shows a typical result obtained; noteworthily, the linear relationship could not be obtained for the Arrhenius plot of cyclization constant Kc which affords the difference in activation energy between intramolecular cyclization and intermolecular propagation, requesting the precise reexamination of the temperature dependency of Kc in the cyclopolymerization of acrylic and methacrylic anhydrides(28) in connection with the results discussed later. Furthermore, the high cyclopolymerization tendency of N-substituted dimethacrylamides(2-5) and diallyl ammonium salts(29) may fall on the same category because the corresponding monoene counterparts show no polymerizability.

Second, the steric factors will govern the addition modes, resulting in the favored five-membered ring formation because the addition selectivity of the uncyclized methacrylyl radical to allyl group leading to five- or six-membered ring formation based on the thermodynamic stability of the resulting radicals is very low as is discussed in the preceding section. That is, the formation of five-membered ring via intramolecular hh addition of uncyclized methacrylyl radical to allyl group is sterically favorable by the introduction of two successive methylene units, which separate adjacent cyclic units, into the crowded polymer chain in comparison with the six-membered ring formation via intramolecular ht addition which introduces only one methylene unit between cyclic units. These expectations were verified clearly for the polymerization of AMA at elevated temperatures(27); the five-membered ring formation, evident from the absorption characteristic of carbonyl group of γ-lactone at 1765 cm^{-1} in the IR spectrum shown in Figure 4, was quite enhanced with temperature and dilution. In this connection, the bulk polymerization of allyl atropate(AAT) was carried out at 80°C since the ceiling temperature in methyl atropate was very low, i.e., -8°C in bulk(30); the resulting poly-AAT(f_c = 0.46, $\overline{P}_n$ = 17) had no pendant atropyl group and its IR spectrum showed a clear absorption at 1765 cm^{-1}(Figure 4-B). These results may be ascribed to the release of steric crowding by the formation of five-membered ring; the details are now in progress. This may be also the case of the preferential five-membered ring formation in the cyclopolymerization of diallyl ammonium salts(8,9,11), in which the electrostatic repulsion between ionized groups is probably more significant than steric factors.

Cyclopolymerization of Acrylic Anhydride

The radical polymerization of acrylic anhydride(AA) as a typical 1,6-diene has been investigated in detail in terms of cyclopolymerization. Thus, in 1958 Crawshaw and Butler(31) and Jones(32) have independently demonstrated that AA could be polymerized in solution by an alternating intramolecular-intermolecular chain propagation or cyclopolymerization, leading to the formation of saturated, linear polymers consisting exclusively of six-membered ring anhydride structure. Later, Mercier and Smets(33)

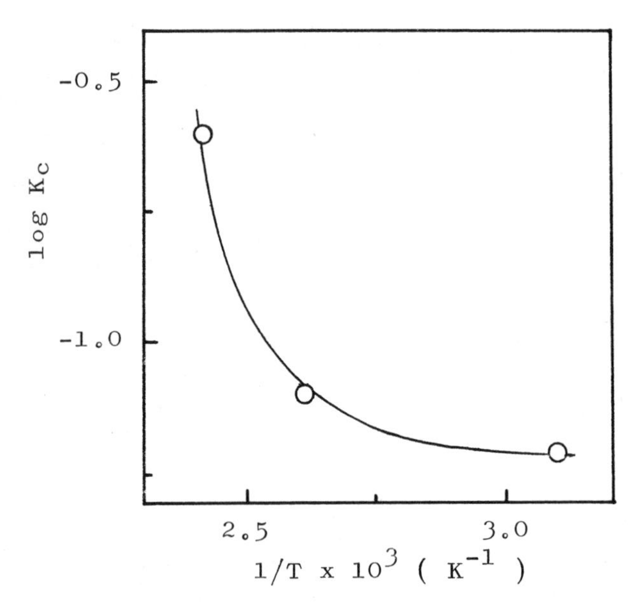

Figure 3. Arrhenius plot of cyclization constant K_c for the cyclopolymerization of allyl methacrylate.

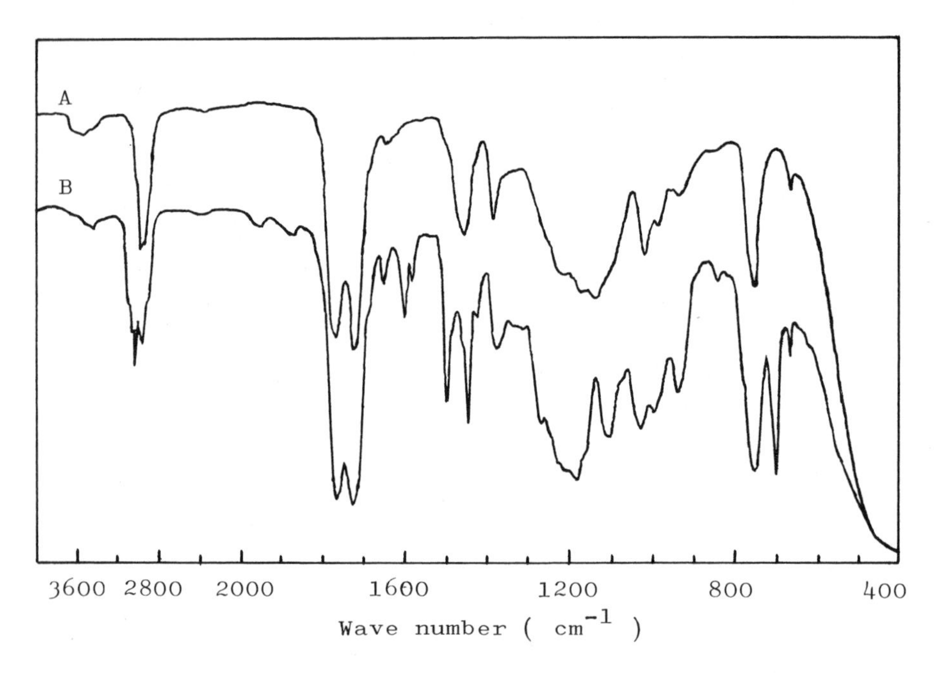

Figure 4. IR spectra of poly(methyl methacrylate) (A) and poly(allyl atropate) (B).

have examined in detail the structure of poly-AA by IR spectroscopy and chemical analysis; thus, two types of six-membered ring anhydrides, different in their stereochemical structure, were shown to be present although a relatively high content of five-membered ring anhydride units was formed only by polymerization at high temperatures, i.e., at 115°C in xylene as solvent.

On the other hand, we reexamined in detail the ring size of the cyclic structural units of poly-AA's by means of IR, ^{1}H-NMR, and ^{13}C-NMR spectroscopy; these analytical procedures were applied to the structural analysis of poly-AA, the poly(acrylic acid) derived from hydrolysis of the poly-AA, and the poly(methyl acrylate) obtained by subsequent esterification of the poly(acrylic acid) in comparison with the corresponding model polymers of five- or six-membered ring structure. Then, we investigated in detail the effects of polymerization conditions on the ring size of poly-AA's, i.e., on the intramolecular addition modes in the cyclopolymerization of AA since five- or six-membered ring anhydride structure can be formed via intramolecular hh or ht addition of the uncyclized radical to the internal double bond(22,23).

The radical polymerization of AA was conducted in various solvents at [M] = 0.167 mole/liter and at -40~130°C in order to determine the solvent and temperature effects on the five-membered ring formation. In this investigation, polymerization conditions were always set up as a highly diluted system in order to obtain highly cyclized polymers because the presence of large amount of uncyclized structural unit carrying unreacted pendant methacrylyl group would complicate the structural analysis of the resulting polymers and, moreover, leads easily to gelation. Thus, the content of the five-membered ring of poly-AA increased with an increase in the polarity of the solvent used and with the elevation of polymerization temperature(22). For example, for the polymerization in toluene at 0°C, cyclic poly-AA consisting predominantly of six-membered ring(99%) was obtained, whereas the content of five-membered ring of cyclic poly-AA was 90% in γ-butyrolactone at 130°C. Furthermore, the five-membered ring formation was highly favored with dilution, i.e., the content of five-membered ring reached 89% in γ-butyrolactone even at 50°C(23).

In summary, the five-membered ring formation, i.e., the occurrence of intramolecular hh addition was favored with the increase of the polarity of solvent, the elevation of polymerization temperature, and the decrease of monomer concentration; thus, the ring size of cyclic structure could be controlled freely by choosing appropriate polymerization conditions. In addition to these results the rate of polymerization and the molecular weight of the polymer were reduced under polymerization conditions favorable to five-membered ring formation(23).

In order to interpret these results we consider the reaction scheme for propagation based on polymerization equilibrium as follows:

$\sim CH_2-CH(CH_2)CH\cdot$ (six-membered anhydride ring, $O=C-O-C=O$) $\underset{}{\overset{ht\ (+M)}{\rightleftharpoons}}$ $\sim CH_2-CH\cdot$ $CH=CH_2$ ($O=C-O-C=O$)

ht ⇅

$\sim CH_2-CH\cdot$ $CH=CH_2$ ($O=C-O-C=O$)

hh ⇅

$\sim CH_2-CH$——$CH-CH_2\cdot$ (five-membered anhydride ring, $O=C-O-C=O$) $\overset{tt(+M)}{\rightleftharpoons}$ $\sim CH_2-CH\cdot$ $CH=CH_2$ ($O=C-O-C=O$)

Thus, if intermolecular ht addition is equilibrated to prevent appreciably ht propagation at a very low monomer concentration as will be expected from the great dependency of five-membered ring formation on monomer concentration, the probability of the occurrence of the intramolecular hh addition will increase correspondingly. Here it should be noted that the energetically unfavorable intramolecular hh addition would be considerably compensated by the subsequently occurring intermolecular tt addition, which is energetically favorable. In these circumstances, the intramolecular hh addition becomes the rate-determining step for the cyclopolymerization of AA; in this connection, the rate of polymerization and the molecular weight of the polymer would be reduced as observed above.

On the other hand, for a reason why the intermolecular ht addition may be equilibrated, the electrostatic repulsion between the highly polar anhydride units may be considered similar to the well known case in the radical copolymerization of vinyl monomers carrying carbonyl(34) or nitrile group(35) in which the penultimate effect is involved. That is, the polymer(D) obtained via intramolecular hh and intermolecular tt additions in which five-membered anhydride units are separated by two methylene units is sterically and/or electrostatically favorable compared with the polymer(E) formed via intramolecular ht and intermolecular ht additions in which six-membered anhydride units are separated by only one methylene unit.

$\sim CH_2-CH$——$CH-CH_2-CH_2-CH$——$CH-CH_2\sim$ (each ring: $O=C-O-C=O$) (D)

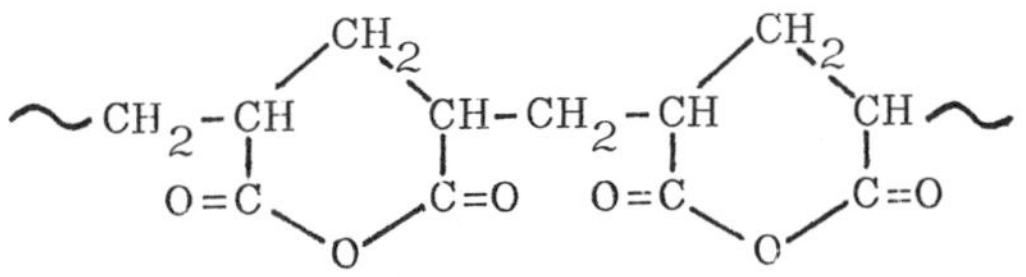

$$\sim CH_2-\underset{\displaystyle O=C\diagdown_{O}\diagup C=O}{CH \quad \overset{CH_2}{} \quad CH}-CH_2-\underset{\displaystyle O=C\diagdown_{O}\diagup C=O}{CH \quad \overset{CH_2}{} \quad CH}\sim \qquad (E)$$

Thus, the solvent effect on five-membered ring formation could be reasonably understood because in polar media the polarization of carbonyl groups would be higher and consequently, the electrostatic repulsion between cyclic anhydride units would be enhanced.

Here it should be remembered that the five-membered ring formation was favored with an increase in solvent polarity for the cyclopolymerization of dimethacrylamide, an explanation being, however, offered in correlation with conformation of imide group(36).

Cyclopolymerization of Methacrylic Anhydride

In the cyclopolymerization of methacrylic anhydride(MA) as a typical 1,6-diene, five- and six-membered ring anhydride structures can be formed, respectively corresponding to intramolecular hh and ht addition of uncyclized radical. The IR spectrum of poly-MA has been tentatively compared earlier with that of poly-MA prepared by dehydration of poly(methacrylic acid)(37); no five-membered cyclized anhydrides were detected, any absorption at higher frequency due to the strain of a five-membered ring being absent. In this connection, we reconfirmed the exclusive formation of six-membered ring(22).

However, the five-membered ring formation should be expected for the cyclopolymerization of MA at a highily diluted monomer concentration and a higher temperature, i.e., above the ceiling temperature in methacrylic isobutyric anhydride which corresponds to a monoene counterpart of MA as an extension of our preceding discussion. Figure 5 shows the comparison of IR spectrum of the cyclic poly-MA(I) obtained in benzene at 50°C, the ring size of the anhydride structure of the polymer which consisted exclusively of six-membered(22), with that of the cyclic poly-MA(II) obtained in benzonitrile at 160°C; the difference in both spectra of I and II was quite remarkable in the C=O and C-O-C absorption bands at 1900∼1700 and 1100∼900 cm^{-1}, respectively. Thus, the new absorptions at 1840 and 1775 cm^{-1} which appeared in II in addition to the spectrum of I are assigned to five-membered ring anhydride structure. This was further supported by the comparison of ^{13}C-NMR spectra; thus, the chemical shift of the carbonyl carbon of the anhydride unit in I was singly placed at 171.9 ppm, whereas in II another new peak appeared at 174.5 ppm. Other supporting evidence was obtained from the fact that poly-MA containing five-membered ring could be reasonably converted to poly(methyl methacrylate) carrying a considerable amount of hh linkage(38).

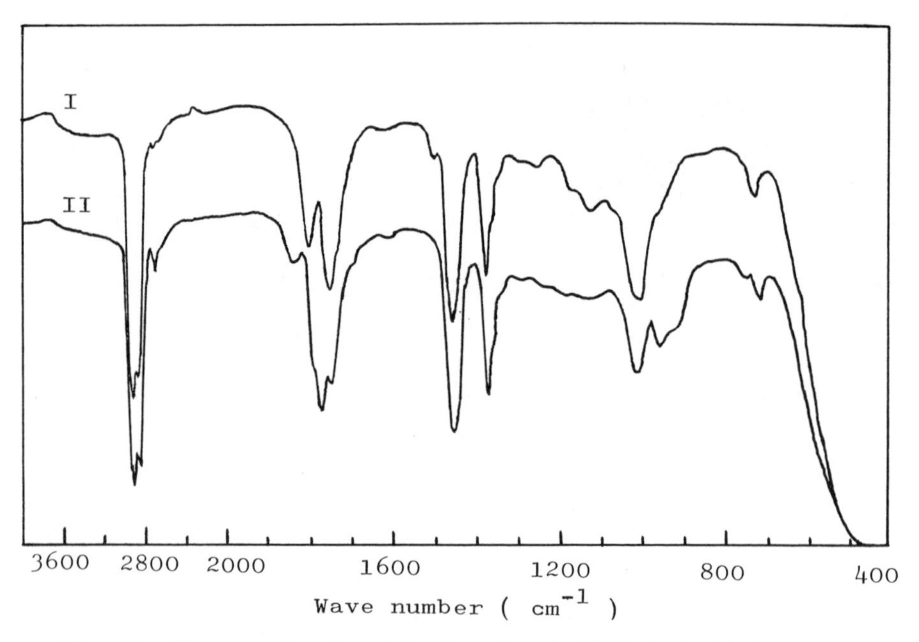

Figure 5. IR spectra of cyclic poly(methacrylic anhydride) obtained in benzene at 50°C (I); and in benzonitrile at 160°C (II).

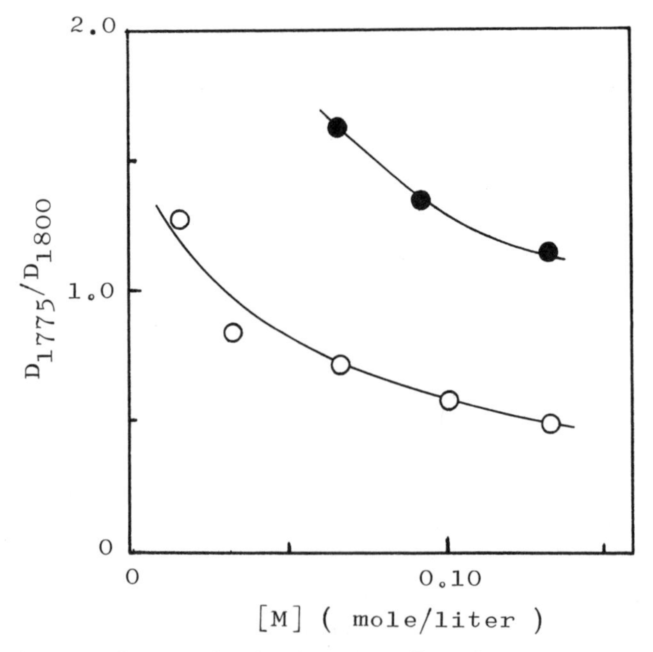

Figure 6. Dependence of the absorbance ratio D_{1775}/D_{1800} on monomer concentration in the cyclopolymerization of methacrylic anhydride in benzonitrile. Key: ○, 80°C; and ●, 140°C.

This novel five-membered ring formation at a higher temperature due to the reluctance of ht propagation was further enhanced by the elevation of temperature and the lowering of monomer concentration and, moreover, in polar media as is consistent with our expectation(39). For example, Figure 6 shows the dependence of five-membered ring formation on monomer concentration in the solution polymerization of MA in benzonitrile at 80 and 140°C, in which the absorbance ratio D_{1775}/D_{1800} estimated from the IR spectrum of cyclic poly-MA was employed as a measure of the content of five-membered anhydride structure units in the polymer since for the cyclopolymerization of AA a reasonable correlation was observed between the absorbance ratio D_{1780}/D_{1805} and the content of five-membered ring in the polymer(22). As the values of D_{1775}/D_{1800} were estimated at 0.60 and 1.44 for the cyclic poly-MA's containing zero and 53% of five-membered ring, respectively, all poly-MA's obtained at 140°C involved five-membered ring anhydride unit and its content increased with a decrease in monomer concentration. In addition, it should be noted that even at a rather lower temperature of 80°C the five-membered ring formation was observed in a very dilute solution, although the poly-MA obtained at a rather higher monomer concentration contained exclusively six-membered ring anhydride structures as was the case in the previous article (22) where the monomer concentration was 0.265 mole/liter, being low enough to be sure to obtain the almost perfectly cyclized poly-MA.

Literature Cited

1. Yonezawa, T.; Hayashi, K.; Nagata, C.; Okamura, S.; Fukui, K. J. Polym. Sci. 1954, 14, 312; 1956, 20, 537.
2. Götzen, F.; Schröder, G. Makromol. Chem. 1965, 88, 133.
3. Sokolova, T. A.; Rudkovskaya, G. D. J. Polym. Sci. 1967, C16, 1157.
4. Butler, G. B.; Myers, G. R. J. Macromol. Sci. Chem. 1971, 5, 135.
5. Kodaira, T.; Aoyama, A. J. Polym. Sci. Polym. Chem. Ed. 1974, 12, 897.
6. Solomon, D. H. J. Polym. Sci. Symp. 1975, 49, 175.
7. Hawthorne, D. G.; Johns, S. R.; Solomon, D. H.; Willings, R. I. J. Chem. Soc. Chem. Commun. 1975, 982.
8. Beckwith, A. L. J.; Ong, A. K.; Solomon, D. H. J. Macromol. Sci. Chem. 1975, 9, 115, 125.
9. Johns, S. R.; Willings, R. I.; Middleton, S.; Ong, A. K. J. Macromol. Sci. Chem. 1976, 10, 875.
10. Hawthorne, D. G.; Solomon, D. H. J. Macromol. Sci. Chem. 1976, 10, 923.
11. Lancaster, J. E.; Baccei, L.; Panzer, H. P. J. Polym. Sci. Polym. Lett. Ed. 1976, 14, 549.
12. Tsukino, M.; Kunitake, T. Polym. J. 1979, 11, 437.
13. Corfield, G. C.; Monks, H. H. J. Macromol. Sci. Chem. 1975, 9, 1113.

14. Matsoyan, S. G. J. Polym. Sci. 1961, 52, 189.
15. Minoura, Y.; Mitoh, M. J. Polym. Sci. 1965, A3, 2149.
16. Butler, G. B.; Crawshaw, A.; Miller, W. L. J. Am. Chem. Soc. 1958, 80, 3615.
17. Butler, G. B. J. Polym. Sci. Polym. Symp. 1978, 64, 71.
18. Matsumoto, A.; Iwanami, K.; Oiwa, M. J. Polym. Sci. Polym. Lett. Ed. 1980, 18, 211.
19. Matsumoto, A.; Iwanami, K.; Oiwa, M. J. Polym. Sci. Polym. Lett. Ed. 1980, 18, 307.
20. Matsumoto, A.; Iwanami, K.;Oiwa, M. J. Polym. Sci. Polym. Chem. Ed. 1981, 19, 213.
21. Matsumoto, A.; Iwanami, K.; Oiwa, M. J. Polym. Sci. Polym. Lett. Ed. in press.
22. Butler, G. B.; Matsumoto, A. J. Polym. Sci. Polym. Lett. Ed. 1981, 19, 167.
23. Matsumoto, A.; Kitamura, T.; Oiwa, M.; Butler, G. B. J. polym. Sci. Polym. Chem. Ed. in press.
24. Oiwa, M.; Matsumoto, A. "Progress in Polymer Science Japan, Vol. 7"; Kodansha Ltd.: Tokyo, 1974; p 107.
25. McCurdy, K. G.; Laidler, K. J. Can. J. Chem. 1964, 42, 818.
26. Sawada, H. "Thermodynamics of Polymerization"; Dekker: New York, 1976; p 31.
27. Ishido, H.; Matsumoto, A.; Oiwa, M. Polymer Preprints, Japan 1981, 30(5), 800.
28. Matsumoto, A.; Kitamura, T.; Oiwa, M. Polymer Preprints, Japan 1981, 30(1), 25.
29. Butler G. B.; Ingley, F. L. J. Am. Chem. Soc. 1951, 73, 895.
30. Hopff, H.; Lüssi, H.; Borla, L. Makromol. Chem. 1965, 81, 268.
31. Crawshaw, A.; Butler, G B. J. Am. Chem. Soc. 1958, 80, 5464.
32. Jones, J. F. J. Polym. Sci. 1958, 33, 15.
33. Mercier, J.; Smets, G. J. Polym. Sci. 1963, A1, 1491.
34. Barb, W. G. J. Polym. Sci. 1953, 11, 117.
35. Ham, G. E. J. Polym. Sci. 1954, 14, 87.
36. Boyarchuk, Y. M. Vysokomol. Soedin. 1969, A11, 2161.
37. Smets, G.; Hous, P.; Deval, N. J. Polym. Sci. 1964, A2, 4825.
38. Matsumoto, A.; Kawaguchi, N.; Oiwa, M. to be submitted for publication in J. Polym. Sci. Polym. Chem. Ed.
39. Matsumoto, A.; Kitamura, T.; Oiwa, M. presented at the 42nd Meeting of the Chemical Society of Japan,Sendai, September 1980.

RECEIVED January 23, 1982.

4

Styrene–Methyl Methacrylate Copolymers Derived from Styrene–Methacrylic Anhydride Copolymers

DANIEL L. NEUMANN and H. JAMES HARWOOD

University of Akron, Institute of Polymer Science, Akron, OH 44325

Mathematical procedures for calculating structural features of cyclocopolymers and of copolymers derived from them are proposed and are used in studies on the ^{1}H-NMR spectra of styrene-methyl methacrylate copolymers derived from styrene-(methacrylic anhydride) copolymers. Reactivity ratios and cyclization constants for styrene-methacrylic anhydride copolymerization were determined from structural features of the derived styrene-methyl methacrylate copolymers. The amount of uncyclized methacrylic anhydride units present in styrene-methacrylic anhydride copolymers having high styrene contents is considerably less than that predicted by these copolymerization parameters. The methoxy proton resonances of the derived copolymers are more intense in the highest field methoxy proton resonance area than would be expected if such resonance were due only to cosyndiotactic SMS triads. Possible explanations for these discrepancies are proposed.

Cyclocopolymerization of methacrylic anhydride or acrylic anhydride with vinyl monomers, followed by chemical conversion of the resulting copolymers into corresponding acrylic acid -, methyl acrylate -, methacrylic acid -, or methyl methacrylate (MMA) - containing copolymers, provides a route to copolymers having structures that differ from those of copolymers prepared by conventional copolymerization methods. For example, in copolymerizations of methacrylic anhydride with styrene, the great tendency of methacrylic anhydride to cyclopolymerize (1-4) causes the derived styrene (S)-MMA (M) copolymers to contain a high proportion of MM placements, high proportions of MMM, MMS and SMM triads and a very low proportion of SMS triads. Furthermore, a high proportion of the MM placements can probably be expected to have meso configurations (1) in the derived S/MMA copolymers, whereas they are mostly racemic in S/MMA copolymers prepared by free radical

0097-6156/82/0195-0043$06.00/0
© 1982 American Chemical Society

initiated polymerizations. These differences cause the derived copolymers to have considerably different chemical characteristics than those of conventional copolymers (5).

$$CH_2{=}CRCO{-}O{-}COCR{=}CH_2 \;+\; CH_2{=}\underset{X}{CH} \longrightarrow \left[{-}CH_2{-}\overset{R}{C}\langle CH_2 \rangle \overset{R}{C}{-}CH_2{-}\underset{X}{CH}{-}\right]_n \text{ (six-membered anhydride ring: } C(=O){-}O{-}C(=O)\text{)}$$

$$\downarrow H_2O$$

$$\left[{-}CH_2{-}\overset{R}{\underset{\underset{OCH_3}{O=C}}{C}}{-}CH_2{-}\overset{R}{\underset{\underset{OCH_3}{C=O}}{C}}{-}CH_2{-}\underset{X}{CH}{-}\right]_n \xleftarrow{CH_2N_2} \left[{-}CH_2{-}\overset{R}{\underset{COOH}{C}}{-}CH_2{-}\overset{R}{\underset{COOH}{C}}{-}CH_2{-}\underset{X}{CH}{-}\right]_n$$

Studies on the properties of the derived copolymers can thus be helpful in understanding structure-property relationships of copolymers. Furthermore, since nuclear magnetic resonance can be used to characterize the structures of S-MMA copolymers (7,8), structural studies on the derived copolymers can provide information about cyclocopolymerization processes. For these reasons, we have developed a procedure for converting styrene-methacrylic anhydride copolymers into styrene-methyl methacrylate copolymers, have developed methods for calculating structural aspects of the derived S-MMA copolymers and, have investigated the ^1H-NMR spectra of S/MMA copolymers derived from styrene-methacrylic anhydride copolymers.

Several theoretical treatments of cyclocopolymerization have been reported previously (8-11). These relate the compositions of cyclocopolymers to monomer feed concentrations and appropriate rate constant ratios. To our knowledge, procedures for calculating sequence distributions for either cyclocopolymers or for copolymers derived from them have not been developed previously. In this paper we show that procedures for calculating sequence distributions of terpolymers can be used for this purpose. Most previous studies on styrene-methacrylic anhydride copolymerizations (10,12,13) have shown that a high proportion of the methacrylic anhydride units are cyclized in these polymers. Cyclization constants were determined from monomer feed concentrations and the content of uncyclized methacrylic anhydride units in the copolymers. These studies invoked simplifying assumptions that enabled the conventional copolymer equation to be used in determinations of monomer reactivity ratios for this copolymerization system.

In the present study, this information was obtained primarily from structural studies on S/MMA copolymers derived from styrene-methacrylic anhydride copolymers. A computer oriented procedure was used to analyze the data and this made it unnecessary to employ simplifying assumptions.

Experimental

Methacrylic anhydride was prepared from methacroyl chloride and either sodium methacrylate (14) or methacrylic acid (15). The relative intensity of resonance observed at δ=11.6 ppm in the nmr spectrum of this monomer and the intensity of carbonyl absorption observed at 1700 cm^{-1}, relative to that observed at 1724 and 1782 cm^{-1} indicated that the monomer had less than 3 mole percent methacrylic acid. For the most part, monomer used for copolymerization experiments, b.p. 50° (1 mm), was prepared from sodium methacrylate. It contained less than one mole percent methacrylic acid.

Copolymerization Experiments. Copolymerizations were conducted in benzene solution under nitrogen at 40° and 60° using AIBN as initiator. The methacrylic anhydride concentration was 0.936M in all experiments. Styrene concentrations ranged from 0.029M to 7.85M. Conversions were below 10 wt. percent. The polymerization mixtures were poured into hexane to precipitate the copolymers. Working carefully to avoid exposure to moisture, these were reprecipitated from benzene into hexane and were then extracted exhaustively with hexane to remove residual monomer. This adhered strongly to the polymers and was detected by resonances at δ=5.66 and 6.03 ppm in nmr spectra of the copolymers in DMSO-d_6 solution (Figure 1). Uncyclized methacrylic anhydride units in the copolymers were detected by resonances at δ=5.72 and 6.15 ppm. This latter assignment was established by recording the spectrum of an 85/15 styrene/methacrylic acid copolymer that had been allowed to equilibrate with methacrylic anhydride (Figure 1).

The ^{1}H-NMR spectra of the copolymers in DMSO solution were recorded and the ratios (X) of uncyclized methacrylic anhydride units to styrene units in the copolymers were determined from the relative intensities of the resonances at δ=5.72, 6.15 and 6.5-7.5 ppm.

Hydrolysis of styrene-methacrylic anhydride copolymers. One gram samples of the copolymers were suspended in distilled water (150 ml.) and the mixtures were refluxed, with stirring, until solutions were obtained that were stable at room temperature. Copolymers with high styrene contents hydrolyzed slowly and required 108 hr. reaction times. These polymers formed soap-like solutions when completely hydrolyzed. The hydrolyzed polymers were isolated by freeze-drying and were examined by infrared spectroscopy to establish the completeness of hydrolysis.

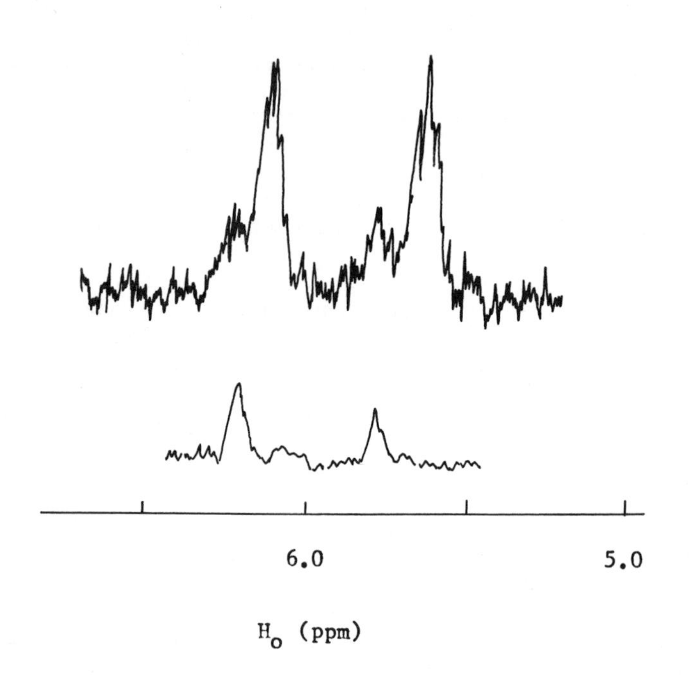

Figure 1. Olefinic proton resonances of styrene–methacrylic anhydride copolymers containing both uncyclized methacrylic anhydride units and absorbed monomer (top), and uncyclized methacrylic anhydride units but little absorbed monomer (bottom). This sample was prepared by reacting a styrene–methacrylic acid copolymer with methacrylic anhydride.

Methanolysis of styrene-methacrylic anhydride copolymers. One gram samples of the copolymers were suspended in a mixture containing methanol (50 ml), sodium methoxide (10 ml of a 20% solution in methanol), benzene (5 ml) and water (1-5 ml). The mixtures were refluxed with stirring for 6-24 hrs, until the copolymers dissolved. The reaction mixtures were then poured slowly into a large excess of acidified water, with rapid stirring, to precipitate the products. These were washed with water and dried. Infrared spectra of the copolymers were examined to ensure the complete decomposition of the anhydride groups.

Methylation of hydrolyzed or methanolized copolymers. One gram samples of hydrolyzed or methanolized copolymers were suspended or dissolved in dry benzene and the mixtures were treated with diazomethane until methylation was complete. This was indicated by the complete dissolution of the copolymers and by the presence of the yellow color of diazomethane in the reaction mixtures for 8 hrs. The reaction mixtures were poured into methanol to precipitate the products, which were reprecipitated twice from benzene solution into methanol before being dried in vacuo. Completely methylated copolymers were soluble in CCl_4, but copolymers containing very small amounts of acid units were not. This simple test was employed routinely to ensure the completeness of methylation. The ^{1}H-NMR spectra of the copolymers in C_6D_6 solution were examined for the presence of carboxylic acid proton resonances at δ=11.5 ppm as an additional criterion for the completeness of methylation.

^{1}H-NMR studies. Varian A-60 and HR-100 NMR spectrometers were used to measure the ^{1}H-NMR spectra of styrene-methacrylic anhydride copolymers in DMSO-d_6 solution at 90° and of the derived styrene-methyl methacrylate copolymers in CCl_4 and C_6D_6 solution at 75-80°C. Solvent resonances interfered with composition determinations in the case of styrene-methacrylic anhydride copolymers, but the ratio of uncyclized methacrylic anhydride to styrene units (X) could be measured from the relative intensities of resonances observed at δ=5.72 and 6.15 ppm (olefinic protons) and at 6.5-7.5 ppm (aromatic protons). The compositions of the derived styrene/-MMA copolymers were calculated from the proportion of aromatic proton resonance observed in the spectra of copolymers dissolved in CCl_4, as was described previously (6). Letting Y represent the ratio of styrene to MMA units in the derived copolymers, the compositions of the parent styrene-methacrylic anhydride copolymers were calculated as follows:

Mole fraction of styrene units = $2Y/(2Y + XY + 1) = F_S$

Mole fraction of uncyclized methacrylic anhydride units
$= 2XY/(2Y + XY + 1) = F_u$

Mole fraction of cyclized methacrylic anhydride units
$= (1-XY)/C2Y + XY + 1) = F_c$

Table I lists compositions determined this way for styrene-methacrylic anhydride copolymers prepared at 40°.

TABLE I

Compositions of Styrene-Methacrylic Anhydride Copolymers Prepared at 40°

Monomer Comp.		Observed Copolymer Comp.			Calculated Copolymer Comp.[c]		
$[S_f]^a$	$[S]^b$	F_s^a	F_u^a	F_c^a	F_s^a	F_u^a	F_c^a
0.855	5.54	0.61	0.02	0.36	0.61	0.14	0.21
0.776	3.23	0.61	0.02	0.37	0.56	0.15	0.29
0.733	2.57	0.61	0.03	0.37	0.55	0.06	0.39
0.597	1.38	0.55	0.03	0.41	0.51	0.07	0.42
0.548	1.13	0.53	0.04	0.44	0.49	0.08	0.43
0.388	0.596	0.49	0.04	0.45	0.45	0.06	0.49
0.270	0.346	0.42	0.03	0.55	0.41	0.04	0.55
0.207	0.244	0.39	0.07	0.54	0.37	0.04	0.59
0.141	0.159	0.31	0.06	0.62	0.32	0.04	0.64
0.029	0.028	0.13	0.06	0.81	0.12	0.03	0.85

a - mole fraction, b- molar concentration, [M]=0.936 moles/liter, c - r_1 = 0.19, r_2 = 0.11, r_3 = 0.19, r_c = 37.4 moles/liter, r_c'= 7.1 moles/liter.

As in the case of S/MMA copolymers prepared by direct copolymerization, the methoxy proton resonances of the derived S/MMA copolymers occurred in three general areas (A - δ = 3.2-3.8 ppm, B - δ = 2.7-3.2 ppm and C - δ = 2.2-2.7 ppm). Figure 2 compares the aliphatic proton resonance patterns of styrene-MMA copolymers prepared by direct polymerization and by modification of styrene-methacrylic anhydride copolymers. The methoxy and α-methyl proton resonance patterns of the copolymers differ considerably even though they have similar compositions. The proportions of MeO resonance occurring in the A- and B- areas (F_A, F_B) were calculated by dividing the A- and B- resonance areas by the total MeO resonance area expected, based on the compositons of the copolymers (i.e., MeO resonance area expected = 3/8 x % MMA/100). The proportion of MeO resonance occurring in the C- area was calculated by subtracting these quantities from one ($F_C = 1 - F_A - F_B$).

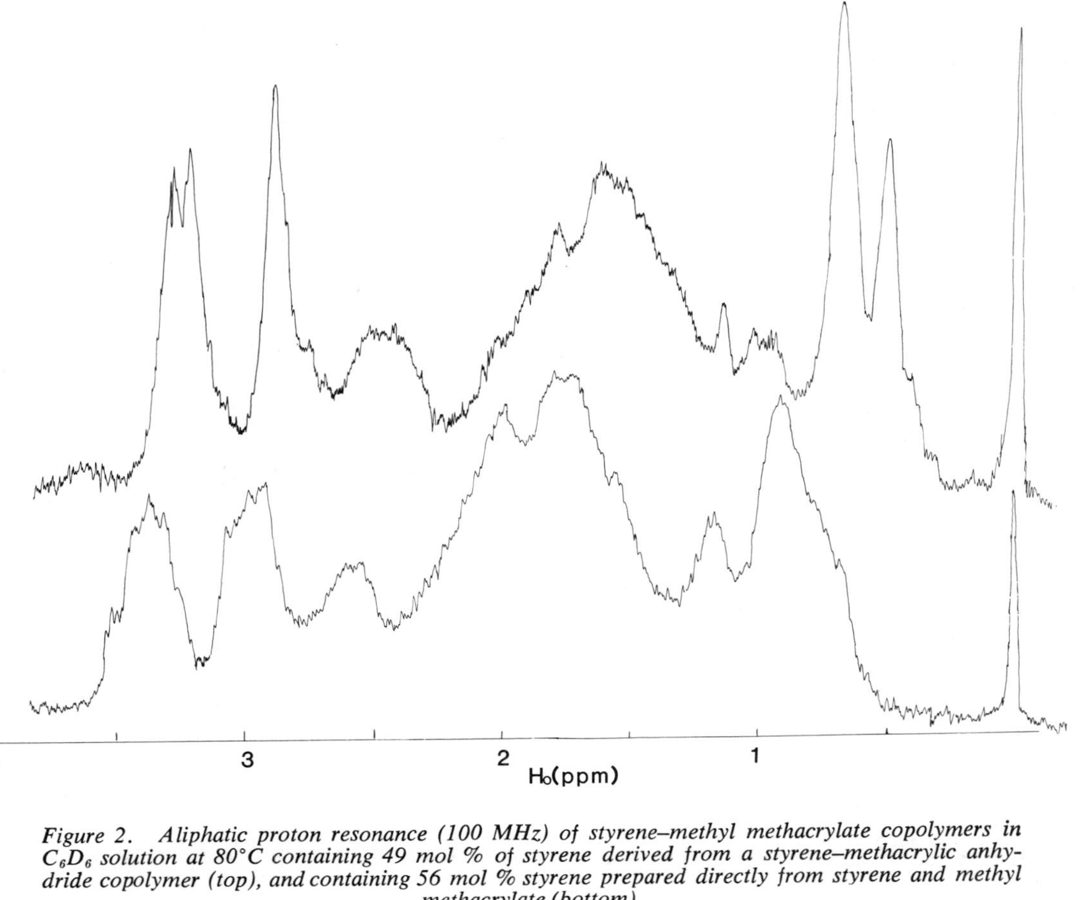

Figure 2. Aliphatic proton resonance (100 MHz) of styrene–methyl methacrylate copolymers in C_6D_6 solution at 80°C containing 49 mol % of styrene derived from a styrene–methacrylic anhydride copolymer (top), and containing 56 mol % styrene prepared directly from styrene and methyl methacrylate (bottom).

Copolymers prepared at 40° and 60° from a given monomer mixture had equivalent F_S, F_A, F_B and F_C values, within the limits of experimental reproducibility, which was approximately ± 0.03.

Theoretical Aspects

In the copolymerization of styrene (S) with methacrylic anhydride (Anh), three structures are incorporated into the resulting copolymers: styrene units (S), uncyclized methacrylic anhydride units (U), and cyclized methacrylic anhydride units (C). For simplicity, possible reactions involving U units will be ignored. This is partially justified by the fact that the copolymerizations were only allowed to proceed to low conversions. In accordance with previous treatments of this copolymerization process, the following reactions and reactivity ratios are believed to be important for determining copolymer compositions and structures. S· refers to a styryl radical, A· refers to an uncyclized methacrylic anhydride radical and B· refers to a completely cyclized methacrylic anhydride radical.

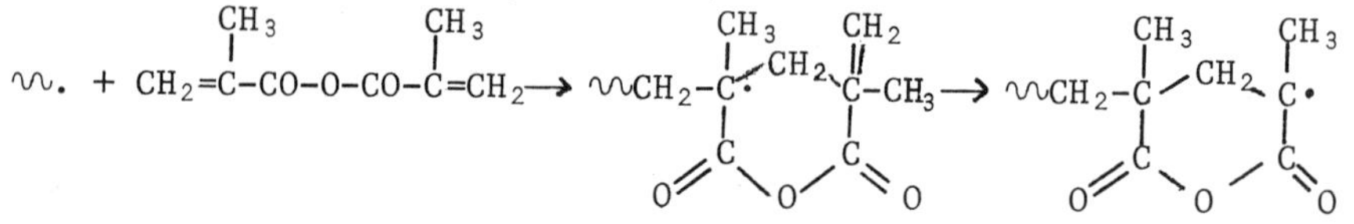

$$\sim\cdot + CH_2{=}C(CH_3){-}CO{-}O{-}CO{-}C(CH_3){=}CH_2 \longrightarrow \text{A}\cdot \longrightarrow \text{B}\cdot$$

Anh. A· B·

$$\sim S\cdot + S \xrightarrow{k_{SS}} \sim S\cdot$$

$$\sim S\cdot + Anh \xrightarrow{k_{SA}} \sim A\cdot$$

$$\sim A\cdot \xrightarrow{k_c} \sim B\cdot$$

$$\sim A\cdot + S \xrightarrow{k_{AS}} \sim S\cdot$$

$$\sim A\cdot + Anh \xrightarrow{k_{AA}} \sim A\cdot$$

$$\sim B\cdot + S \xrightarrow{k_{BS}} \sim S\cdot$$

$$\sim B\cdot + Anh \xrightarrow{k_{BA}} \sim A\cdot$$

$$r_1 = k_{AA}/k_{AS}$$

$$r_2 = k_{SS}/k_{SA}$$

$$r_c = k_c/k_{AA}$$

$$r_c' = k_c/k_{AS}$$

$$r_3 = k_{BA}/k_{BS}$$

The copolymerization process can thus be considered to be a Markoffian process involving three states, in which the following transition probabilities prevail.

$$P(A/A) = (Anh)/((Anh)+r_c + (S)/r_1)$$

$$P(B/A) = r_c/((Anh + r_c + (S)/r_1)$$

$$P(S/A) = 1 - P(A/A) - P(B/A)$$

$$P(S/B) = (S)/((S) + r_3(Anh))$$

$$P(B/B = 0$$

$$P(A/B) = 1 - P(S/B)$$

$$P(A/S) = (Anh)/((Anh) + r_3(S))$$

$$P(B/S) = 0$$

$$P(S/S) = 1 - P(A/S)$$

Following methods reviewed previously (16), these transition probabilities can be used to calculate the compositions of styrene-methacrylic anhydride copolymers. Thus, the method of Price (17) yields the following results, where P(S), P(A) and P(B) represent the relative concentrations of the following groups.

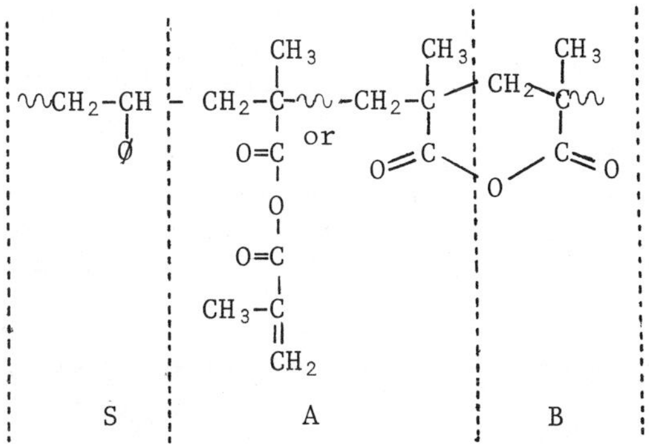

S: ∿CH₂–CH(Ø)

A: CH₂–C(CH₃)(C(=O)–O–C(=O)–C(CH₃)=CH₂)∿ or

B: –CH₂–C(CH₃)–CH₂–C(CH₃)∿ with C=O, O, C=O (cyclic anhydride)

$$P(S) = \frac{(P(B/A)P(S/B) + P(S/A))}{1 + P(A/B) + P(S/A) - P(S/S) - P(S/S)P(B/A) + P(B/A)P(S/B)}$$

$$P(A) = \frac{1 - P(S)}{1 + P(B/A)}$$

$$P(B) = P(A)P(B/A)$$

Knowing these relative concentrations and transition probabilities, probabilities for various sequences of S, A and B entities present in styrene-methacrylic anhydride copolymers can be calculated in the usual way, viz.,

$$P(SAB) = P(S)P(A/S)P(B/A)$$

$$P(AABS) = P(A)P(A/A)P(B/A)P(S/B)$$

etc.

The results of such calculations need to be "translated" to obtain the relative concentrations of styrene units, F_s, cyclized methacrylic anhydride units, F_c, and uncyclized methacrylic anhydride units, F_u, present in styrene-methacrylic anhydride copolymers. To do this, it must be realized that each A-entity corresponds to a methacrylic anhydride unit and that each A-entity which is followed by a B-entity is cyclized. Such considerations lead to the following results.

$$F_s = P(S)/(P(S) + P(A))$$

$$F_c = P(A)\cdot P(B/A)/CP(S) + P(S))$$

$$F_u = 1 - F_s - F_c$$

Sequence probabilities may also be calculated, *viz.*,

$$P(SUC) = \frac{P(S)\cdot P(A/S)\cdot P(A/A)\cdot P(B/A)}{P(S) + P(A)}$$

$$P(SUU) = \frac{P(S)\cdot P(A/S)P(A/A)(1 - P(B/A))}{P(S) + P(A)}$$

Of special interest to the present study is the calculation of the compositions and sequence distributions of styrene-MMA copolymers derived from styrene-methacrylic anhydride copolymers. A "translation" technique based on the fact that A- and B- entities *both* lead to MMA units, leads to the following results:

$$F_S = P(S)/(P(A) + P(B) + P(S)) = P(S)$$

$$F_M = 1 - F_S = P(A) + P(B)$$

Since A- and B- entities can both become MMA-units in the derived copolymers, the calculation of sequence probabilities is more difficult than for the case of styrene-methacrylic anhydride copolymers; there are often many combinations of A-, B- and S- entities that can yield a given sequence of styrene and MMA units. For example, an MMM triad can result from the following sequences: AAA, (A)BAA, (A)BAB, ABA and AAB. Thus,

$$P(MMM) = P(A)\cdot P(A/A)^2 + P(B)P(A/A)^2 + P(B)P(A/B)P(B/A) + P(A)P(B/A)P(A/B) + P(A)P(A/A)P(B/A)$$

Similarly,

$$P(MMS) = P(B)P(A/B)P(S/B) + P(A)P(A/A)P(S/A) + P(A)P(B/A)P(S/B)$$

$$P(SMM) = P(S)P(A/S)P(A/A) + P(S)P(A/S)P(B/A)$$

$$P(SMS) = P(S)P(A/S)P(S/A)$$

In a similar manner, probabilities of other sequences can also be calculated. For analyses of the methoxy proton resonance patterns of the derived S/MMA copolymers, M-centered triad fractions are needed. These can be calculated in the usual way, viz.,

$$f_{MMM} = \frac{P(MMM)}{P(M)} ; \quad f_{(MMS+SMM)} = \frac{P(MMS)+P(SMM)}{P(M)} ; \quad f_{SMS} = \frac{P(SMS)}{P(M)}$$

On the basis of the above considerations, two computer programs have been written to calculate MMA-centered triad and pentad distributions for S/MMA copolymers derived from styrene-methacrylic anhydride copolymers. One program, written specifically for this application, is programmed with the results of the translations described above. The second program is a combination of one of our programs for calculating structural features of terpolymers (18), and a subroutine to perform the necessary translations. Both programs yield the same results, but the second one enables calculations to be performed for polymers prepared in high conversions.

Results and Discussion

The copolymerization of methacrylic anhydride with styrene has been investigated by several groups. With the exception of Smets, et al. (12) it has been reported that the methacrylic anhydride units in the polymers are almost completely cyclized. This is also in accord with our experience. It proved difficult to separate unpolymerized methacrylic anhydride from the copolymers, and considerable effort was made to remove unreacted monomer from the polymers. ^{1}H-NMR spectroscopy proved to be an effective method for distinguishing uncyclized methacrylic anhydride units present on the polymers from adsorbed monomer (see Experimental). In many samples, no resonances due to uncyclized anhydride units were detected. This was particularly true of copolymers with high styrene contents. Table I lists the compositions determined for styrene-methacrylic anhydride copolymers prepared at 40°. The styrene contents of the copolymers are in good agreement with those reported by Smets, et al., for copolymers prepared from comparable monomer ratios, but where the anhydride concentration was $\sim$2M. However, Smets, et al. report that 30-50 percent of the anhydride units were uncyclized in their copolymers and it appears that the extent of cyclization is better than 90 percent, generally about

95 percent, in our samples. It is doubtful that the use of a lower anhydride concentration (∿1M) in our preparations is responsible for this difference.

Roovers and Smets have derived the following equation, which relates the fraction of anhydride units cyclized (f_c) to the cyclization constants r_c and r_c' and to monomer concentrations. Since

$$\frac{1}{f_c} = 1 + \frac{(M)}{r_c} + \frac{(S)}{r_c'}$$

[M] was 0.936M in our work, a plot of $\frac{1}{f_c}$ vs (S) should yield a straight line having $1/r_c'$ as slope and $1 + 0.936/r_c$ as intercept. Construction of such a plot from the data in Table I yields the surprising result that $r_c' > 100M > (r_c \sim 17M)$. This result is probably in conflict with the kinetic model (9) described earlier in this paper, since $r_c' = r_c \cdot r_1$ and r_1 must clearly be less than one. Perhaps a penultimate effect makes it easier for an anhydride radical to cyclize if it is attached to a styrene unit than if it is attached to a cyclized anhydride unit. This would cause the extent of cyclization to be higher in copolymers with high styrene contents than in those with low styrene contents. This point merits additional study. For the present, we will utilize an r_c value of ∿40M for calculating structural features of S/MMA copolymers derived from styrene-methacrylic anhydride copolymers.

Smets, et al. (12) noted that the compositions of styrene-methacrylic anhydride copolymers prepared from given styrene-methacrylic anhydride mixtures were independent of the amount of solvent present in the system and concluded that r_1 and r_3 in the kinetic scheme outlined above must be equal. This conclusion, which was also accepted by Baines and Bevington (13), enabled reactivity ratios for this copolymerization system to be calculated by use of the standard copolymer equation. Unfortunately, this is not a valid conclusion; the results in Table II show that for conditions similar to those employed in the previous work, the styrene contents of the copolymers are imperceptibly affected by dilution of the system, even when r_3 is five times greater or less than r_1.

Due to the difficulty of working with styrene-methacrylic anhydride copolymers, we have elected to determine reactivity ratios and cyclization constants from the compositions and structures of styrene-MMA copolymers derived from these copolymers. As is discussed in the experimental section it is possible to measure the styrene contents and the proportions of methoxy proton resonance occurring in three different areas (designated A, B and C) from the ^{1}H-NMR spectra of S/MMA copolymers. The proportions of methoxy proton resonance observed in the A (F_A), B (F_B) and C (F_C) areas obey the following relationships in conventional styrene-MMA copolymers (6,7).

$$F_A = P(MMM) + 0.5\,P(MMS + SMM) + 0.25P(SMS)$$
$$F_B = 0.5\,P(MMS + SMM) + 0.50P(SMS)$$
$$F_C = 0.25P(SMS)$$

TABLE II

Effect of Dilution on the Compositions of Styrene-Methacrylic Anhydride Copolymers when r_1=0.26, r_2 = 0.12, r_c = 5 or 45 moles/ℓ. and r_3 is varied

r_c (moles/ℓ.)	Monomer Concentrations (moles/ℓ.)		Calculated Mole Fraction of Styrene in Copolymer				
	(S)	(Anh)	r_3 = 4.0	1.0	0.26	0.10	0.05
5	2.21	2.21	0.316	0.404	0.471	0.493	0.501
5	1.77	1.77	0.301	0.399	0.471	0.495	0.503
5	1.41	1.41	0.285	0.393	0.471	0.497	0.505
5	0.90	0.90	0.268	0.383	0.471	0.499	0.508
45	2.21	2.21	0.208	0.367	0.471	0.503	0.514
45	1.41	1.41	0.199	0.364	0.471	0.503	0.514
45	0.90	0.90	0.194	0.362	0.471	0.504	0.515

The coefficients in the above equations are related to the probability (σMS = 0.5) that M-S or S-M placements in the copolymers have meso (erythro) configurations. It was assumed that this probability would also prevail in the case of S/MMA copolymers derived from styrene-methacrylic anhydride copolymers. The optimization program STEPIT (19) was then used in conjunction with the programs described earlier in this paper to find the set of reactivity ratios and cyclization constants that provided the best agreement between observed F_S, F_A and F_B values, and those calculated from monomer feed concentrations. In a preliminary optimization, r_1 and r_3 were maintained equal. This led to the results shown in Table III. The optimization was then repeated with r_2, r_c and r_c' maintained at the values shown in Table III and with r_1 and r_3 being evaluated independently. Using ten different starting points, STEPIT converged on r_1 = 0.184 and r_3 = 0.189 in all cases. It seems that Smets' assumption that $r_1 \cong r_3$ is correct.

TABLE III

Reactivity Ratios and Cyclization Constants Estimated for Styrene-Methacrylic Anhydride Copolymerization

	This work[a]	Smets (12)	Bevington (13)
$r_1 = r_3$	0.19	0.26	0.33
r_2	0.11	0.12	0.10
r_c	37 moles/ℓ.	45 moles/ℓ.	-
r_c'	7.1 moles/ℓ.	5 moles/ℓ.	-
Polymer Temp.	60°	36.6°	60°
Solvent	Benzene	Cyclohexanone	Benzene

a) These constants have not been corrected for the fact that two double bonds are present in methacrylic anhydride.

Figure 3 and Table I compare the compositions of styrene-methacrylic anhydride and styrene/MMA copolymers prepared in this study with those calculated from monomer feed concentrations and the reactivity ratios and cyclization constants determined in this study. It should be remembered that the F_u contents of the parent copolymers may be high since some of the copolymers did not seem to contain uncyclized units. It can be seen that the calculated compositions are in good agreement with observed compositions for polymers with low styrene contents, but that the agreement is not very good for polymers with high styrene contents. In particular, calculated F_u contents are much higher than observed ones. This suggests that it may be necessary to adopt a modified kinetic model to bring everything into agreement, as is mentioned earlier in this paper.

In analyzing methoxy proton resonance patterns of S/MMA copolymers we have found it convenient to construct HR plots (6). These are plots of the left-hand parts (L.H.P.) of the following equations vs P(SMM+MMS)/P(SMS).

$$\frac{F_A - P(MMM)}{P(SMS)} = \frac{X\ P(SMM+MMS) + Y}{P(SMS)}$$

$$\frac{F_B}{P(SMS)} = \frac{X'\ P(SMM+MMS) + Y'}{P(SMS)}$$

$$\frac{F_C}{P(SMS)} = \frac{X''\ P(SMM+MMS) + Y''}{P(SMS)}$$

These equations are based on the assumption that the methoxy proton resonance associated with MMM triads occurs entirely in the A-area. The slopes and intercepts of HR plots indicate the distribution of (MMS+SMM)- and (SMS)- methoxy proton resonances among the A, B and C areas. The reactivity ratios determined in this work are based on the assumption that X = X' = 0.5, X" = 0., Y = Y" = 0.25 and Y' = 0.5. It was of interest to construct HR plots from the results obtained in this study to see how well this assumption holds up and access the quality of the fit obtained by use of the STEPIT optimization routine. As can be seen in Figure 4, the HR plots are linear. The coefficients of these plots indicate that the methoxy proton resonances can be described by the following equations.

$$F_A = P(MMM) + 0.48\ P(SMM+MMS) + 0.25\ P(SMS)$$

$$F_B = 0.44\ P(SMM+MMS) + 0.50\ P(SMS)$$

$$F_C = 0.08\ P(SMM+MMS) + 0.25\ P(SMS)$$

These equations are similar to those assumed for the reactivity ratio determination. In contrast to what has been observed for conventional styrene-MMA copolymers, however, these equations indicate that a substantial proportion of the (SMM+MMS)-type resonance appears to occur in the C-area. The proportion of methoxy resonance observed in the C-area, in fact, exceeds P(SMS) by a substantial amount for many of the copolymers. This can be due to the assumption of an inadequate model for the copolymerization reaction, to the use of incorrect reactivity ratios and cyclization constants for the calculations or to an inadequate understanding of the methoxy proton resonance patterns of S/MMA copolymers. It is possible that intramolecular reactions between propagating radicals and uncyclized methacrylic anhydride units present on propagating chains result in the formation of macrocycles. Failure to account for the formation of macrocycles would result in overestimation of r_c and r_c' and in underestimation of the proportions of MMA units in SMS triads in the derived S/MMA copolymers. This might account for the results obtained. An alternate possibility is that a high proportion (>50%) of the M-M placements in the copolymers studied in this work can be expected to have meso placements (1,2), whereas only a small proportion of such placements (∿20%) are meso in conventional S/MMA copolymers. Studies with molecular models (20) have indicated that the methoxy protons on MMA units centered in structures such as the following can experience appreciable shielding by next nearest styrene units.

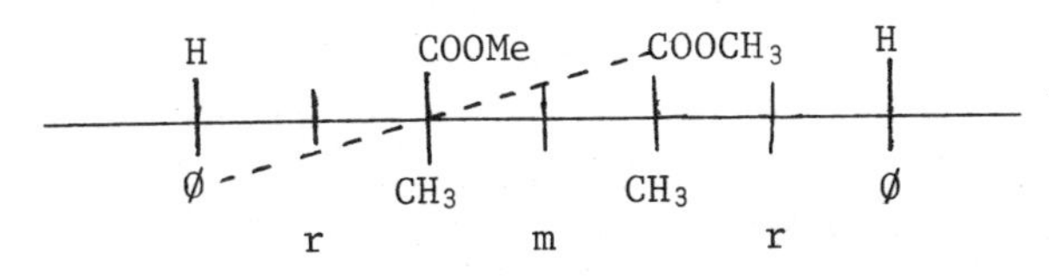

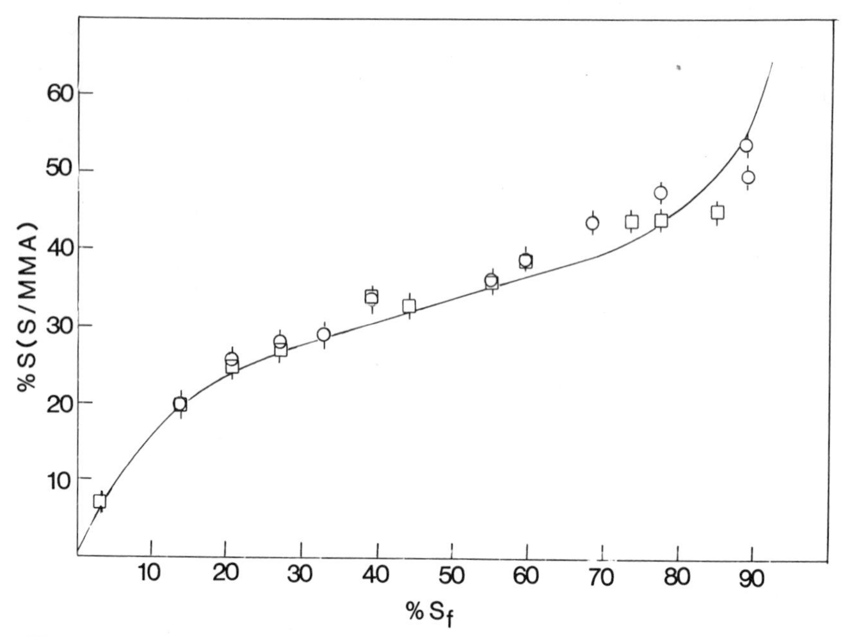

Figure 3. Mole fraction of styrene units in styrene–methyl methacrylate copolymers %S (S/MMA) derived from styrene–methacrylic anhydride copolymers, as a function of the molar percentage of styrene (% S_f) in the monomer mixture used to prepare the styrene–methacrylic anhydride copolymers. The solid line was calculated using $r_1 = r_3 = 0.19$, $r_2 = 0.11$, $r_c = 37$ and $r_c' = 7.7$. Key: ○, 60°C polymerization; and □, 40°C polymerization.

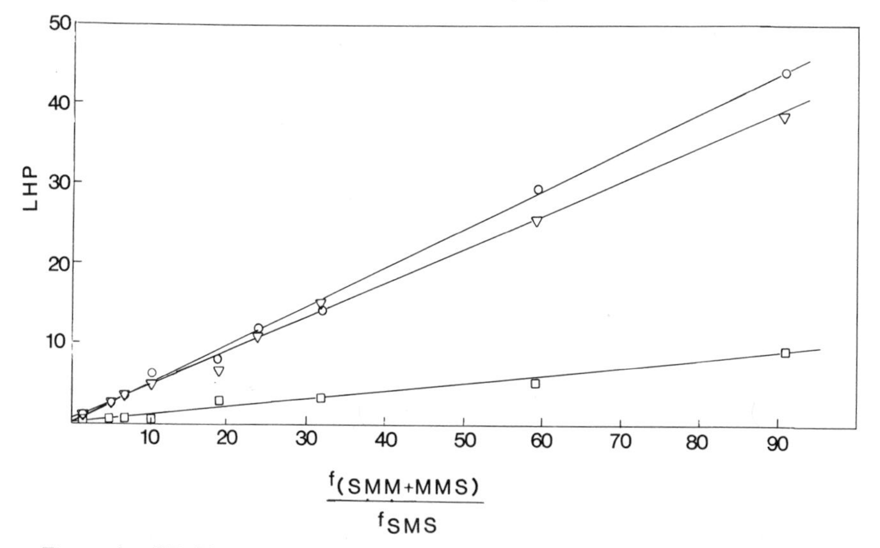

Figure 4. HR-Plots of methoxy proton resonance data for styrene–methyl methacrylate copolymers derived from styrene–methacrylic anhydride copolymers prepared at 60°C. Key: ○, (F_A-P(MMM))/P(SMS) vs. P(SMM + MMS)/P(SMS); ▽, F_B/P(CMS) vs. P(SMM + MMS)/P(SMS); □, F_C/P(SMS) vs. P(SMM + MMS)P(SMS).

Should the resonance of such protons occur in the C-area, the apparent observation of (SMM+MMS) resonance in this area in the present study, might be explained. Thus, the concentration of such structures can be expected to be higher in S/MMA copolymers derived from styrene-methacrylic anhydride copolymers than in conventional S/MMA copolymers.

We hope to do additional work on this problem using 300 MHz-^{1}H-NMR and ^{13}C-NMR spectroscopies.

Conclusion

Procedures for converting styrene-methacrylic anhydride copolymers into styrene-methacrylic acid and styrene-methyl methacrylate copolymers have been developed. Mathematical procedures and appropriate computer programs have been developed for calculating the compositions and structural features of cyclocopolymers and of products derived from them. By fitting the compositions and methoxy proton resonance patterns of styrene-MMA copolymers derived from styrene-methacrylic anhydride copolymers to compositions and methoxy proton resonance patterns calculated using these computation methods, reactivity ratios and cyclization constants for the styrene-methacrylic anhydride copolymerization system were derived. The methoxy proton resonance patterns of the derived copolymers indicate that a higher proportion of this resonance occurs in the highest field methoxy proton resonance area (C) than would be expected based on our present understanding of these patterns. This point merits further investigation.

Acknowledgement

The authors are indebted to the National Science Foundation for partial support of this study, to Dr. Ronald E. Bockrath for helpful discussions, and to Mr. Everett R. Santee for experimental assistance.

Literature Cited

1. Miller, W.L.; Brey, Jr., W.S.; Butler, G.B. J. Polym. Sci. 1961, 54, 329.
2. Hwa, J.C.H. J. Polym. Sci. 1962, 60, S12.
3. Smets, G.; Hous, P.; Deval, N. J. Polym. Sci., Part A 1964, 2, 4825.
4. Gibbs, W.E.; Murray, J.T. J. Polym. Sci. 1962, 58, 1211.
5. Thall, E., Ph.D. Dissertation, University of Akron, 1972.
6. Harwood, H.J.; Ritchey, W.M. J. Polym. Sci., Part B 1965, 3, 419.
7. Ito, K.; Yamashita, Y. J. Polym. Sci., Part B 1965, 3, 625.
8. Gibbs, W.E.; McHenry, A.J. J. Polym. Sci., Part A 1964, 2, 5277.

9. Roovers, J.; Smets, G. Makromol. Chem. 1963, 60, 89.
10. Hwa, J.C.H.; Miller, L. J. Polym. Sci. 1961, 55, 197.
11. Aso, C. J. Polym. Sci. 1959, 39, 475.
12. Smets, G.; Deval, N; Hous, P. J. Polym. Sci., Part A 1964, 2, 4835.
13. Baines, F.C.; Bevington, J.C. Polymer 1970, 11, 647.
14. British Patent 538,319 (July 29, 1941); Chem. Abstr., 36, 1619 (1942).
15. Yakubovich, Y.; Muler, J.; Bayelevski, D. J. Gen. Chem. (U.S.S.R) 1960, 30, 1299.
16. Harwood, H.J.; Kodaira, Y.; Neumann, D.L. in "Computers in Polymer Science", ed. by Mattson, J.S., Mark, Jr., H.B., and MacDonald, Jr., H.C., M. Dekker, Inc., New York, N.Y., 1977, Chapter 2.
17. Price, F.P. J. Chem. Phys. 1962 36, 209.
18. Kodaira, Y.; Harwood, H.J. Amer. Chem. Soc., Div. Org. Coat. Plast. Chem., Preprints, 1981, 45, 305.
19. Chandler, J. P. Subroutine STEPIT, Program Number 66.1, Quantum Chemistry Program Exchange, Indiana University, Bloomington, Indiana
20. Bockrath, R.E. Ph.D. Dissertation, University of Akron, 1971.

RECEIVED February 1, 1982.

5

N,N-Dimethyl-3,4-Dimethylenepyrrolidinium Bromide Polymer

RAPHAEL M. OTTENBRITE

Virginia Commonwealth University, Department of Chemistry, Richmond, VA 23284

The formation of cyclic structures during the polymerization processes had led to many interesting and useful polymers. Aqueous solution of N,N-dimethyl-3,4-dimethylenepyrrolidinium bromide were found to polymerize spontaneously at 60°C. The polymerization occurs by a 1,4-addition and the polymer product has properties very similar to that of poly(N,N-dimethyldiallylammonium chloride). High molecular weight polymers were obtained with monomer concentrations of 40-50%. Higher monomer concentrations yielded insoluble gels.

Polyelectrolytes are one of the most interesting and utilizable polymer materials presently under development. These water soluble polymers may be classified into three main groups: nonionic polyelectrolytes, such as polyols, polyethers and polyamides; anionic polyelectrolytes, such as carboxylates, sulfonates, and phosphates; and cationic polyelectrolytes, such as ammonium, quaternary amines, sulfoniums and phosphoniums (1). These materials have wide and varied industrial applications. The cationic polyelectrolytes, in particular, are versatile and useful materials with applications being found in water treatment, waste treatment, preparation of papers, textiles, oil drilling, plastics and biomedical areas (2).

One of the prime environmental objectives is the removal of suspended contaminants from water and waste streams. Water turbidity in nature is the result of colloidal clay dispersion and the color is from decayed wood and leaves (tannins and lignins) and organic soil matter. In addition to these contaminants, there are viruses, algae, bacteria, metal oxides, oils and other pollutants. In recent years, synthetic organic polyelectrolyes, in particular the cationic polymers, have been used very effectively in water treatment (3). These polyelectrolytes are high

0097-6156/82/0195-0061$06.00/0
© 1982 American Chemical Society

performance materials and can be used in very low dosages. The amount of polymer used is only 2 or 3 percent of the amount of the inorganic materials necessary for coagulation and flocculation. This results in reduced bulk of treatment chemicals, increased capacity of the plant and less bulk for disposal after treatment. Another benefit is better sludge conditioning and caking which results in easier dewatering and the filter runs cleaner, longer and more efficiently than with inorganic materials. Polyelectrolytes are much more effective in flocculation than inorganic polymetal hydrate complexes because their effective chain length is much longer. Consequently, bridging between particles can occur more readily and result in more microflocs.

Cyclic Quaternary Ammonium Polymers

Attempts have been made to prepare high molecular weight "ionene" cationic polyelectrolytes without success (the highest molecular weight was about 40,000 (4, 5, 6). In fact, the molecular weight was reported to be almost inversely proportional to the charge density of the polymer (5).

Butler (7), however, has developed one of the most interesting and effective industrial cationic polymers by free radical polymerization of N,N-dimethyldiallylammonium chloride (DMDAC). The polymerization process involves an intra-intermolecular cyclopolymerization which results in a polymer which was originally reported to have a 6-membered piperidinium ring as a repeating unit in the polymer chain (8).

$$\text{(CH}_2\text{=CHCH}_2)_2\text{N}^+(\text{CH}_3)_2\ \text{Cl}^- \xrightarrow{R^\circ} \sim\text{CH}_2\text{–[piperidinium ring, }\text{N}^+(\text{CH}_3)_2]\sim\ \text{Cl}^-$$

1 2

However, two cyclic structures are possible: a 6-membered ring 3 and 5-membered ring 4. The 6-membered ring structure is thermodynamically more stable than the 5-membered due to relief of 1,2-interactions in the chair conformation. The six-membered ring also gives rise to a 2° radical in the propagation sequence whereas the 5-membered-ring gives a less stable 1° radical. The 6-membered ring is usually regarded as the thermodynamic product because of its greater apparent stability, and the 5-membered ring is regarded as the kinetic product.

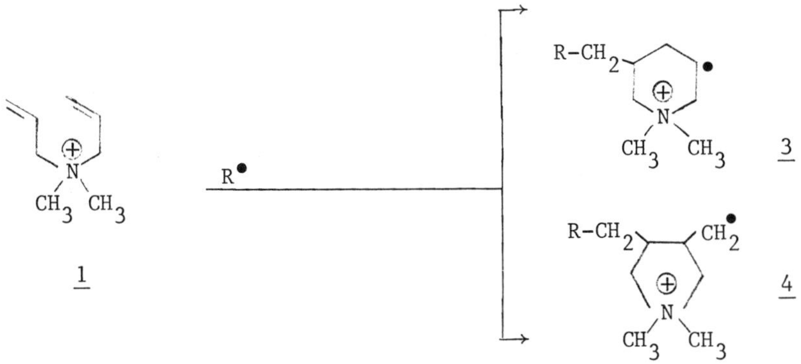

Subsequently, several studies over the years have been carried out to determine whether 1,6-diene systems undergo cyclization in an intramolecular mode by forming a 5-membered ring or 6-membered ring (9). Since polymer systems are structurally difficult to evaluate some very interesting studies on monomer substrates were carried out by Brace (10), Julia (11), Solomon (12), Lancaster (13), and Ottenbrite (9). These studies have shown radical addition to a 1,6-diene system occurs readily in an intramolecular mode to preferentially form a 5-membered ring. There are several factors that influence this cyclization process: (a) the relative conformational stability of a 6-membered ring system over a 5-membered ring system (9, 10, 11, 12, 14), (b) steric influences of 2,5-substituents (12), (c) radical stabilization of the initial site (11) and (d) electron density differences between C-5 and C-6 causing coulombic interactions (9).

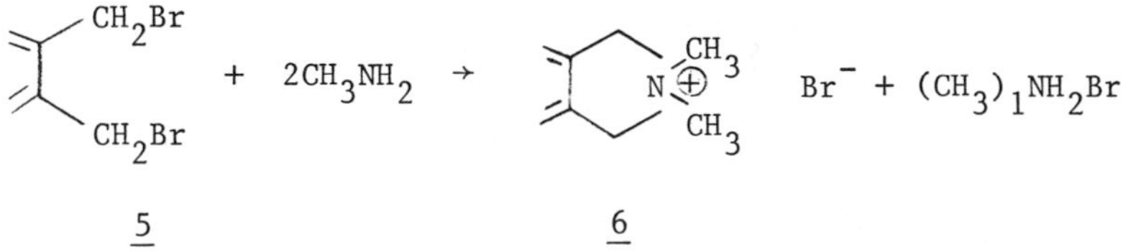

Although several systems have been evaluated to determine the structure of the ring formed in cyclopolymerization, very little work has been reported on the original cyclopolymer reported by Butler (7) that was obtained from N,N-dimethyldiallylammonium bromide or the chloride. Consequently, we undertook a study of a five-membered quaternary ammonium polymer, poly(1,1,3,4-tetramethyl-3-pyrrolinium bromide) as a model system. N,N-dimethyl-3,4-dimethylenepyrrolidinium bromide (6) was prepared by reacting dimethylamine with 2,3-bis(bromomethyl)-1,3-butadiene (5). In addition to the bromide salt (6), the iodide salt was also prepared by reacting 2,3-bis(bromomethyl)-1,3-butadiene with methylamine in acetonitrile to form a solution of N-methyl-3,4-dimethylene pyrrolidine 8. Methyl iodide was added to a solution of 8 to form the iodide salt 9.

Aqueous solutions of the exocyclic diene 6 readily poly-

$$\underset{7}{CH_2Br,\ CH_2Br} + 2CH_3NH_2 \rightarrow \underset{8}{N\text{-}CH_3} + CH_3\overset{\oplus}{N}H_3Br$$

$$8 + CH_3I \rightarrow \underset{9}{\overset{\oplus}{N}(CH_3)(CH_3)}\ I^-$$

merized to form the cationic polyelectrolyte containing a 3-pyrrolinium ring 6 as part of the polymer chain. The NMR spectra

$$\underset{6}{\overset{\oplus}{N}(CH_3)(CH_3)\ Br^-} \rightarrow \underset{10}{\sim CH_2\ \ CH_2\ \ \overset{\oplus}{N}(CH_3)(CH_3)\ \ Br^-}$$

of the polymer obtained from 6 are consistent with the proposed structure 10 which would result from 1,4-polymerization of the diene system.

We conducted a ^{13}C NMR study (9) of a 6-membered quaternary ammonium monomer model, a 5-membered quaternary ammonium model (a hydrogen reduced derivative of N,N,3,4-tetramethyl-3-pyrrolidinium bromide) and poly(N,N-dimethyldiallylammonium chloride). The spectral data indicated that poly(N,N-dimethyldiallylammonium chloride) is composed of 5-membered rings rather than the 6-membered system. A similar study was carried out by Lancaster et. al. (13) and analogous results were obtained.

Subsequently, we were able to completely reduce the unsaturated 5-membered poly(N,N-dimethyl-3-pyrrolidinium bromide) to yield completely saturated 5-membered rings. The ^{13}C NMR of this reduced polymer material gave a spectrum that was the same as that obtained from poly(N,N-dimethyldiallylammonium chloride) which unequivocally shows that the cyclopolymerization produces exclusively 5-membered rings along the polymer backbone in a predominently cis configuration (9).

Polymerization Studies of N,N-dimethyl-3,4-dimethylenepyrrolidinium Bromide

It was found that N,N-dimethyl-3,4-dimethylenepyrrolidinium bromide undergoes spontaneous polymerization in aqueous solutions when the concentration of monomer is greater than 10% at 60°C. Conversion studies by ^{1}H NMR show that rate and degree of conversion increase with increasing concentration (Table I).

Table I: Concentration Study on the Polymerization of N,N-Dimethyl-3,4-Dimethylenepyrrolidinium Bromide

No.	Monomer Concentration %	Temperature °C	Time hr	Yield %	Intrinsic Viscosity of polymer[a]
1	10	60	24	45	1.37
2	20	60	24	81	1.70
3	40	60	5	100[b]	3.90
4	50	60	5	100[b]	4.05
5	60	60	5	100[b]	Gel

a) Measured in 2.0M KBr at 30°C.
b) Contains 5-10% of a dimer formed by a self Diels-Alder Reaction

At low monomer concentrations (10-20%), the polymerization mixture after 24 hrs contained polymer 45% and 81%, respectively, and unreacted monomer. At higher concentration (>40%), all the monomer appeared to have been totally reacted in less than 5 hrs. The reaction mixture, however, was found to contain 5-10% of a dimer 11 formed by a self Diels-Alder reaction of the monomer.

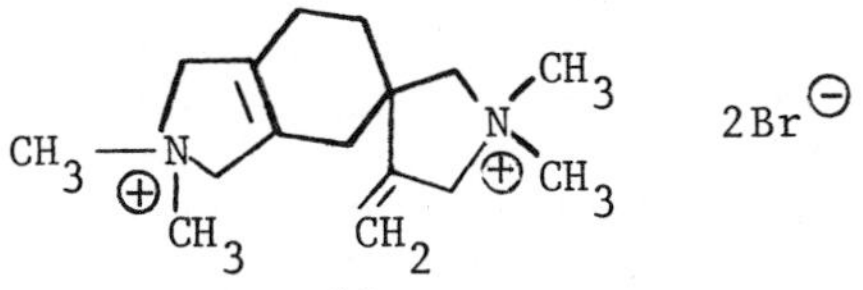

11

It was observed that the molecular weight of the polymer increased significantly with increased monomer concentration. For example, at 10% monomer concentration, the intrinsic viscosity was 1.37 dl/g while the viscosity was 4.05 dl/g at 50% monomer concentration. At 60% monomer concentration, an insoluble gel was formed which appeared to have considerable crosslinking. It was also observed from rate conversion studies that the rates of polymerization were larger at higher concentration of monomer (Figure 1).

The spontaneous polymerization in aqueous solution experienced in this polymer system is not unique to quarternary systems. Salomone (15) extensively studied the spontaneous polymerization of 4-vinylpyridinium salts. Although there are several similarities in the two systems, it appears that the initiating species may be different and will be the subject of future studies. For example, 20% monomer solutions in methanol at 60° for 24 hrs showed no polymerization thus indicating that the spontaneous initiation may be dependent on the aqueous media. Other initiators (Table II) have varying effects (Figure 2); t-butyl hydroperoxide

Table II: Conversion Studies on the Polymerization of N,N-Dimethyl-3,4-dimethylenepyrrolidinium Bromide

No.	Monomer Concentration %	Initiator % of Monomer	Temperature °C	Time hr	Conversion %
1	20	None	60	24	93
2	20	$(NH_4)_2S_2O_8$ 1%	60	24	77
3	20	$(NH_4)_2S_2O_8$ 1%	25	24	50
4	20	t-Butyl Hydroperoxide 1%	60	5	100
5	20	None in methanol	60	24	0

produced 100% conversion in 5 hrs, while ammonium persulfate appeared to inhibit the process with only 77% conversion after 24 hrs. The latter result may be due to bromide ion oxidation by the persulfate ions to generate bromine which can act as an inhibitor. A similar observation by Negi and coworkers (16) was made with diallyldimethylammonium bromide salt which does not polymerize as readily in the presence of ammonium persulfate as did the corresponding chloride salt.

The addition of other salts to the polymerization system was also evaluated Figure 3. Ammonium bromide and sodium bromide showed a slight increase in conversion rate which was attributed to an increase in the dielectric constant of media. It was noted that the ammonium ion does not appear to have any effect on the initiation process. A large rate increase with 10%

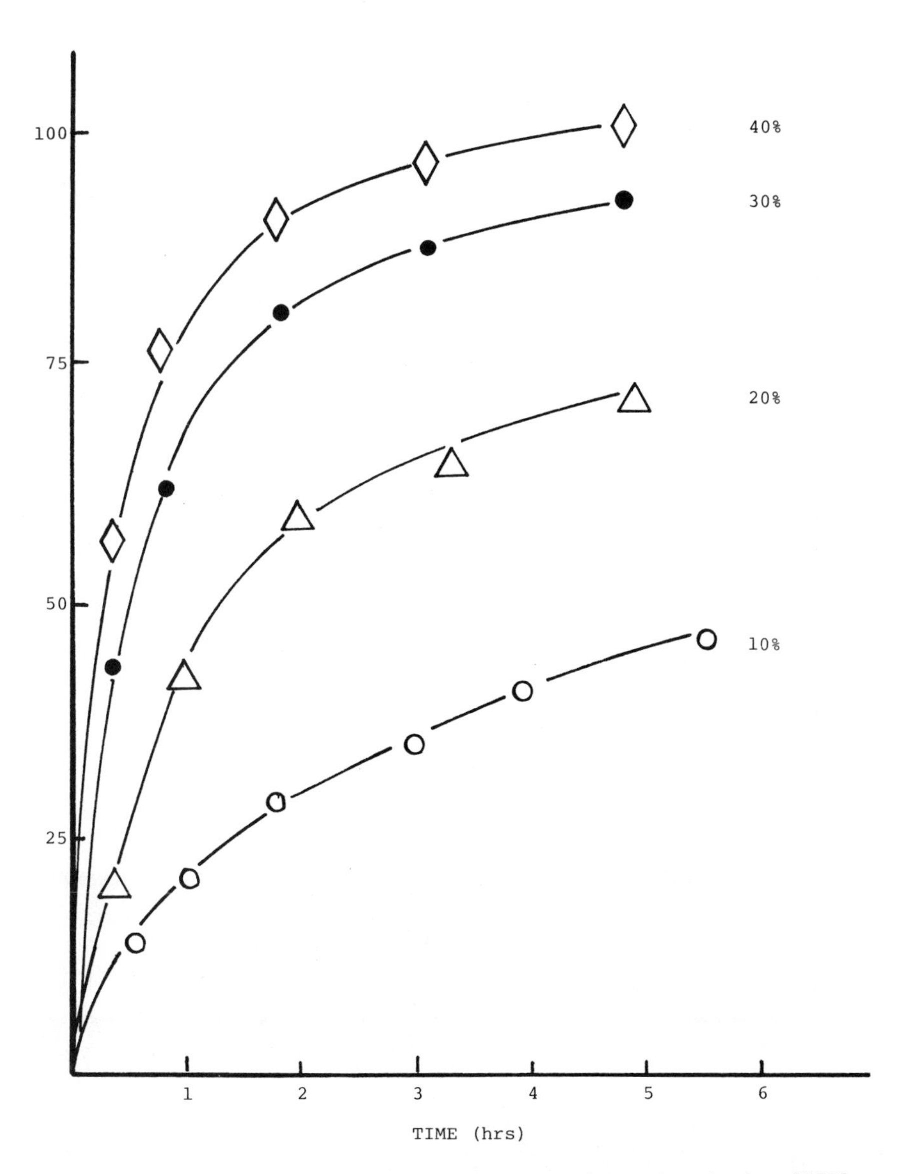

Figure 1. Effect of monomer concentration on the rate of polymerization of N,N-*dimethyl-3,4-dimethylenepyrrolidinium bromide.*

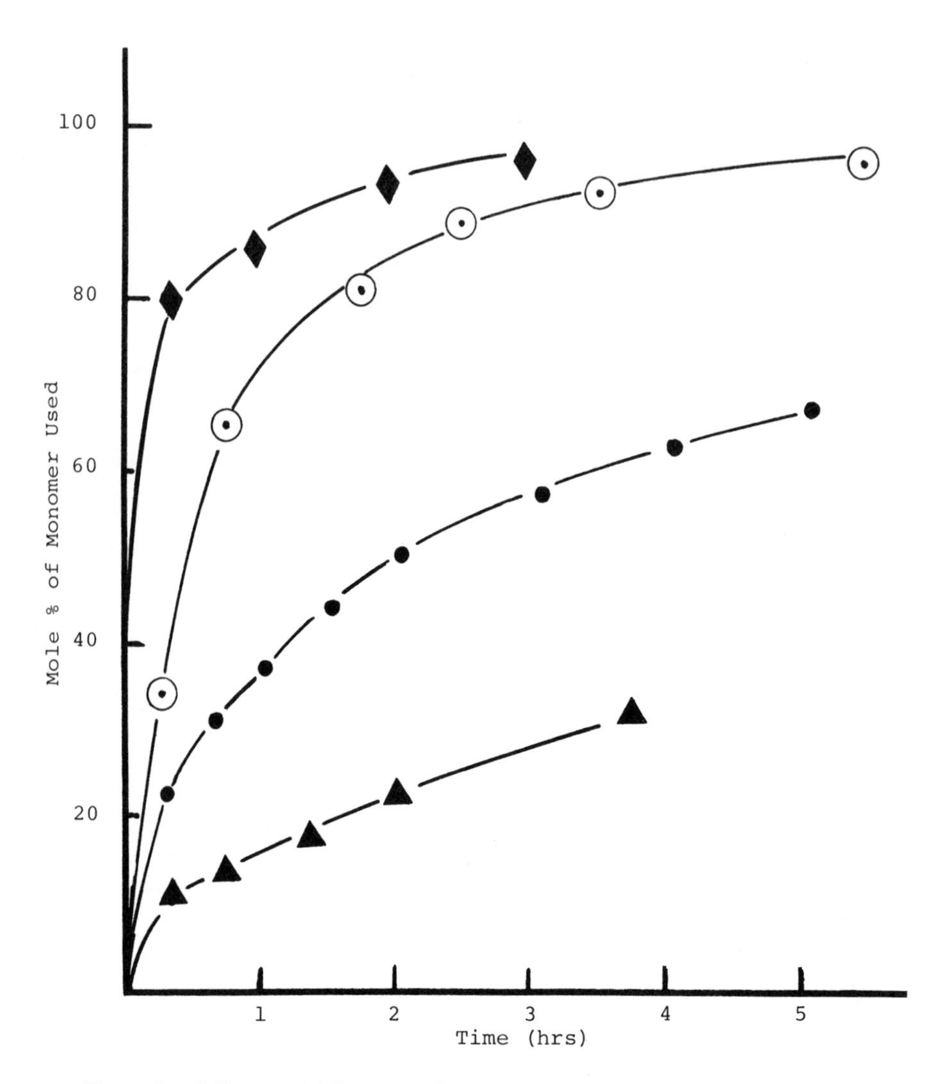

Figure 2. Influence of initiators and concentration on the rate of polymerization of N,N-*dimethyl-3,4-dimethylenepyrrolidinium bromide. Key:* ♦, *40% monomer;* ⊙, *20% monomer with* t-*butyl hydroperoxide;* ●, *20% monomer;* ▲, *20% monomer with ammonium persulfate.*

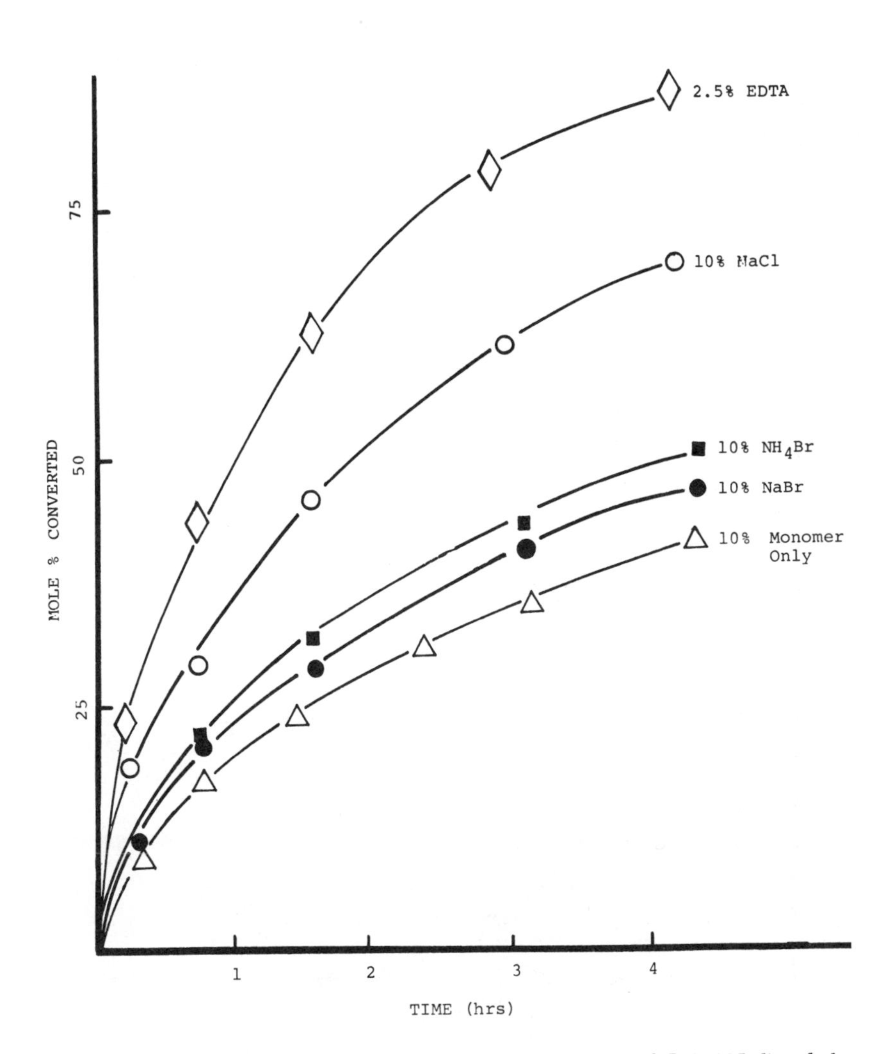

Figure 3. Effect of additives on the rate of polymerization of 0.5M N,N-*dimethyl-3,4-dimethylenepyrrolidinium bromide.*

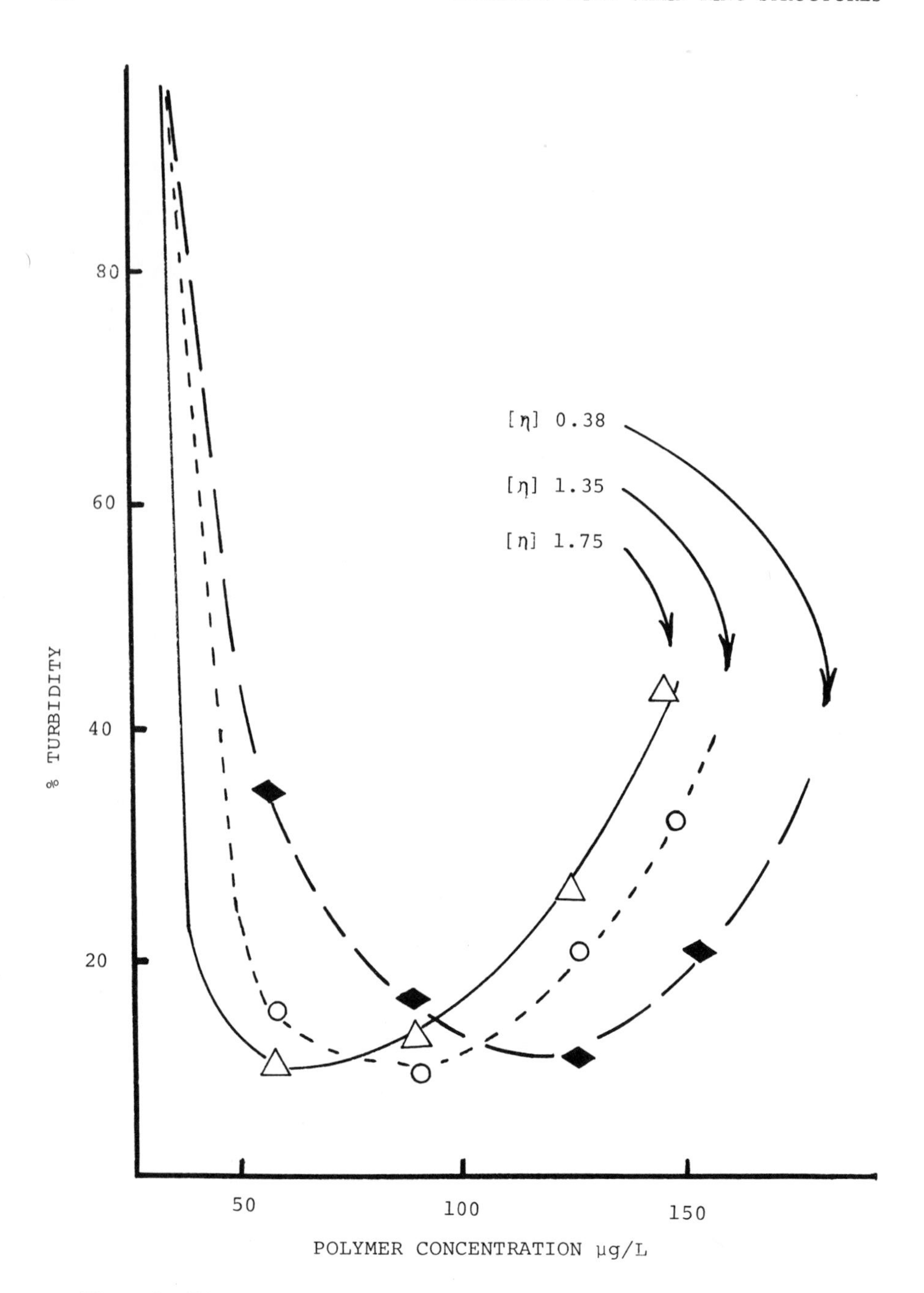

Figure 4. Effect of polymer molecular weight on clarification of clay suspensions.

NaCl was experienced. This observation was similar to the findings (16) in the case of diallyldimethylammonium chloride salt which polymerizes more readily than the bromide salt. Another large increase in rate of conversion was obtained with 2.5% EDTA. The reason for this great increase is not evident but was also observed for the diallyldimethylammonium chloride polymerization by Hoover (17).

Initial studies showed that this polymer has the potential of being an excellent flocculating agent. Performance evaluations were made by comparing flocculating data to commerical Cat-Floc (18) on a 50 ppm Montmorillonite clay suspension using the method of Black and Vilaret (19). Floc times and residual turbilities of low molecule weight samples (η = 1.35 and 1.75) were comparable to Cat-Floc. It is anticipated the higher molecular weight polymer will give better performance. Buchner funnel tests (21,20) to determine activated sludge dewatering performance was better than those obtained with Cat-Floc.

Acknowledgement

I wish to thank Dr. Rodney Meyers for his efforts in these studies as well as the cooperation of the personnel at Calgon Corporation. I also wish to thank the National Science Foundation for support in part for this study.

Literature Cited

1. Eisenberg, A. and Hoover, M.F., "Ion Containing Polymers", American Chemical Society Publications, Washington, D.C., 1972.
2. Rembaum, A. and Selegny, Z., "Polyelectrolytes and Their Applications", Vol. 1 and 2, D. Reidel Publishing Company, Holland (1975).
3. Ottenbrite, R.M. and Ryan, W.S., I & EC Prod. Res. and Devel., 19, 528 (1980).
4. Ottenbrite, R.M. and Myers, G.R., J. Poly. Sci., 11, 1443 (1973).
5. Noguchi, H. and Rembaum, A., J. Polym. Sci. B, 1, 385 (1969).
6. Rembaum, A., Baumgartnes, W. and Eisenberg, W., J. Polym. Sci. B, 6, 159 (1968).
7. Butler, G.B. and Ingley, F., J. Amer. Chem. Soc., 73, 895 (1951); Butler, G.B. and Angelo, R.J., ibid., 79, 3128 (1957); Butler, G.B., U.S. Patent 3,288,770, 1966.
8. Butler, G.B., Gropp, A.H., Angelo, R.J., Husa, W.J. and Jorolan, E.P., 5th Quaterly Report. Atomic Energy Commission Contract AT(40-1)-1353 (1953).
9. Ottenbrite, R.M. and Shillady, D.D., IUPAC Symposia: Quaternary Ammonium and Amine Polymers, Goethals, E.J., Ed., p. 143 Pergamon Press, New York, 1980.

10. Brace, N.O., J. Poly. Sci., 8, 2091 (1970); J. Org. Chem., 44, 212 (1978).
11. Julia, M., Chem. Eng. News, 44, 100 (1966); Julia, M. and Manny, M., Bull. Soc. Chem., 434 (1966).
12. Hawthorne, D.G. and Solomon, D.H., J. Macromol. Sci. Chem., A10, 923 (1976).
Beckwith, A.J., Ong, A.K. and Solomon, D.H., J. Macromol. Sci. Chem., A9, 115 (1975); ibid., A9, 125 (1975); Hawthorne, D.G., Johns, S., Solomon, D. and Willings, R., Chem. Commun., 982 (1975).
13. Lancaster, J., Baccei, L. and Panzer, H., Polymer Letters, 14, 549 (1976).
14. Smith, T.W. and Butler, G.B., J. Org. Chem., 43, 6 (1978).
15. Salomone, J.
16. Negi, Y., Haradu, S. and Ishizuka, O., J. Polym. Sci. A, 5, 1951 (1967).
17. Boothe, J.E., Flock, H.G. and Hoover, M.F.
18. Commercially known as Cat-Floc produced and marketed by Calgon Corporation, Pittsburg, Pa.
19. Black, A.P. and Vileret, M.R., Jour. A.W.W.A., 61, 209 (1969).
20. Water Pollution Control Federation Manual Practice No. 20, "Sludge Dewatering", 3900 Wisconsin Avenue, Washington, D.C.
21. Calgon Technical Information Bullentin No. 12-1034 "Buchner Funnel Test", 1968.

RECEIVED February 17, 1982.

6

Radical Cyclopolymerization of Divinyl Formal: Polymerization Conditions and Polymer Structure

MITSUO TSUKINO

Kitakyushu Technical College, Department of Chemical Engineering,
Kokura-minami, Kitakyushu 803, Japan

TOYOKI KUNITAKE

Kyushu University, Department of Organic Synthesis, Faculty of Engineering,
Fukuoka 812, Japan

Radical polymerization of divinyl formal was conducted in benzene and the polymer structure examined by ^1H- and ^{13}C-NMR spectroscopy. The polymer contains the cis-dioxolane ring as the major structure, along with the trans-ring and the branched structure. The major structures are connected in the meso and racemic fashions. The content of the minor structures becomes negligible by lowering polymerization temperature and by increasing monomer concentration. The unsaturated unit and the six-membered unit are not formed under any condition.

We have been conducting detailed structural studies on the cyclopolymers of several unconjugated dienes. Radical cyclopolymerization of divinyl ether had been known to give polymers with the unsaturated monocyclic unit and the bicyclic unit(1, 2). Our initial ^{13}C-NMR study indicated that the polymer was composed of the five-membered monocyclic unit with the pendent unsaturation and the bicyclic unit with the bicyclo[3,3.0]octane skeleton(3). According to the stereochemistry of the polymer established by more recent NMR examination(4), the polymerization process was shown to be highly stereoselective.

$CH_2{=}CH{-}O{-}CH{=}CH_2 \xrightarrow{R\cdot}$ —CH_2-(monocyclic unit with CH_2— and O-$CH{=}CH_2$) + —CH_2-(bicyclic unit)-CH_2—

or —CH_2-(monocyclic unit with CH_2— and O-$CH{=}CH_2$) + —CH_2-(bicyclic unit)-CH_2—

0097-6156/82/0195-0073$06.00/0
© 1982 American Chemical Society

Similar studies were performed for the radical polymers of cis-propenyl vinyl ether and 2-methylpropenyl vinyl ether, with the conclusion that the polymer structures were basically the same as that of poly(divinyl ether)(5).

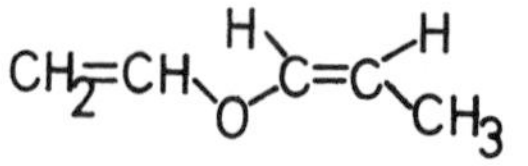

$CH_2{=}CH{-}O{-}CH{=}CH{-}CH_3$	$CH_2{=}CH{-}O{-}CH{=}C(CH_3)_2$
cis-propenyl vinyl ether 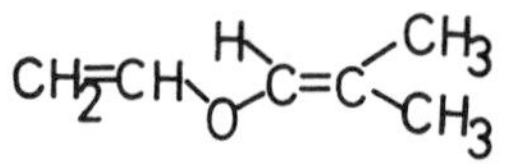	2-methylpropenyl vinyl ether

Divinyl acetals are also known to give soluble cyclopolymers by radical polymerization. Although the polymer structure had been studied by a variety of spectroscopic and chemical methods (6-9), the results were inconclusive and no information was obtained as to the stereochemistry. Our ^{13}C-NMR methodology was thus applied to polymers of divinyl acetal and its derivatives (acetaldehyde divinyl acetal and acetone divinyl acetal). Comparison of ^{13}C-NMR spectra of the polymers with those of 1,3-dioxolanes and 1,3-dioxanes which are model compounds of the cyclic units indicates that these polymers have essentially the same structure(10). The propagation process of these monomers is shown in Scheme I with divinyl formal as an example. The intramolecular cyclization produces cis-closed five-membered cyclic radical I predominantly. The radical either reacts with monomer or abstracts hydrogen from the neighboring exocyclic methylene. The newly-formed radical propagates subsequently to produce branched unit III. On the other hand, NMR evidence indicates the absence of the uncyclized structure and the trans-closed rings IV and V. Furthermore, it is indicated that the meso and racemic modes exist in equal amounts for the connection between the major structure II.

Following these previous results, we investigated in this study the influence of the polymerization conditions(monomer concentration and polymerization temperature) on the structure of poly(divinyl formal).

Experimental

Materials. Divinyl formal was prepared as described before (10). Azobis(isobutyronitrile)(AIBN) was recrystallized from methanol. Benzene and methanol were purified by the conventional procedure.

Polymerization. The polymerization was conducted in benzene with AIBN initiator at 50 and 70°C. The initiation was accelerated at 10 and 30°C by irradiation with a high-pressure Hg lamp. Required amounts of divinyl formal, benzene and AIBN were placed in ampoules, subjected to the freeze-pump-thaw cycle, and the

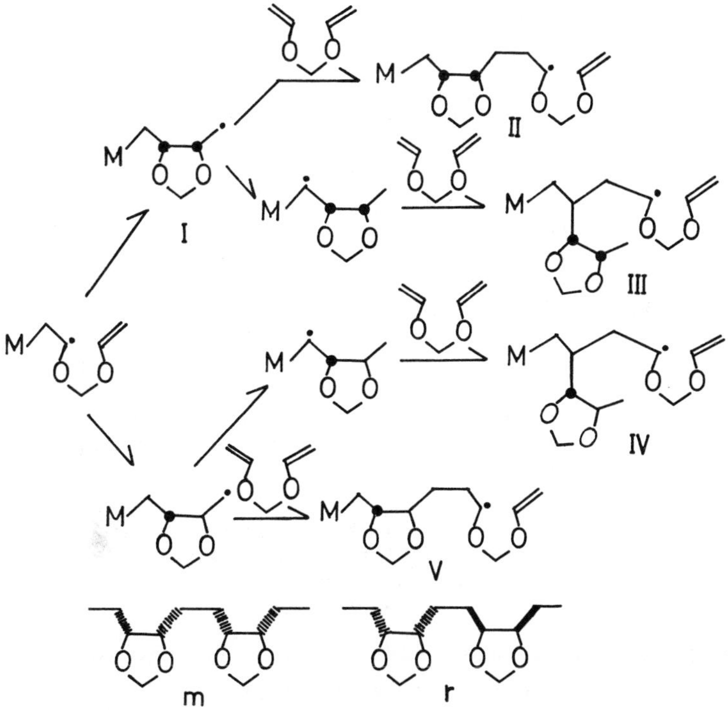

M
I
II
III
IV
V
m
r

Scheme 1

ampoules sealed. The polymers were precipitated by methanol, reprecipitated from benzene and methanol, and dried. They were freeze-dried from benzene, when necessary.

Miscellaneous. NMR spectra of 1,3-dioxolanes were obtained with a Varian A-60 spectrometer. ^{1}H- and ^{13}C-NMR spectra of the polymers were obtained with a JEOL FX-100 spectrometer. The molecular weight was determined by gel permeation chromatography (Toyo Soda HLC-802UR) using monodisperse polystyrene as reference.

Results and Discussion

It was previously confirmed that the six-membered ring and the unsaturated unit were not present in the polymer(10). The five-membered ring unit may possess the structures shown in Figure 1. Structure A is the cis-closed ring in the main chain and B is the branched cis-closed ring. C and D are the trans-closed rings in the main chain and in the side chain, respectively. E and F are the chain end rings with the cis and trans stereochemistries.

Figure 2 shows a ^{1}H-NMR spectrum of the polymer obtained at low monomer concentration and high polymerization temperature. The carbon chemical shifts of the trans rings are close to those of structure B, and the acetal carbons of structures A to E in Figure 1 possess almost the same chemical shift. From these data we tentatively concluded in our previous study that the trans ring units were absent(10). However, the stereochemical difference of the ring is clearly reflected in the chemical shift of the acetal hydrogen in the ^{1}H-NMR spectrum of improved resolution(Figure 2). The peak assignment is performed on the basis of the chemical shift of 1,3-dioxolanes: peaks at ca. 1 ppm are attributed to the methyl proton of structures B and D, and peaks at 1.5-2.0 ppm attributed to the exocyclic methylene proton of structures A and C. The molecular weight of this polymer is relatively small(MW = 2800), and a methyl proton peak of the initiator fragment is observable at 1.4 ppm. The peaks of the ring methine protons are located at 3.3-4.3 ppm, and the peaks of the acetal methylene proton are separately found at 4.5-5.0 ppm, in correspondence to the structures of Figure 1. The total amount of the trans ring unit(C + D + F) is 20 %, as estimated from the relative peak area of the acetal proton.

Figure 3 gives ^{13}C-NMR spectra of the polymer prepared at 70°C with different monomer concentrations. These spectra are assigned as described before(10) by referring to ^{13}C-NMR data of 1,3-dioxolanes. In Figure 3a, a peak at 14 ppm suggests the presence of the methyl carbon of structures B and E. Peaks at 25 ppm are ascribed to the exocyclic methylene carbon of structures A and E, and they are split due to the two modes of ring connection(meso and racemic). The methine carbon peaks are located at 72-82 ppm. The acetal carbon peak of all the ring

Figure 1. Conceivable five-membered ring units for poly(divinyl formal).

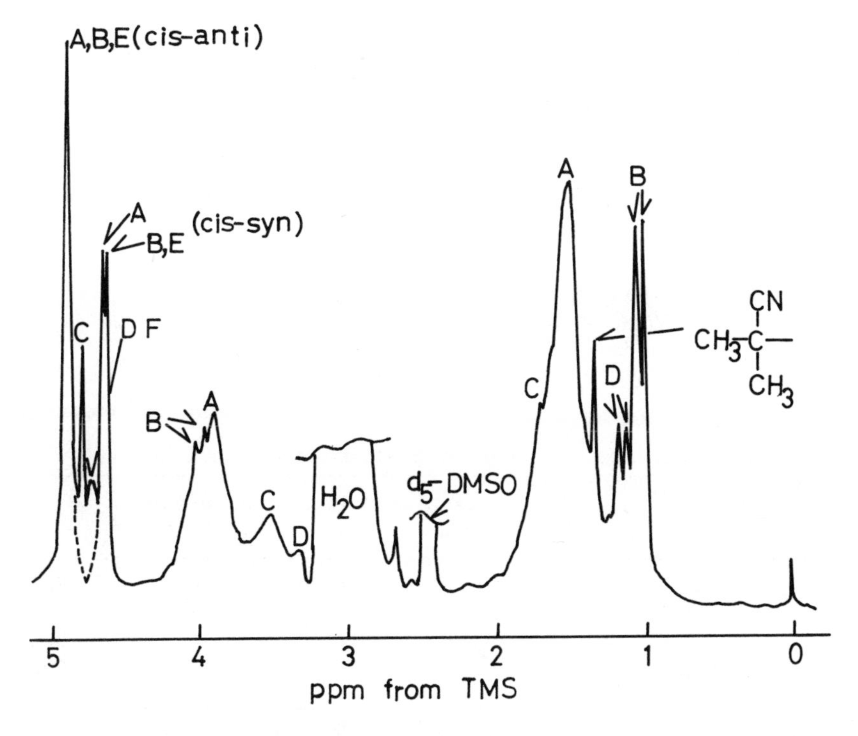

Figure 2. ^1H NMR spectrum of poly(divinyl formal). Table I, run 3: 3 w/v % in d_6-DMSO at 150°C; accumulation, 32 scans.

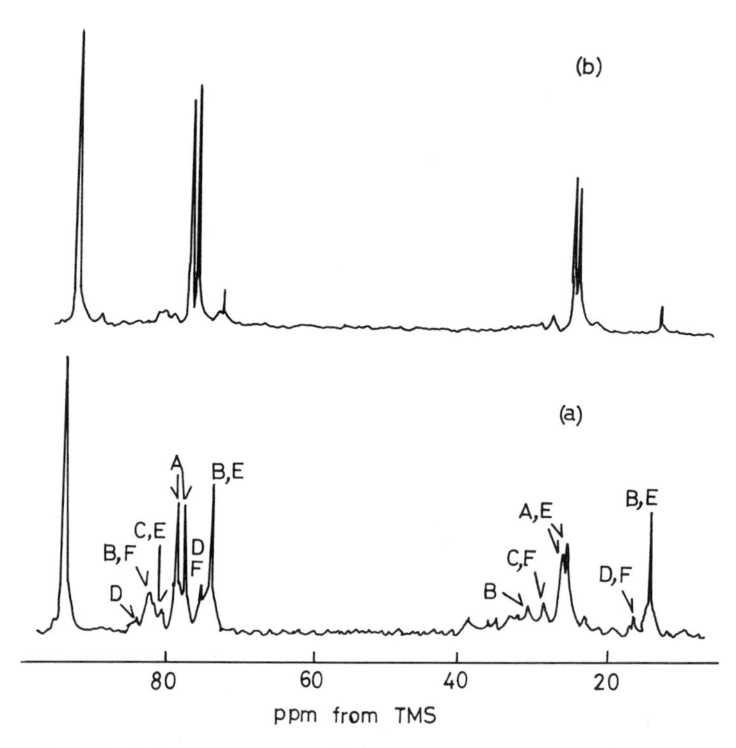

Figure 3. ^{13}C *NMR spectra of poly(divinyl formal). Key: a, Table I, run 3, 28 wt % in* C_6D_6 *at 75°C, accumulation, 2000 scans; and b, Table I, run 4, 28 wt % in* C_6D_6 *at 75°C, accumulation, 1600 scans.*

Table I. Radical Polymerization of Divinyl Formal[a)]

Run	Monomer mol/l	AIBN mol/l	Temp °C	Time hr	Conversion %	$\bar{M}_n$
1	2.5	0.1[b)]	10	8.0	8.6	5,500
2	2.5	0.02	70	1.5	8.1	8,100
3	1.0	0.02	70	6.0	8.6	2,800
4	5.0	0.02	70	0.75	8.6	20,000

a. in benzene,
b. irradiated with a high-pressure Hg lamp

structures(A to F) is at 93.6 ppm. Thus, the branched structure and the trans rings are included in the polymer apart from the main structure(A), when the monomer concentration is low.

In contrast, the formation of the minor structures are suppressed by conducting polymerization at higher monomer concentrations, as shown by Figure 3b. The fraction of structure A is estimated from the relative peak area of the acetal carbon (total structures) and the ring methine carbon at 77-78 ppm (structure A), by assuming that NOE's are the same for these peaks. At 70°C, the fraction of A increased from 47 to 78 % upon increase in the monomer concentration from 1.0 to 5.0 mol/l.

Figure 4 is a ^{1}H-NMR spectrum of the polymer obtained under the polymerization conditions of relatively low monomer concentration and low polymerization temperature. The formation of structures other than A is again suppressed. The content of the trans rings(C + D + F) is 5 % in this case.

Figure 5 compares ^{13}C-NMR spectra of the polymers prepared at different temperatures. The main structure A becomes predominant at a low polymerization temperature as shown in Figure 5a, but the content of the other structures, particularly the branched cis ring B, increases at a higher temperature(Figure 5b). The fraction of structure A increased from 66 to 89 % by lowering the polymerization temperature from 70 to 10°C at a fixed monomer concentration of 2.5 mol/l.

The mode of connection of the main structure is constant (meso: racemic = 1:1), in spite of the change in the polymerization conditions. The equal amounts of the meso and racemic connections might be randomly distributed. It is interesting, however, that the sequence of type c appears least sterically demanding among the four types of the connections on the basis of the CPK molecular model.

(a) ...rrrrrrrr...
(b) ...mmmmmmmm...
(c) ...mmrrmmrr...
(d) ...mrmrmrmr...

In conclusion, the structure of poly(divinyl formal) was shown to vary with the polymerization conditions (Table I). It is expected that highly stereoregular polymers are obtainable by selecting proper polymerization comditions. These results will be described elsewhere, together with the kinetic analysis of the cyclopolymerization process.

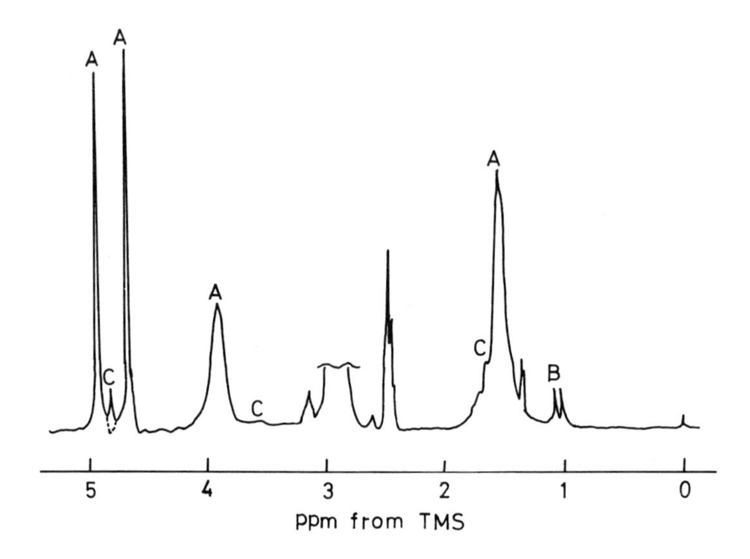

Figure 4. ^{1}*H NMR spectra of poly(divinyl formal) from Table I, run 1: 3 w/v % in* d_6*-DMSO at 150°C; accumulation, 16 scans.*

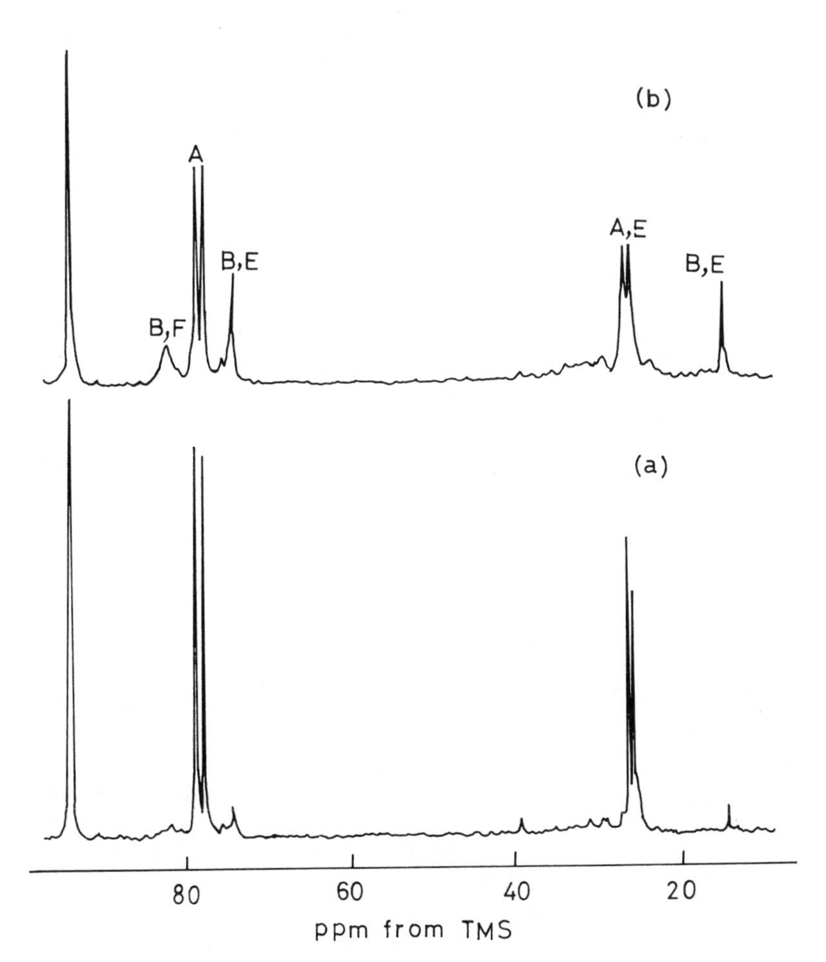

Figure 5. ^{13}C NMR spectra of poly(divinyl formal). Key: a, Table I, run 1, 28 wt % in C_6D_6 at 75°C, accumulation, 2500 scans; and b, Table I, run 2, 28 wt % in C_6D_6 at 75°C; accumulation, 3000 scans.

Literature Cited
1. Aso, C.; Ushio, S. Makromol Chem. 1967, 100, 100.
2. Guaita, M,; Camino, C.; Trossarelli, L. ibid. 1970, 131, 237.
3. Kunitake, T.; Tsukino, M. ibid. 1976, 177, 303.
4. Tsukino, M.; Kunitake, T. Macromolecules 1979, 12, 387.
5. Tsukino, M.; Kunitake, T. Polym. J. 1981, 13, 657.
6. Matsoyan, S. G. J. Polym. Sci. 1961, 52, 189.
7. Minoura, Y.; Mitoh, M. ibid. 1965, A-3, 2149.
8. Aso, C.; Kunitake, T.; Ando, S. J. Macromol. Sci. Chem. 1971, A5, 167.
9. Aso, C.; Kunitake, T.; Tsutsumi, F. Kogyo Kagaku Zasshi 1967, 70, 2043.
10. Tsukino, M.; Kunitake, T. Polym. J. 1979, 11, 437.

RECEIVED March 8, 1982.

7

Divinyl Ether–Maleic Anhydride Copolymer and Its Derivatives: Toxicity and Immunological and Endocytic Behavior

L. GROS, H. RINGSDORF, and R. SCHNEE[1]—Institut für Organische Chemie, Universität Mainz, Johann-Joachim-Becher-Weg 20, D-6500 Mainz, Federal Republic of Germany

J. B. LLOYD—Biochemistry Research Laboratory, Department of Biological Sciences, University of Keele, Staffs, United Kingdom

E. RÜDE, H. U. SCHORLEMMER[2], and H. STÖTTER[3]—Institut für Mikrobiologie und Immunologie, Universität Mainz, Federal Republic of Germany

DIVEMA and some DIVEMA derivatives were synthesized and tested for toxicity and immunological behaviour.
Toxicity can be varied by introducing different side groups into parent DIVEMA molecules by polymer analogous reactions.
The compounds tested showed mitogenic effects, i.e. they stimulated lymphocyte proliferation. Astonishingly, a DIVEMA-DNP conjugate did not induce the production of DNP-specific antibodies. Our results suggest that this is due to the formation of DNP-specific active suppressor cells.
Using serum free culture media, we could show for the first time that DIVEMA and derivatives stimulate macrophages to release cytotoxic factors into the supernatant. However, no significant release of lytic enzymes was detected.
DIVEMA and the derivatives tested failed to stimulate endocytosis by macrophages of ^{125}I-PVP and ^{198}Au (colloidal); higher mol. wt. DIVEMA inhibited pinocytosis. We conclude that the only effect of the compounds tested on macrophages is the release of cytotoxic factors into the supernatant.

[1] Current address: Röhm GmbH, Kirschenallee, D-6100 Darmstadt, Federal Republic of Germany.
[2] Current address: Behringwerke, Postfach 1140, D-3550 Marburg/Lahn 1, Federal Republic of Germany.
[3] Current address: Zentrum f. Innere Medizin der Universität, Steinhövelstr. 9, D-7900 Ulm, Federal Republic of Germany.

0097-6156/82/0195-0083$06.00/0
© 1982 American Chemical Society

Pharmacologically active polymers, and especially polymeric antitumour agents, have been widely discussed, synthesized and tested during the past decade (1, 2, 3). Among the many anionic polymers tested for pharmacological activity (4), DIVEMA, or pyran copolymer, a 1:2 alternating cyclocopolymer of divinyl-ether and maleic anhydride in its hydrolyzed form (5), has been one of the most intensively studied compounds. DIVEMA has been shown to have immune stimulating, antiviral and antitumour activities (4, 6, 7). These properties suggested that DIVEMA might be an appropriate carrier candidate for drug molecules, especially for antitumour agents. Thus, the immunosuppressive folate-antagonists methotrexate (MTX) (8) and Cytoxan [R] (cyclophosphamide) (9, 10) were linked to pyran copolymer (cf. 3).

There are, however, many unsolved problems concerning toxicity, endocytic uptake and immunogenicity of DIVEMA. In order to learn more about the structure dependency of these properties, we sythesized some DIVEMA derivatives and investigated DIVEMA and the new derivatives with appropriate routine biological test systems.

Compounds Tested

DIVEMA samples (1a - 1d) were kindly supplied by Dr.D. Breslow, HERCULES Inc. The synthesis of the derivatives (11) will be published elsewhere. The compounds investigated are listed in Table I.

Acute Toxicity of DIVEMA and two DIVEMA Derivatives

One disadvantage of DIVEMA is its relatively high toxicity which is mainly caused by high molecular weight fractions (7). In an attempt to lower the toxicity of DIVEMA, derivatives (3) and (4) were synthesized.

The pyrrolidone side chain is known to be non-toxic e.g. in poly(vinylpyrrolidone). Incorporation of the dimethylamino-moiety leads to zwitterionic structures; such structures are known to lower the toxicity of the polycationic poly(ethyleneimine) (12).

Acute toxicity of compounds (1d), (3b) and (4b), all of mol. wt. 24,000, was tested *in vivo*. Male C 57/BL.6 mice were given a single i.p. dose. Table II shows the results. Toxicity is expressed as LD 50/21, the dose lethal for 50% of the animals by day 21 after injection of the polymer.

Table I. Compounds synthesized and tested with R groups distributed between ring and chain carboxyls.

NR	-R	X	MW
1a	$-OH$	–	4 400
b			11 000
c			16 800
d			24 000
2	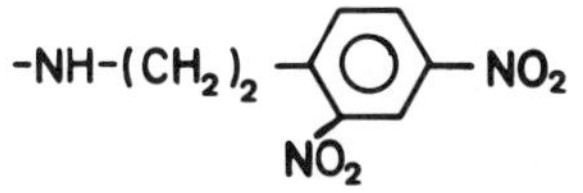 $-NH-(CH_2)_2-C_6H_3(NO_2)_2$ (2,4-dinitrophenyl)	5,3	16 800
3a	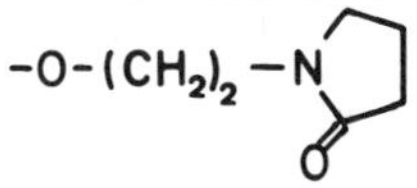 $-O-(CH_2)_2-N$ (2-pyrrolidone)	12	16 800
b			24 000
4a	$-O-(CH_2)_2-N(CH_3)_2$	20	16 800
b			24000
5	$-O-(CH_2)_{17}-CH_3$	4	16 800

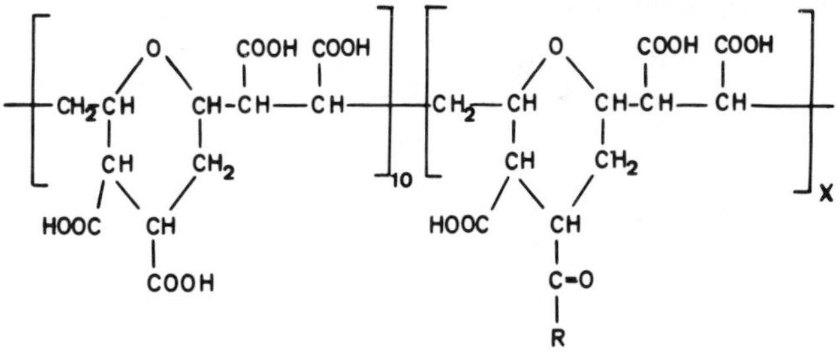

Table II. Acute toxicity of DIVEMA and some derivatives

Substance Nr.	Dose (mg/kg)	Survivors (day 21)	LD50/21*
1d	300	6/10	300-400
	400	4/10	
3b	300	4/8	ca. 300
	400	0/8	
4b	300	8/8	500-600
	400	7/8	
	500	4/8	
	600	4/8	
	700	4/8	

*LD50/21 = lethal dose for 50% of test animals by day 21.

The introduction of the potentially biocompatible pyrrolidone unit (polymer (3b)) unexpectedly leads to a slightly higher toxicity, as compared to the starting material (1d). Decrease of the polyanion character in the zwitterionic polymer (4b), however, substantially lowers toxicity.

Effects of DIVEMA and DIVEMA Derivatives on the Immune System.

DIVEMA is known to act upon different types of immune cells (13) such as the antibody producing B cells (bone marrow derived lymphocytes); on T-cells (thymus processed lymphocytes) and on macrophages (cells which phagocytose foreign material). The mechanisms of interaction between polymers and the immune system are very complex. In order to investigate them, many different experimental approaches can be used.

Our approach is based on the following questions:

- Are the DIVEMA derivatives mitogenic, i.e. do they stimulate lymphocyte proliferation?
- Are the DIVEMA derivatives antigenic, i.e. do they induce production of specific antibodies?
- Do DIVEMA derivatives stimulate macrophages in vitro, i.e. do they induce the release of lytic enzymes or cytotoxic factors?

Mitogenic Activity of DIVEMA Derivatives. A first approach to the effects of a compound on the immune system is to test its ability to stimulate lymphocyte proliferation. An appropriate test is to measure the incorporation of ^{3}H-thymidine into DNA of dividing cells:

Substances (1a) - (1d), (3a) and (4a) are incubated with spleen cell cultures of untreated mice in a serum containing culture medium. A defined amount of radioactive ^{3}H-thymidine is added. After incubation, the cultures are deep frozen in order to stop cell proliferation and DNA is isolated by filtration through glass fibre filters. Radioactivity is counted in a liquid scintillator. Control cultures are incubated without antigen (polymer) added and with a known potent mitogen (Concanavalin A.)

All substances showed significant mitogenic activity. High molecular weight DIVEMA stimulated lymphocyte proliferation at lower concentration than the corresponding low molecular weight samples. Derivatives (3a) and (4a) were slightly less active. It seems that the more the polyanionic character is reduced by substitution, the more the mitogenic activity is lowered.

Antigenicity of a DIVEMA-Hapten-Conjugate (2). The results of the ^{3}H-thymidine incorporation experiments give no information about the type of lymphocytes stimulated by DIVEMA. It is likely, though not formally proven, that the enhanced ^{3}H-thymidine uptake is due to proliferation of B lymphocytes. Whether the polymer furthermore is antigenic and leads to an increase in the number of antibody forming cells can be tested using either the polymer itself or a polymer-hapten conjugate. In the latter case a low molecular weight compound (hapten, which is recognized by the immune system only when bound to an immunogenic carrier) is linked to the polymer. This has the advantage that a standardized test-system common for all substances is available.

Experimental results known so far (e.g. 14,15,16) indicate that the amount of antibodies produced against the polymer-hapten conjugate is closely related to the antigenicity of the polymer itself; the more antigenic the polymer, the higher the antibody production against a hapten fixed to it.

Polymer (2) was used as a test substance. It contains the dinitrophenyl-moiety, a classical hapten (DNP). Production of DNP-specific antibodies was measured by a radio-immunoassay (antigen-binding test).

Female CBA mice were immunized with different doses of (2) (day 0) and reimmunized (boostered) with the same dose (day 21). On day 31, a blood sample was taken and the antibodies in the serum were precipitated by adding a ^{125}I-labelled vinyl polymer containing DNP-moieties (DNP-VINAM),

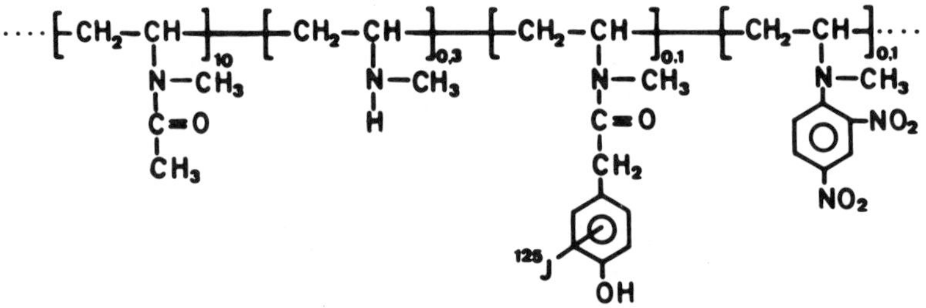

DNP-VINAM

a well established experimental method (17). The radioactivity remaining in the supernatant was measured and the percent binding of hapten to the antibodies was calculated.

No significant antibody production was found. Re-investigation of the number of cells producing DNP-specific antibodies with the plaque assay (which is much more sensitive than the antigen binding test) confirmed this result. These findings would not be too surprising if a weakly immunogenic carrier like poly-(vinylpyrrolidone) (PVP) was used. In the case of DIVEMA, it is unexpected, since this substance - in contrast to PVP - shows strong influences on the immune system, e.g.mitogenicity for lymphocytes. One possible explanation is that (2) induces tolerance against DNP.

In order to verify this assumption, female CBA mice were given a strongly immunogenic DNP protein conjugate called DNP-KLH (keyhole limpet hemocyanine) (18). One group of animals had been pre-treated ("vaccinated") with different doses of (2), a control group had only been given phosphate buffer injections on days 10,6 and 1 before injection of DNP-KLH. Table III shows the results of the antigen-binding test of blood samples from the two groups of mice and of two control groups.

These measurements show that small doses of DNP-DIVEMA (2), given to the mice before the immune system is challenged with the potent antigen DNP-KLH, significantly suppress the formation of DNP-specific antibodies: (2) induces tolerance against DNP-KLH.

Taking into account that (2) is mitogenic but still has no antigenic effect, we suggest that induction of tolerance by this polymer is due to formation of DNP-specific suppressor cells. This would explain why DIVEMA is immune stimulating (induces proliferation of lymphocytes) without causing production of antibodies: (2) induces active suppression and thus leads to the apparent lack of immunogenicity.

Effects of DIVEMA and DIVEMA Derivatives on Mouse Macrophages in Vitro. Polyanions act not only upon lymphocytes, but also on macrophages (13, 19). Macrophages are immune cells which take up foreign material, help lymphocytes recognize and attack antigens and play an important role in the defense of the body against tumour cells (20).

Schultz et al. (21, 22) found that macrophages of DIVEMA-treated mice inhibited tumour cell proliferation in vitro; this means that DIVEMA activates macrophages. The authors did not find cytotoxic factors in the supernatant of these macrophage cultures.

As polyanions can stimulate many different macrophage functions, we tried to find out which functions

Table III

Induction of tolerance by DNP-DIVEMA (2) against DNP-KLH. Mean values of 5 animals. Antibody production is expressed as % hapten (^{125}I-DNP-VINAM) precipitated by the antibodies in the serum (groups A and B) or as titer 33%, that is the reciprocal of the serum dilution at which 33% of ^{125}I-DNP-VINAM are bound and precipitated (groups C,D). Titer 33% was reached in groups C and D only.

PBS = phosphate buffered saline
CFA = complete Freund's Adjuvant

Group	Treatment	Dose of (2),μg mouse	antibodies in serum day 31
A (background control)	PBS days -10,-6,-1 CFA day 0 NaCl day 21		-2.1 ± 1.9
B (effect of DNP-DIVEMA itself)	(2) days -10,-6,-1 CFA day 0 NaCl day 21	0,1 1 10 100	8.6 ± 11.1 7.7 ± 6.0 6.4 ± 5.0 2.2 ± 4.6
C DNP-KLH alone	PBS days -10,-6,-1 DNP-KLH (100 μg in CFA) day 0 DNP-KLH (10 μg in NaCl) day 21		12,340
D (2)+DNP-KLH	(2) days -10,-6,-1 DNP-KLH days 0 and 21	0,1 1 10 100	2,380 6,880 7,220 12,400

are activated. We chose in vitro tests in order to avoid influence of metabolism. Two macrophage functions were tested:

- secretion of lysosomal enzymes and
- release of cytotoxic factors into the supernatant by DIVEMA-treated macrophage cultures.

In order to measure secretion of lysosomal enzymes, macrophages of normal, untreated NMRI mice were incubated in serum-free culture medium (RPMI) and treated with different amounts of compounds (1c), (3a), (4a) (mol.wt. 16,800), and (1d), (3b), (3c) (mol.wt. 24,000). Secretion of the following enzymes was measured after 24 h of incubation using established biochemical test systems:
N-Acetyl-β-D-glucosaminidase (N-Ac-Glu) (23), β-Glucuronidase (β-Glu) (24) and lactate dehydrogenase (LDH) (25). The latter enzyme is located in the cytoplasm and is not actively secreted (as the other two enzymes are) but only set free if the cell membrane is destroyed: the LDH liberated informs about viability of cell cultures or cytotoxic effects on macrophages of the substance tested.

None of the substances caused significant enzyme secretion by macrophages. Only the pyrrolidone derivative (3a) caused a slight increase of LDH with increasing dose of the polymer. This finding corresponds to the higher acute toxicity of (3b) as compared to unsubstituted DIVEMA derivative (1b).

These results are unexpected, since dextran sulfate with similar mol.wt. does induce secretion of enzymes (26).

Release of cytotoxic factors into the supernatant was tested by adding the supernatant of DIVEMA-treated macrophages to cultures of defined amounts of P815 tumour cells and measuring LDH from lysed tumour cells after 24 hr. of incubation.
Figure 1 shows schematically how the experiment was done. All substances tested, (1c), (1d), (3a), (3b), (4a), (4b), caused increasing cytotoxic activity of the supernatant with increasing doses of the polymers. Though the absolute values of LDH-liberation differ greatly from one experiment to another, this qualitative result is the same in all tests (11). One typical diagram showing the dose dependency of supernatant mediated toxicity caused by polymer (3b) is shown in Fig. 2. Relatively low concentrations (50 - 100 μg/ml) were sufficient to cause substantial cytotoxicity. As far as we know, this is the first example of supernatant-mediated toxicity caused by DIVEMA.

It is contradictory to the results of Schultz et al.

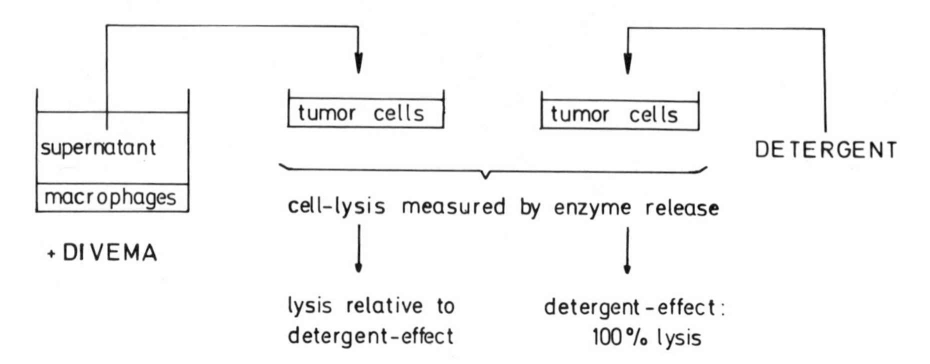

Figure 1. Test system for measuring cytotoxicity in the supernatant of macrophage cultures.

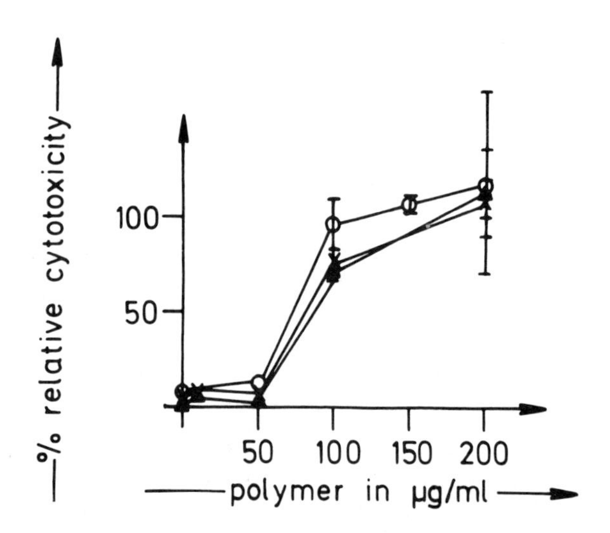

Figure 2. Supernatant-mediated cytotoxicity caused by DIVEMA derivative (3b in Table II) (three independent runs).

(21, 22); this may be due to the fact that these authors used serum-containing culture media. Our results demonstrate that even chemically modified DIVEMA causes - at least *in vitro* - the same biological effect (release of cytotoxic factors) as DIVEMA itself. This aspect is especially important if one uses DIVEMA as a carrier polymer for antitumour agents hoping to retain immune stimulating properties.

Endocytic Behaviour of DIVEMA and Derivatives

Macrophages have one more important property: they ingest foreign material via endocytosis, a process which resembles the ingestion of material by amoebae.

Polymers may either be taken up themselves or may stimulate endocytosis of other substances. We tested the latter possibility with two established test systems (27), (28):

- Uptake of ^{125}I-labelled poly(vinylpyrrolidone) (^{125}I-PVP) by rat macrophages. (This polymer is taken up by pinocytosis in the fluid phase).
- Uptake of ^{198}Au (colloidal) by mouse macrophages. (This colloid is mainly taken up after adsorption at the cell membrane).

Polymers (1a), (1b), (1d), (3a), (4a) and (5) were tested. Figure 3 shows the Endocytic Index (28) measured. The DIVEMA derivatives (1a), (3a) and (4a) have no significant effect on the pinocytosis of ^{125}I-PVP (they do not influence the rate of formation of pinocytic vesicles). Polymer (1d) shows marked inhibition of pinocytosis at the highest dose. (This latter finding *in vitro* corresponds to *in vivo* results of Breslow et al. (29)). They found that the removal of colloidal carbon from the blood of test animals is inhibited by high molecular weight DIVEMA).

Polymer (5) causes a decrease of pinocytosis with increasing dose.

The failure of derivatives (3a) and (4a) to influence pinocytosis may be due to their reduced polyanion-character. The inhibitory effect of the hydrophobic derivative (5) may be attributed to interactions of alkyl chains with the lipid membrane.

These results obtained with the ^{125}I-PVP-test system, as well as experiments using the colloidal gold test system, are discussed in more detail in ref. (27).

This work is part of the Ph.D. thesis of R. Schnee (11).

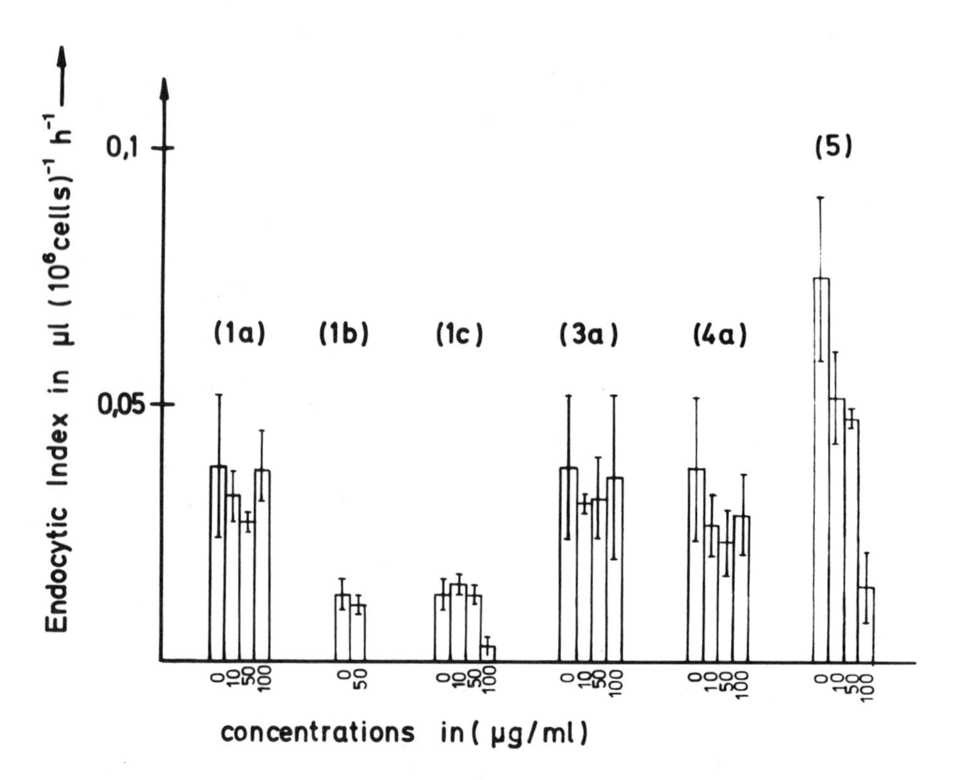

Figure 3. Stimulation of endocytosis by DIVEMA and DIVEMA derivatives. Endocytic index in the macrophage test system and ^{125}I–PVP used as standard substance.

Acknowledgements

We are indebted to Prof. Dr.D. Breslow for the DIVEMA samples.

This work was supported in part by the British Science Research Council and the Deutsche Forschungsgemeinschaft. R.S. thanks the Fonds der Chemie for a Ph.D. fellowship.

Literature Cited

1. Kostelnik, R.J. (Ed.), "Polymeric Delivery Systems", Midland Macromolecular Monographs Vol. 5, Gordon & Breach, N.Y. 1975
2. Donaruma, G.L., Vogl, O. (Eds.), "Polymeric Drugs" Academic Press, N.Y. 1978
3. Gros, L., Ringsdorf, H., Schupp, H., Angew.Chem.Int. Ed., 1981, 93, 305 - 325
4. Donaruma, L.G., Ottenbrite, R.M., Vogl, O. (Eds.), "Anionic Polymeric Drugs", John Wiley & Sons, N.Y. 1980
5. Butler, G.B., see 4., p. 49 - 142
6. Breslow, D.S., Pure Appl.Chem. 1976, 46, 103
7. Regelson, W., Morahan, P., Kaplan, A. in "Polyelectrolytes and Their Applications" (A. Rembaum, E. Sélégny, Eds.) D. Reidel Publ. Comp., Dordrecht+ Boston 1975, p. 131 - 144
8. Przybylski, M., Zaharko, D.S., Chirigos, M.A., Adamson, R.H., Schultz, R.M., Ringsdorf, H. Cancer Treat.Rep. 1978, 62, 1837
9. Hirano, T., Klesse, W., Ringsdorf, H., Makromol. Chem. 1979, 180, 1123
10. Hirano, R. Ringsdorf, H., Zaharko, D.S., Cancer Res. 1980, 40, 2263
11. Schnee, R., Ph.D. Thesis, Mainz 1980
12. Kobayashi, S., Ringsdorf, H., in preparation (unpubl. results)
13. Butler, G.B., see 4., p. 185 - 210
14. Snippe, H., Nab, J., van Eyk, R.V.W., Immunology 1974, 27, 761
15. Golan, D.R., Borel, Y., J.Exp.Med. 1971, 134, 1046
16. Schmidtke, J.R., Dixon, R.J., J.Exp.Med. 1972, 136 392
17. Wrede, J., Rüde, E., Thumb, R., Meyer-Debus, M. Eur.J.Immunol. 1973, 3, 798
18. Mishell, B.B., Shiigi, S.M., "Selected Methods in Cellular Immunology", W.H. Freeman & Co. San Francisco 1980
19. see 4., p. 234 - 236 and p. 196 - 198

20. James, K., McBride, B. and Stuart, H. (Eds.) "The Macrophage and Cancer" Proceedings of the European Reticuloendothelial Society, Symposium Edinburgh Sept. 12 - 14th, 1977, Edinburgh 1977
21. Schultz, R.M., Papamatheakis, J.B., Luetzler, J., Chirigos, M.A., Cancer Res. 1977, 37, 3338
22. Schultz, R.M., Papamatheakis, J.P., Chirigos, M.A., Cell Immunol. 1977, 29, 403
23. Woolen, J.P., Heyworth, R., Walker, P.G., Biochem.J. 1961, 78, 111
24. Talalay, P., Fishman, W.H., Huggins, C., J.Biol. Chem. 1946, 166, 757
25. Bergmeyer, H.U. (Ed.) "Methoden der enzymatischen Analyse" Verlag Chemie, Weinheim FRG, 1970, p. 533
26. Schorlemmer, H.U., Burger, R., Hylton, W., Allison, A.C., Clin.Immunol.Immunopathol. 1977, 7, 88
27. Pratten, M.K., Duncan, R., Cable, H.C., Schnee, R., Ringsdorf, H., Lloyd, J.B., Chemico-Biological Interactions, 1981, 35, 319
28. Pratten, M.K., Lloyd, J.B., Biochem.J. 1979, 180, 567
29. Breslow, D.S., Edwards, E.F., Newburg, N.R., Nature 1973, 246, 160

RECEIVED February 1, 1982.

8

Radical Polymerization of *exo* and *endo* *N*-Phenyl-5-norbornene-2,3-dicarboximides as Models for Addition-Curing Norbornene End-Capped Polyimides[1]

NORMAN G. GAYLORD and MICHAEL MARTAN

Gaylord Research Institute Inc., New Providence, NJ 07974

Homopolymerization of exo and endo cyclopentadiene-N-phenylmaleimide Diels-Alder adducts in the melt at 150-260°C and in chlorobenzene at 120°C was promoted in the presence of radical catalysts having half-lives of less than 2 hr at reaction temperature. The low molecular weight (< 2000), saturated (^{1}H-NMR in $CDCl_3$) homopolymer retained the configuration of the adduct when the polymerization temperature was below 200°C but contained both endo and exo configurations, due to isomerization, when prepared from either isomer at 260°C. The proposed saturated structure with N-phenylnorbornane-2,3-dicarboximide repeating units with 5,7-linkage, is presumably present in the cross-linked products from the high temperature curing of norbornene end-capped addition-type polyimides.

Aromatic polyimides have achieved considerable success as high temperature-resistant polymers. Since the desired high molecular weight polyimides are not readily processable, they are generated in situ by the thermal ring closure of the processable precursor poly(amic-acids) (2). The use of low molecular weight prepolymers, i.e. poly(amic-acids), as laminating resins is limited by their conversion to an intractable state before complete elimination of volatile by-products.

This problem has been alleviated by the use of low molecular weight polyimide prepolymers end-capped with norbornene rings (3, 4, 5)

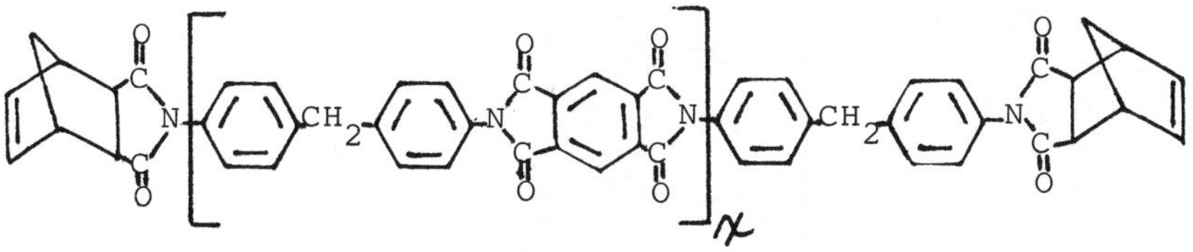

[1] See Reference 1.

0097-6156/82/0195-0097$06.00/0
© 1982 American Chemical Society

or monomeric reactants polymerizable thereto (5-8)

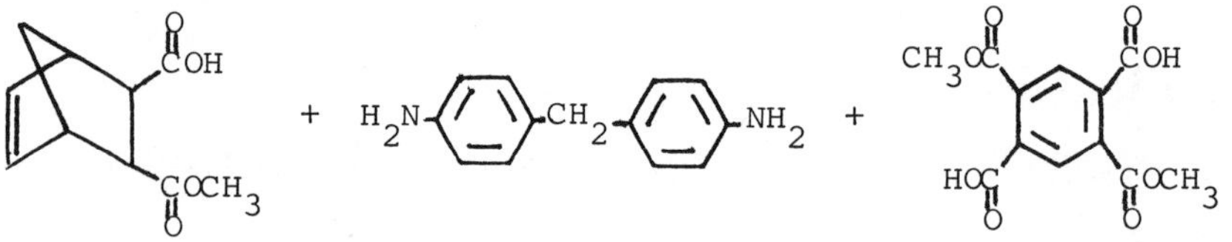

At elevated temperatures the terminal norbornene rings presumably undergo addition polymerization to promote crosslinking with minimal formation of volatile by-products.

The mechanism of the crosslinking reaction has been postulated as (a) dissociation of the terminal cyclopentadiene-N-arylmaleimide Diels-Alder adduct to the monomeric precursors, which immediately react to form an adduct which initiates the homopolymerization of the undissociated terminal norbornene rings to form a saturated polymer (5),

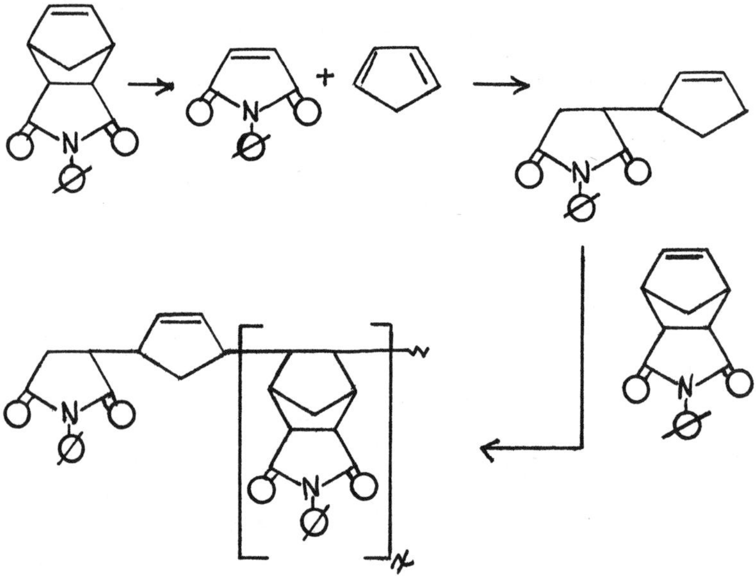

and as (b) homopolymerization of the norbornene ring in the terminal Diels-Alder adduct through an unspecified mechanism, although presumably involving ring opening and/or dissociation, to form an unsaturated alternating copolymer (8, 9).

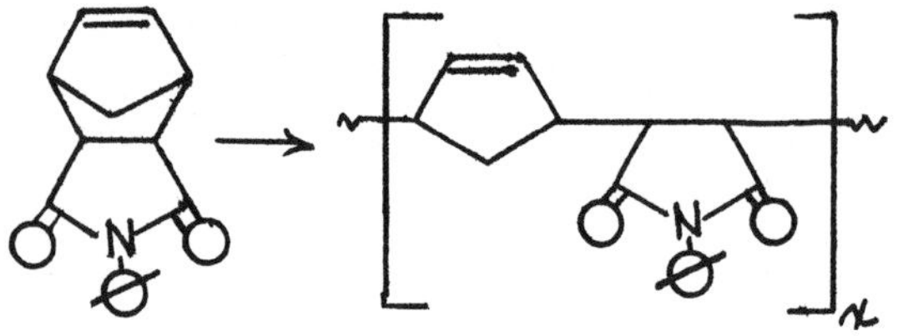

The radical catalyzed homopolymerization of the furan-maleic anhydride (F-MAH) Diels-Alder adduct yields a saturated homopolymer at temperatures below 60°C, and an unsaturated equimolar alternating copolymer at elevated temperatures, due to retrograde dissociation of the adduct (10, 11). The copolymerization of monomeric furan and maleic anhydride yields the same unsaturated alternating copolymer, independent of temperature (10).

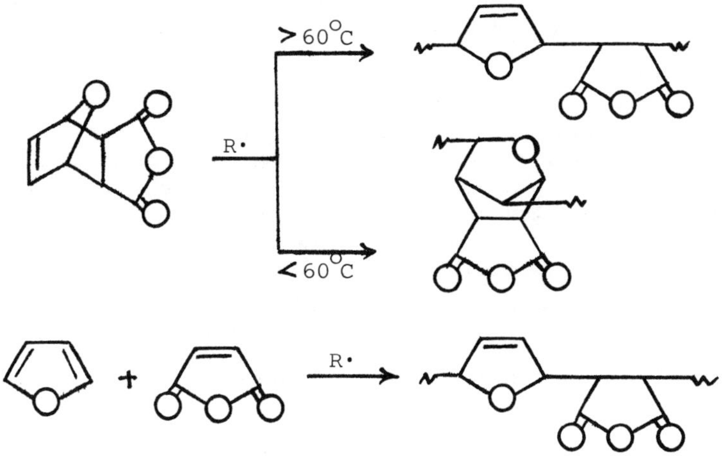

>60°C

R·

<60°C

R·

In contrast, the radical catalyzed homopolymerization of the cyclopentadiene-maleic anhydride (CPD-MAH) Diels-Alder adduct yields a saturated homopolymer at temperatures as high as 220°C, while retrograde dissociation occurs at even higher temperatures. Nevertheless, the copolymerization of monomeric cyclopentadiene and maleic anhydride yields a saturated 1:2 copolymer (12-15).

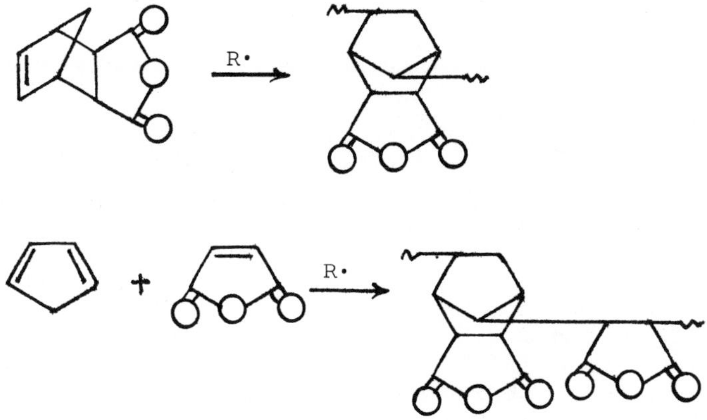

R·

R·

Since the F-MAH and CPD-MAH Diels-Alder adducts yield unsaturated and saturated polymers, respectively, as a result of radical catalyzed homopolymerization at elevated temperatures, it was of interest to investigate the radical catalyzed polymerization of the isomeric exo- and endo-cyclopentadiene-N-phenylmaleimide adducts, models for the norbornene end-capped polyimides whose thermal polymerization products have structures which have been diversely depicted as saturated and unsaturated, without experimental verification.

Experimental

The recrystallized CPD-MAH Diels-Alder adduct, endo-cis-norbornene-2,3-dicarboxylic anhydride, mp 165°C, was maintained at 220°C for 4 hr to yield a mixture of endo and exo anhydrides, mp 101°C. The crude product was recrystallized 3 times from benzene to isolate the exo isomer, mp 141°C. The endo- and exo-cis-N-phenyl-5-norbornene-2,3-dicarboximides, mp 144° and 197°C, respectively, were prepared by reaction of the endo- and exo-CPD-MAH adducts with aniline.

The polymerization of the adducts was carried out by adding 0.2ml of either t-butyl peracetate (tBPA), t-butyl perbenzoate (tBPB) or t-butyl hydroperoxide (tBHP-70 containing 70% tBHP and about 30% di-t-butyl peroxide or tBHP-90 containing 90% tBHP and about 10% t-butyl alcohol), by syringe in four equal portions over a 20 min period to 2g of the adduct in a rubber-capped tube immersed in a constant temperature bath. The mixture was then maintained at temperature for an additional 40 min. Reactions at 120°C were conducted in 3ml chlorobenzene. The reaction product was dissolved in acetone and precipitated with methanol twice.

IR spectra were recorded from films cast on NaCl plates from acetone solution using a Perkin Elmer Model 21 spectrophotometer. NMR spectra were obtained at 60MHz in $CDCl_3$ at 25°C using tetramethylsilane as internal standard.

Results and Discussion

After 5 hr at 170°C, endo-N-phenyl-5-norbornene-2,3-dicarboximide failed to undergo isomerization and remained unchanged. Similarly, no isomerization of the exo adduct was detectible by IR analysis after 5 hr at 170°C. However, after 30 min at 260°C, the exo or endo adduct was converted to a 40/60 endo/exo equilibrium mixture, which remained unchanged after 2 hr at 260°C.

The homopolymerization of the endo and exo adducts was carried out in the melt at 150° to 260°C and in chlorobenzene at 120°C (Table I). Polymer was obtained when the catalyst was used at a temperature where the half-life was short, e.g. less than 2 hr, conditions shown to be effective in the homopolymerization of maleic anhydride (16), norbornene (17, 18) and 5-norbornene-2,3-dicarboxylic anhydride (CPD-MAH adduct) (12-15), as well as the copolymerization of acyclic dienes (19, 20) and cyclopentadiene (12-15) with maleic anhydride.

TABLE I

HOMOPOLYMERIZATION OF CYCLOPENTADIENE-N-PHENYLMALEIMIDE DIELS-ALDER ADDUCTS

Catalyst	$t_{1/2}$, min	Solvent	Temp, °C	Yield, %
		endo adduct		
tBPA	66	CB	120	60
tBHP-70	720	none	150	0
tBPB	4	none	150	20
tBHP-70	0.8	none	210	20
tBHP-70	0.01	none	260	50
		exo adduct		
tBPA	66	CB	120	46
tBPA	2	CB	155	60
tBHP-90	1	none	260	30

Elemental and NMR analyses indicated that the homopolymers contained equimolar amounts of cyclopentadiene and N-phenylmaleimide, irrespective of polymerization temperature within the 120-260°C range. The homopolymers from the exo and endo adducts had softening points of 240° and 265-280°C, respectively. The polymer from the polymerization of the endo adduct at 120°C had a molecular weight of 1645 (vapor pressure osmometry).

The NMR spectra of the homopolymers obtained from the endo and exo adducts at temperatures below 200°C indicated that the polymers retained the configuration of the monomeric adducts. The spectrum of the endo adduct homopolymer contained a peak centered at 6.8τ, not present in the spectrum of the exo adduct homopolymer. The endo adduct homopolymer spectrum contained broad peaks at 7.0-7.7 and 7.8-8.6τ, while the exo adduct homopolymer spectrum had a narrow peak at 7.0-7.4τ, centered at 7.25τ, and a broader peak at 8.3-8.9τ. Both spectra had the peaks of aromatic hydrogens at 2.4-3.0τ. Neither spectrum contained peaks at about 4.0τ attributable to vinylic hydrogens, indicating the absence of unsaturation (Figure 1).

The NMR spectra of the homopolymers obtained from the endo and exo adducts at 260°C were similar to each other and different from those obtained from the homopolymers of either adduct polymerized at lower temperatures. The spectra contained peaks arising from both endo and exo isomers, indicating that the adducts had undergone some endo-exo isomerization. No unsaturation was shown in the spectrum of either homopolymer obtained at 260°C (Figure 2).

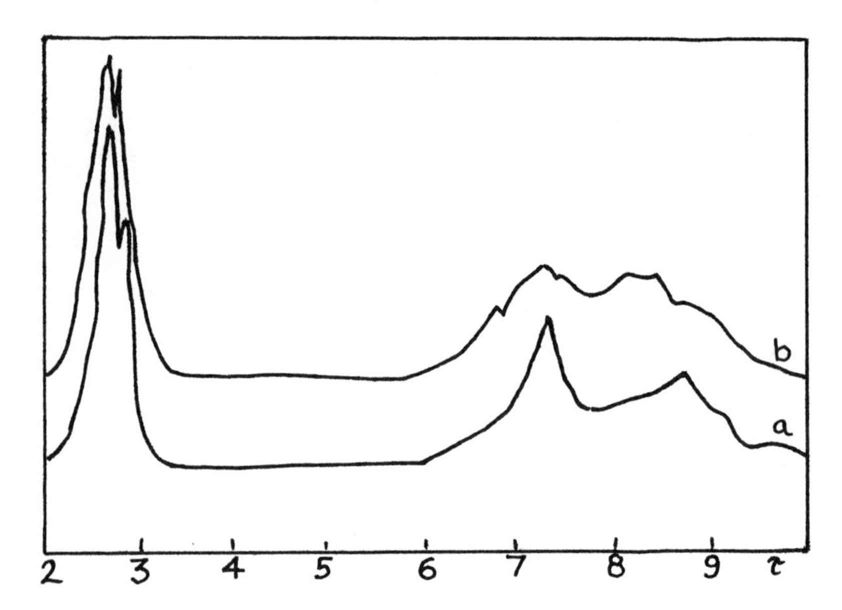

Figure 1. NMR spectra of homopolymer from peroxide catalyzed polymerization of the exo adduct at 155°C (a), and the endo adduct at 120°C (b).

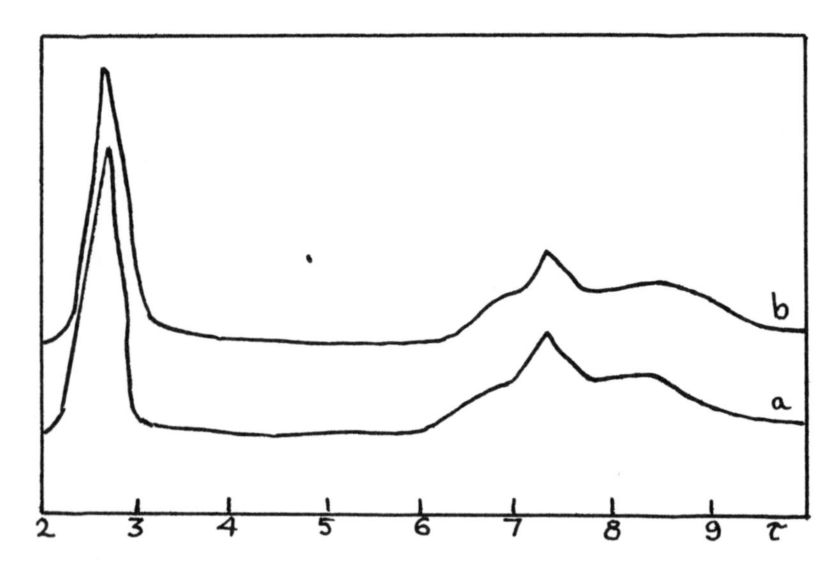

Figure 2. NMR spectra of homopolymer from peroxide catalyzed polymerization of the exo adduct at 260°C (a), and the endo adduct at 260°C (b).

The saturated structures of the homopolymers obtained both at temperatures below 200°C with retention of the endo or exo configuration of the monomeric adduct, and at 260°C with both endo and exo configurations, suggests a simple double bond addition.

It has previously been proposed that the saturated products from the homopolymerization of norbornene (17) and the CPD-MAH Diels-Alder adduct (12-15), and probably from the furan-maleic anhydride Diels-Alder adduct (11), have rearranged structures. An analogous structure would arise from the homopolymerization of the cyclopentadiene-N-phenylmaleimide CPD-NPMI adduct, as follows:

The retrograde dissociation of the Diels-Alder adduct to generate the diene and dienophile, followed by the polymerization of the comonomer charge transfer complex, has previously been proposed (12-15) in the polymerization of the CPD-MAH adduct, in the presence of peroxides having short half-lives at the elevated polymerization temperatures. Under these conditions, the ground state complex, presumably involved in endo-exo isomerization, is converted to the excited state complex which undergoes the indicated polymerization.

The crosslinked products resulting from the high temperature "pyrolytic polymerization" (5) curing of the norbornene end-capped "addition-type" polyimide prepolymers, such as P13N, P10P, etc., as well as those prepared by means of in situ polymerization of monomeric reactants (PMR), at least those obtained at temperatures up to 260°C, may be presumed to have saturated structures similar to those obtained from the polymerization of the isomeric N-phenyl-5-norbornene-2,3-dicarboximides at elevated temperatures. The weight loss or volatilization noted during curing has been attributed to the escape of volatile by-products, e.g. cyclopentadiene, which are formed by dissociation of the norbornene end-caps (3, 8). This is consistent with our proposed polymerization mechanism and polymer structure.

Subsequent to the presentation of the results reported herein (21), ^{13}C and 270 MHz ^{1}H-NMR studies of the thermally induced polymerization of the N-phenyl-5-norbornene-2,3-dicarboximides (22) and ^{13}C-NMR studies of norbornene end-capped polyimide prepolymers (23) were reported. Endo-exo isomerization of the adducts takes place at 200°C and above. Retrograde dissociation of the CPD-NPMI adduct occurs above 275°C and the saturated polymers obtained at temperatures of 285°C and above contain units derived from the exo and endo CPD-NPMI adducts as well as units derived from N-phenylmaleimide and the Diels-Alder adducts of liberated cyclopentadiene and the CPD-NPMI adducts. The proposed saturated structures suggest simple addition across the norbornene (CPD-NPMI and CPD-CPD-NPMI) and maleimide (NPMI) double bonds, in the absence of concrete evidence to support the rearranged structures postulated herein and in the products of earlier catalytic polymerizations. The proposed copolymerization of the norbornene and maleimide double bonds is in agreement with our finding that the peroxyester-catalyzed copolymerization of monomeric cyclopentadiene and N-phenylmaleimide at 85°C and 155°C yields saturated copolymers containing the NPMI and CPD in 2/1 mole ratio (24).

Literature Cited

1. Donor-Acceptor Complexes in Copolymerization. LXI.
2. Sroog, C. E. Macromol. Revs. 1976, 11, 161.
3. Lubowitz, H. R. (to TRW Inc.) 1970, U. S. Patent 3,528,950.
4. Lubowitz, H. R. Polymer Preprints 1971, 12, 329.
5. Serafini, T. T.; Delvigs, P.; Lightsey, G. R. Applied Polymer Symposium 1973, 22, 89; and other references cited therein.
6. Serafini, T. T.; Delvigs, P.; Lightsey, G. R. J. Appl. Polym. Sci. 1972, 16, 905.
7. Serafini, T. T. in May, C. A., Ed. "Resins for Aerospace"; ACS Symposium Series 1980, 132, 15.
8. Dynes, P. J.; Panos, R. M.; Hamermesh, C. L. J. Appl. Polym. Sci. 1980, 25, 1059.
9. Dokoshi, N. Kobunshi 1974, 23, 125.
10. Gaylord, N. G.; Maiti, S.; Patnaik, B. K.; Takahashi, A. J. Macromol. Sci.-Chem. 1972, A6, 1459.

11. Gaylord, N. G.; Martan, M.; Deshpande, A. B. J. Polym. Sci., Polym. Chem. Ed. 1978, 16, 1527.
12. Gaylord, N. G.; Solomon, O.; Stolka, M.; Patnaik, B. K. J. Polym. Sci., Polym. Lett. Ed. 1974, 12, 261.
13. Gaylord, N. G.; Solomon, O.; Stolka, M.; Patnaik, B. K. J. Macromol. Sci.-Chem. 1974, A8, 981.
14. Gaylord, N. G. Polymer Preprints 1976, 17, 666.
15. Gaylord, N. G.; Deshpande, A. B.; Martan, M. J. Polym. Sci., Polym. Lett. Ed. 1976, 14, 679.
16. Gaylord, N. G.; Maiti, S. J. Polym. Sci., Polym. Lett. Ed. 1973, 11, 253.
17. Gaylord, N. G.; Mandal, B. M.; Martan, M. J. Polym. Sci., Polym. Lett. Ed. 1976, 14, 555.
18. Gaylord, N. G.; Deshpande, A. B.; Mandal, B. M.; Martan, M. J. Macromol. Sci.-Chem. 1977, A11, 1053.
19. Gaylord, N. G.; Stolka, M.; Takahashi, A.; Maiti, S. J. Macromol. Sci.-Chem. 1971, A5, 867.
20. Gaylord, N. G.; Stolka, M.; Patnaik, B. K. J. Macromol. Sci.-Chem. 1972, A6, 1435.
21. Gaylord, N. G.; Martan, M. Polymer Preprints 1981, 22, 11.
22. Wong, A. C.; Ritchey, W. M. Macromolecules 1981, 14, 825.
23. Wong, A. C.; Garroway, A. N.; Ritchey, W. M. Macromolecules 1981, 14, 832.
24. Gaylord, N. G.; Martan, M.; Deshpande, A. B. Polym. Bulletin 1981, 5, 623.

RECEIVED February 25, 1982.

9

Cyclization Reaction of *N*-Substituted Dimethacrylamides in the Crystalline and Glassy States at 77 K

T. KODAIRA, M. TANIGUCHI, and M. SAKAI

Fukui Universiy, Department of Industrial Chemistry, Faculty of Engineering, Fukui 910, Japan

N-Methyldimethacrylamide(MDMA) cyclizes to give a 5-membered cyclic radical at 77 K. This conclusion has been deduced based upon the ESR measurements of MDMA irradiated at 77 K. ESR studies of MDMA deuterated at its four methylene protons of the methacryl groups supported above conclusion. Crystal data of MDMA tell us that the formation of 5-membered ring is the most favorable reaction in the crystalline state. It was found that N-benzyl and N-propyl substituents can also form 5-membered ring at 77 K in both their crystalline and glassy states. Molecular structures of these monomers in these states are not available, but these results suggest that BDMA and PDMA have favorable conformations for the formation of 5-membered ring.

Cyclopolymerization of N-substituted dimethacrylamide(RDMA) has been reported by several research groups(1-7). The repeating units which are expected in the polymers prepared from RDMA are 5-membered ring, 6-membered ring, and uncyclic pendant groups. Structural investigations have shown that their main repeating unit is 5-membered ring. They contain small amounts of 6-membered ring but do not contain any detectable pendant double bond(4,5). Propyl(PDMA) and benzyl(BDMA) substituents form glassy and supercooled liquid states when they are quenched rapidly from above their melting points. The polymerization at the supercooled liquid state yields polymers with essentially the same structure as those obtained in the liquid state(7). The contents of 5-membered ring are almost the same for both polymers formed in the liquid and supercooled liquid states while the fraction of 5-membered ring decreases on their polymerization in the solid state. However, the polymers formed in the solid state were found to be amorphous and the mechanism of the polymerization could be explained fundamentally based upon that of liquid state(7). The detailed investigation on the cyclopolymerization of RDMA has led to

0097-6156/82/0195-0107$06.00/0
© 1982 American Chemical Society

the conclusion that the high tendency of RDMA to cyclization was due to the non-polymerizability of its monofunctional counterpart (5). N-Alkyl-N-isobutyrylmethacrylamide(I) just corresponds to the monofunctional counterpart of RDMA and it would not polymerize to give a high polymer(4,5). It has been proposed that the lower the polymerizability of the monofunctional counterparts of bifunctional monomers, the higher their cyclopolymerizability(5) and the validity of the hypothesis has been proven in several monomers (8-11). Some of the reported results on the cyclopolymerization of bifunctional monomers are considered to be the additional evidences which support above hypothesis, judging from their high tendency to cyclization and low polymerizability of their monofunctional counterparts(12,13,14).

During these studies it was found that the rate-determining step of the cyclopolymerization of RDMA in the solid and supercooled liquid states is the intramolecular cyclization reaction (5,7,15). This conclusion has been deduced because only an un-

I: $CH_2=C(CH_3)-C(=O)-N(R)-C(=O)-CH(CH_3)-CH_3$

RDMA: $CH_2=C(CH_3)-C(=O)-N(R)-C(=O)-C(CH_3)=CH_2$ $\xrightarrow{H\bullet}$ II: $CH_3-\dot{C}(CH_3)-C(=O)-N(R)-C(=O)-C(CH_3)=CH_2$

II $\rightarrow$ III: $CH_3-C(CH_3)$—$C(CH_3)-CH_2\bullet$ (ring: $O=C-N(R)-C=O$) $\xrightarrow{RDMA}$ IV: $-\overset{\beta}{CH_2}-\overset{\alpha}{\dot{C}}(CH_3)-C(=O)-N(R)-C(=O)-C(CH_3)=CH_2$

$CH_2=C(CH_3)-X$ $\xrightarrow{H\bullet}$ V: $CH_3-\dot{C}(CH_3)-X$

cyclic propagating radical(IV) was observed on the measurements of ESR spectra of irradiated RDMA at polymerization temperature. Radiation polymerization of unsaturated monomers in the solid state proceeds through the initiation radical(V) which is formed by the additiom of hydrogen atom to their double bond(16). Based upon these facts the formation of an initiation radical(II) in RDMA by the reaction between a hydrogen atom and one of the double bonds of RDMA is reasonably assumed. This initiation radical cyclizes to give the 5-membered cyclic radical(III) which reacts with other monomer to yield the uncyclic propagating radical(IV). At higher temperature where polymerization proceeds, only the uncyclic propagating radical was detected as mentioned previously. Decrease in temperature reduces molecular motion and the polymerization does not proceed. These considerations suggest that we might be able to observe the initiation radical(II) and the 5-membered cyclic radical(III) at lower temperature. The initiation radical(II) affords a seven line spectrum if two methyl groups are rotating freely, and the 5-membered cyclic radical gives a 3-line spectrum if the rotation abount the C-C bond of $>C-CH_2\cdot$ group is free. A 12-line spectrum has already been assigned to the uncyclic propagating radical(15). Therefore, we can easily distinguish these three radicals and accordingly, the two reaction steps, i.e., an intramolecular cyclization and intermolecular propagation between the radical III and a RDMA molecule, by measuring the ESR spectra of irradiated RDMA.

These reactions should be strongly influenced by the phases where they are carried out. The effect of the phases on these reactions can be studied by using RDMA because PDMA and BDMA can form glassy, supercooled liquid and crystalline states. The purpose of the present investigation is to identify the initiation(II) and cyclic(III) raicals, and to see the effect of the phases on these intramolecular cyclization and intermolecular propagation reactions. Crystalline structure of MDMA has recently been determined(17). Therefore, these reaction procedures can be studied in connection with the crystalline structure in the case of MDMA.

Experimental

RDMA was prepared according to the procedure described previously(6,18). MDMA and PDMA deuterated at their $CH_2=C<$ protons (MDMA-d_4 and PDMA-d_4, respectively) were synthesized starting from deuterated methacrylic acid which was obtained by the hydrolysis of deuterated ethylmethacrylate prepared based upon the reported procedure using CD_2O(19). The degree of deuteration was 100% according to NMR measurements. Single crystals of MDMA and MDMA-d_4 were grown by slow evaporation of benzene solution at room temperature. The crystals obtained from both the monomers had form shown in Figure 1. The coordinate axis system employed for the ESR measurements is also given in Figure 1.

Samples for ESR spectra were prepared in Suprasil tubes. The shape of the bottom of the sample tubes was modified to accomodate the single crystal according to Kurita(20). However an error of about 10° in determining the orientation of crystal was inevitable because of the difficulty in settling the crystal. ESR measurements were made with JEOL JES-FE-1X or Varian E-3 EPR X-band spectrometer. γ-Ray irradiation was carried out by using ^{60}Co source.

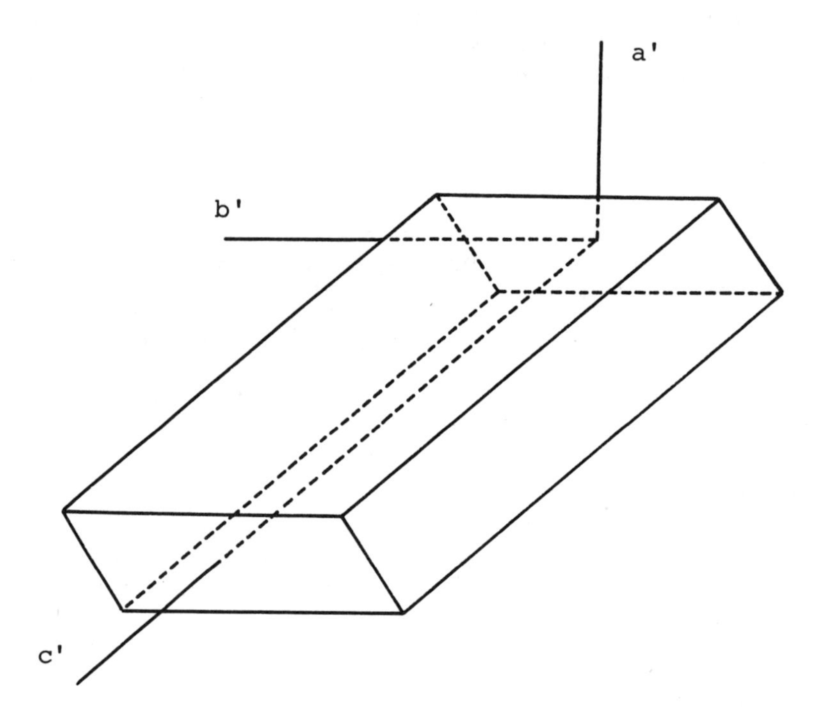

Figure 1. The crystal form of MDMA and MDMA-d_4 *and the axes employed for ESR measurements.*

Results and Discussion

ESR Studies of MDMA and MDMA-d_4. An isolated radical and radical pairs are trapped in MDMA when it is subjected to inonizing radiation at 77 K. The ESR measurements of MDMA at 77 K after warming at 195 K for 5 min showed that the radical pairs disappear during the heat-treatment but the spectral pattern of the isolated radical does not change though its intensity decreases slightly (15). The ESR spectrum illustrated in Figure 2a is the one obtained on measurements at 77 K after the heat-treatment at 195 K. With increasing temperature from 77 K to 193 K the pattern changes gradually and at 193 K it became the triplet with an intensity of 1:2:1 and with a coupling constant of 26.5 gauss(Figure 2b). The spectral pattern changed when the ESR spectra were recorded at different orientations of the single crystal which suggests that the hyperfine splitting is due to α-protons. When temperature was lowered again to 77 K the spectral pattern returned to that of Figure 2a. These phenomena were precisely reproduced when temperature was increased and decreased repeatedly. The triplet changed gradually to the 12-line spectrum ascribed to the uncyclic propagating radical(IV) on further increase in temperature as shown in Figures 2c and 2d. The 3-line spectrum is due to the interaction between an unpaired electron and two protons, and accordingly, is ascribed to the 5-membered cyclic radical(III). The temperature dependence of the spectral pattern is interpreted as restricted rotation about the C-C bond of $>C-CH_2\bullet$ group at lower temperature and its free rotation at higher temperature. To confirm this assignment ESR spectra of irradiated MDMA-d_4 were measured. If the 3-line spectrum is due to the 5-membered cyclic radical(III), the radical(VI) should be observed in the case of MDMA-d_4, and its gives a 5-line spectrum based upon the interaction between an un-

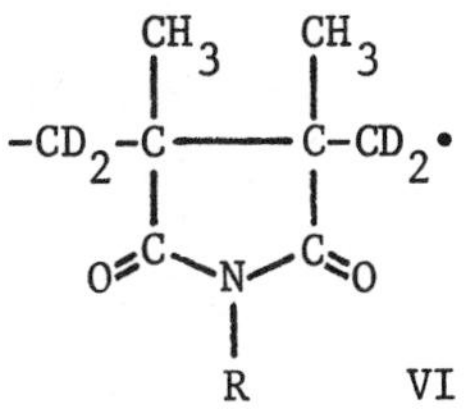

CH_3 CH_3
$-CD_2-C$——$C-CD_2\bullet$
$O{=}C-N-C{=}O$
R
VI

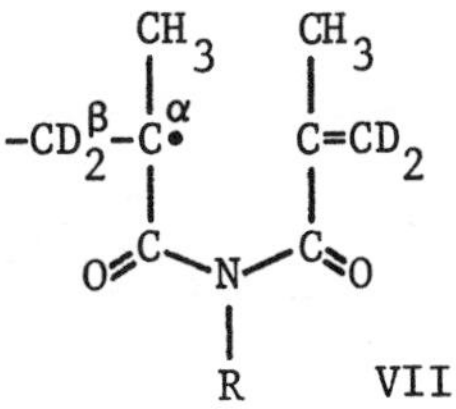

CH_3 CH_3
$-CD_2^{\beta}-C^{\alpha}\bullet$ $C{=}CD_2$
$O{=}C-N-C{=}O$
R
VII

paired electron and two deuteriums as shown in Figure 3c. The coupling constant is predicted to be 4.1 gauss. The results obtained by the ESR measurements at the same orientation of crystal as that of MDMA are illustrated in Figure 3. The big difference between the ESR spectra of MDMA(Figure 2) and MDMA-d_4(Figure 3) indicates that the methylene protons of the methacryl groups are responsible for the spectra in Figure 2. In accordance with our consideration, the quintet with a splitting constant of 4.0 gauss was detected at higher temperature(Figure 3b). The spectral change between a and b in Figure 3 indicates the free rotation about the C-C bond of $>C-CD_2\bullet$ group at higher temperature and re-

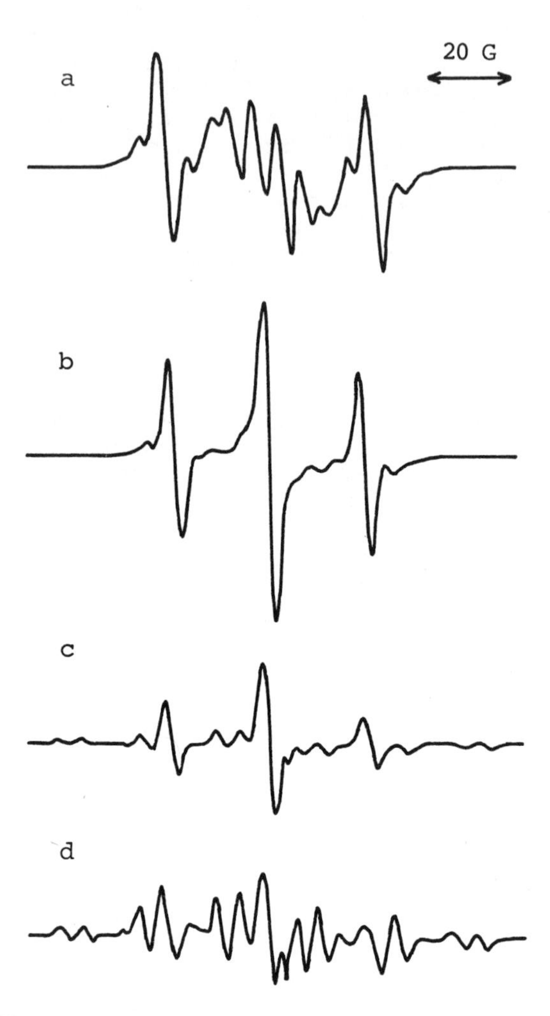

Figure 2. ESR spectra of MDMA single crystal irradiated at 77 K to 5.34 Mrad and heat-treated for 5 min at 195 K. The magnetic field is parallel to the c′ axis in the b′c′ plane, and recorded at 77 K (a), 183 K (b), 213 K (c), and 243 K (d). Spectra b, c, and d were measured after heating to the respective temperature for 20 min. Sensitivity of the instrument was kept constant except for the measurement of spectrum d.

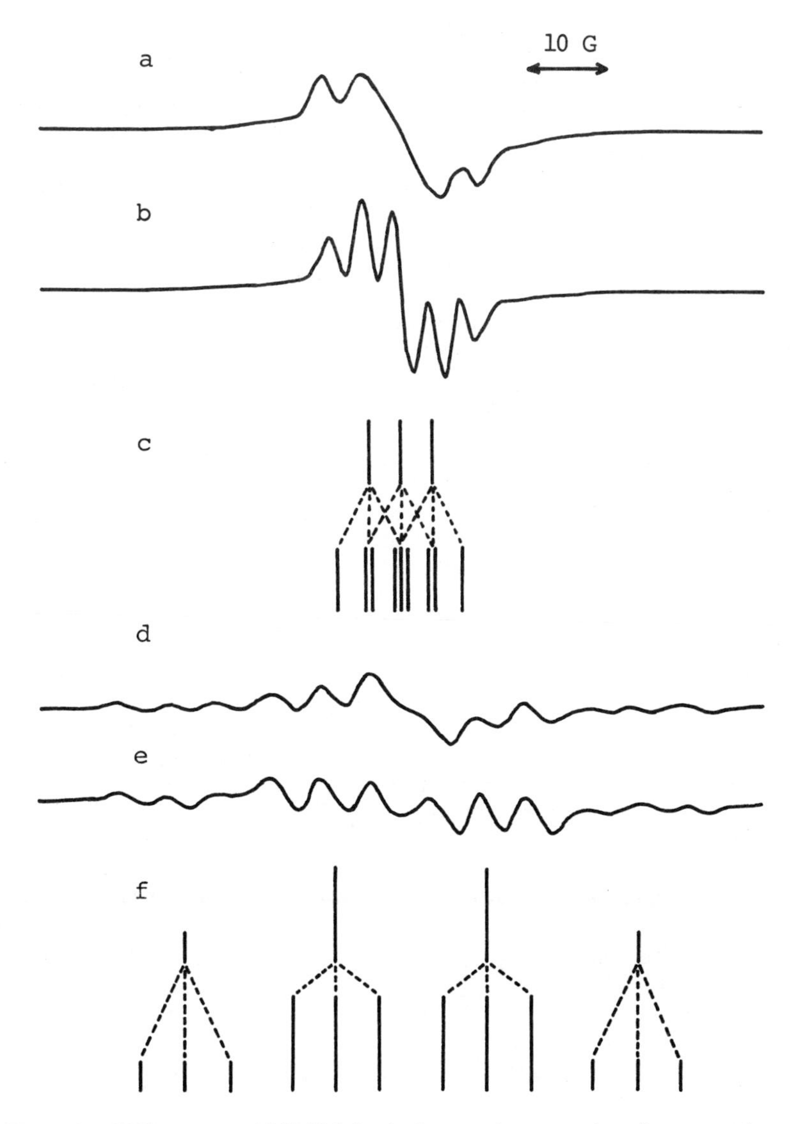

Figure 3. ESR spectra of MDMA-d_4 *single crystal measured at the same orientation as that of MDMA in Figure 2 after the irradiation at 77 K to 5.34 Mrad and subsequent warming to 195 K for 5 min, and recorded at 77 K (a) and 183 K (b). Spectra d and e were observed at 77 K after heat-treatment at 213 K and 243 K, respectively, for 5 min. Diagrams c and f are predicted patterns for b and e, respectively. Sensitivity of the instrument was kept constant except for the measurement of spectrum e.*

stricted rotation at lower temperature. On further increase in temperature the quintet changed gradually to a 12-line spectrum (Figures 3d and 3e). If we assume that the two β-deuteriums of the methylene group of radical VII take the same conformations as those of two β-methylene protons of the uncyclic propagating radical(IV) of MDMA, the 12-line spectrum is attributable to the uncyclic propagating radical(VII) based upon the following consideration. ESR studies of irradiated MDMA showed that the angles between the π-orbital and the projection of the two C_β-H bonds along the C_α-C_β bond of radical(IV) are 0° and 120°, respectively, and accordingly, each deuterium should give rise to coupling constants of 6.1 and 1.5 gauss. The latter splitting is considered to be too small to be observed due to line broadening in present experimental conditions. Therefore, each component of the quartet produced by the freely rotating methyl group bonded to C_α should splits into a triplet with a coupling constant of 6.1 gauss and with an equal intensity. The 12-line spectrum illustrated as a stick diagram in Figure 3f should be observed. It agrees well with 12-line spectrum obtained from irradiated MDMA-d_4 as shown in Figure 3e.

The 12-line spectra in Figures 2 and 3 are not seen at lower temperature although sensitivity of the instrument was kept constant during these measurements. These facts suggest that uncyclic propagating radicals were formed during the heat-treatment. The intensities of the 12-line and 3-line spectra of Figure 2 were plotted against temperature from 183 K to 263 K in Figure 4. The total amount of radicals and the quantity of the 5-membered cyclic radical decrease with increasing temperature, but the 12-line which was not observed at 183 K appears, and it increases its intensity and then decreases. Figure 4 shows that the intermolecular reaction between the 5-membered cyclic radical(III) and a MDMA molecule occurs at the temperature range from 183 K to 263 K.

MDMA molecule can take various conformations. The distance between the two intramolecular double bonds is rather long if it takes an extended structure. Such a conformation is unfavorable

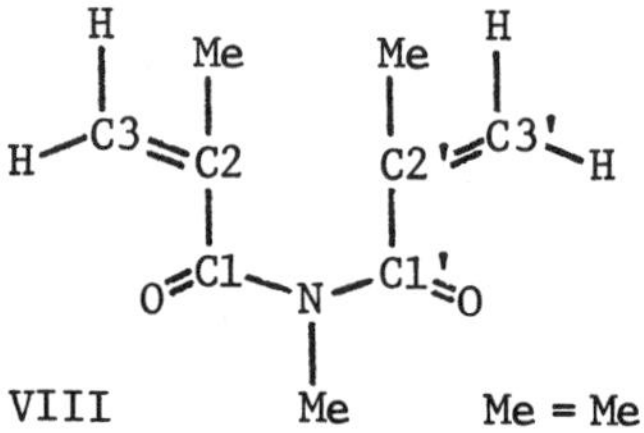

H, Me, Me, H; H–C3=C2, C2'=C3'–H; O=C1–N–C1'=O; Me

VIII Me = Methyl group

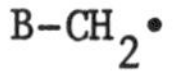

$B\text{-}CH_2\cdot$

IX

for cyclization reaction at 77 K because of the restricted rotation about chemical bonds. Therefore, it is interesting to know the molecular structure of MDMA in crystalline lattice. A crystallographic study of MDMA revealed that the distance between the carbon atoms C2 and C2'(VIII) is 2.909 Å, and it is the shortest one among the distances between possible reaction sites(17). This result

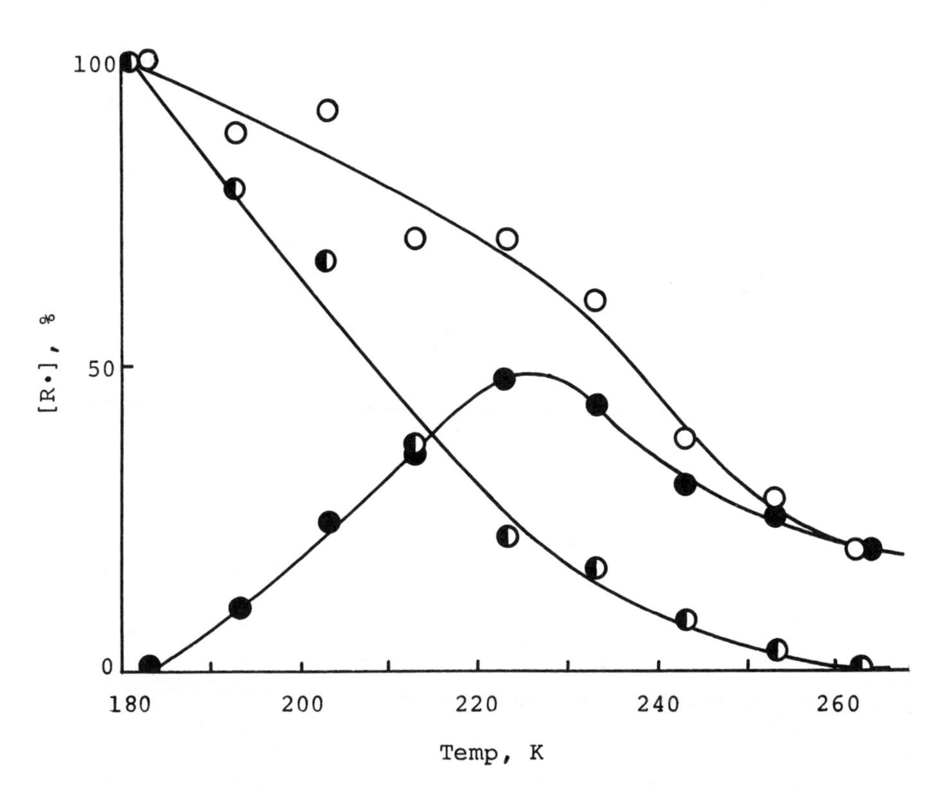

Figure 4. Variation of radical concentration [R·] trapped in MDMA irradiated at 77 K to 5.34 Mrad, and recorded at each temperature after heat-treatment for 20 min. Key: ◐, *3-line;* ●, *12-line; and* ○, *total concentration.*

suggests that the formation of 5-membered ring is the most favorable reaction in the solid state of MDMA. It can not be said anything from the ESR spectra of Figures 2a and 2b on the structure of the part written as B in Scheme IX except the fact that it does not contribute to the hyperfine structure of ESR. However, this crystallographic study supports the assignment of the triplet to the monomeric 5-membered cyclic radical(III). It appeared that temperature lower than 77 K is required to observe initiation radical in the solid state cyclopolymerization of MDMA.

ESR Studies of PDMA, PDMA-d_4, and BDMA. To see the effect of the substituents on nitrogen of RDMA and the phases where the reactions are carried out, ESR spectra of irradiated PDMA and BDMA were investigated. These monomers are favorable for this purpose because they form glassy, supercooled liquid, and solid states, depending on the conditions by which they are treated. Irradiated polycrystalline PDMA gives a broad 3-line spectrum with a coupling constant of about 20 gauss as shown in Figure 5a. With increasing temperature from 77 K to 193 K the central line of the triplet sharpens strongly, leaving the side bands almost unaltered(Figure 5b). When temperature was lowered to 77 K the spectral pattern returned to that shown in Figure 5a. These phenomena were essentially the same as those observed in the irradiated single crystal of MDMA. Therefore, the 3-line spectrum in Figure 5a is attributable to the 5-membered cyclic radical(III). ESR spectrum of irradiated PDMA-d_4 is shown in Figure 5c. It does not show hyperfine structure, but the large difference between Figures 5a and 5c indicates that the methylene protons of methacryl groups are closely related to the active species. When temperature was increased to 193 K the quintet with a coupling constant of 4.0 gauss was detected(Figure 5d). The hyperfine splitting is considered to be due to the interaction between an unpaired electron and two deuteriums. The spectrum illustrated in Figure 5c reappeared when ESR was measured again at 77 K. These spectral change just corresponds to what has been observed in irradiated MDMA-d_4 and the active species which yields these spectra is ascribed to the 5-membered cyclic radical(VI).

Glassy PDMA irradiated at 77 K contains the anion radical of methacryl group which affords 3-line spectrum with a coupling constant of 11 gauss. This anion radical could be bleached at 77 K by UV-light(21). ESR spectrum obtained after photo-bleach is shown in Figure 6a. When temperature was raised, its pattern changed to that depicted in Figure 6b. Central part decreases its intensity as temperature is lowered(Figure 6c). This temperature dependence of spectral pattern is very similar to that observed in crystalline MDMA and PDMA. Thus, the radical which yields these ESR spectra is ascribed to the 5-membered cyclic radical(III). PDMA-d_4 in glassy state yields spectra shown in Figures 6d-6f. These spectra were recorded after photo-bleach at 77 K. It is difficult to identify these spectra, but they clearly show that

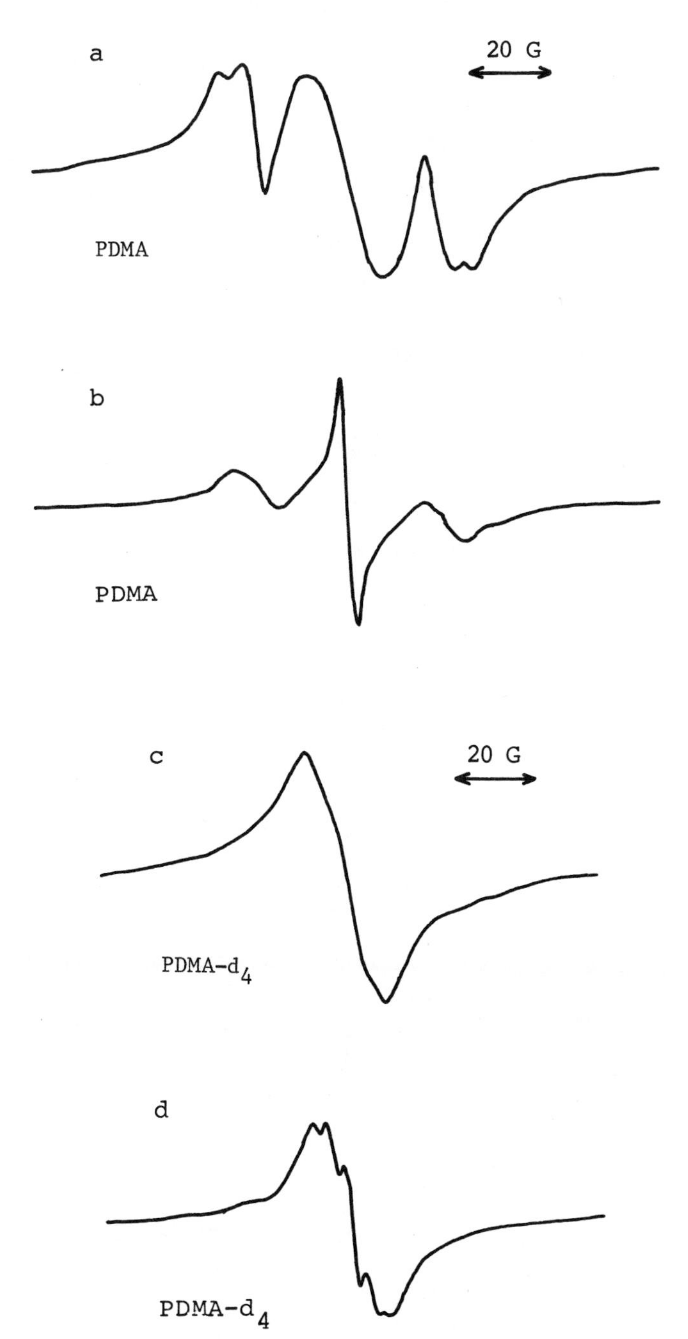

Figure 5. ESR spectra of polycrystalline PDMA and PDMA-d_4 *irradiated at 77 K to 5.34 Mrad. Recorded at 77 K (a,c) and 193 K (b,d).*

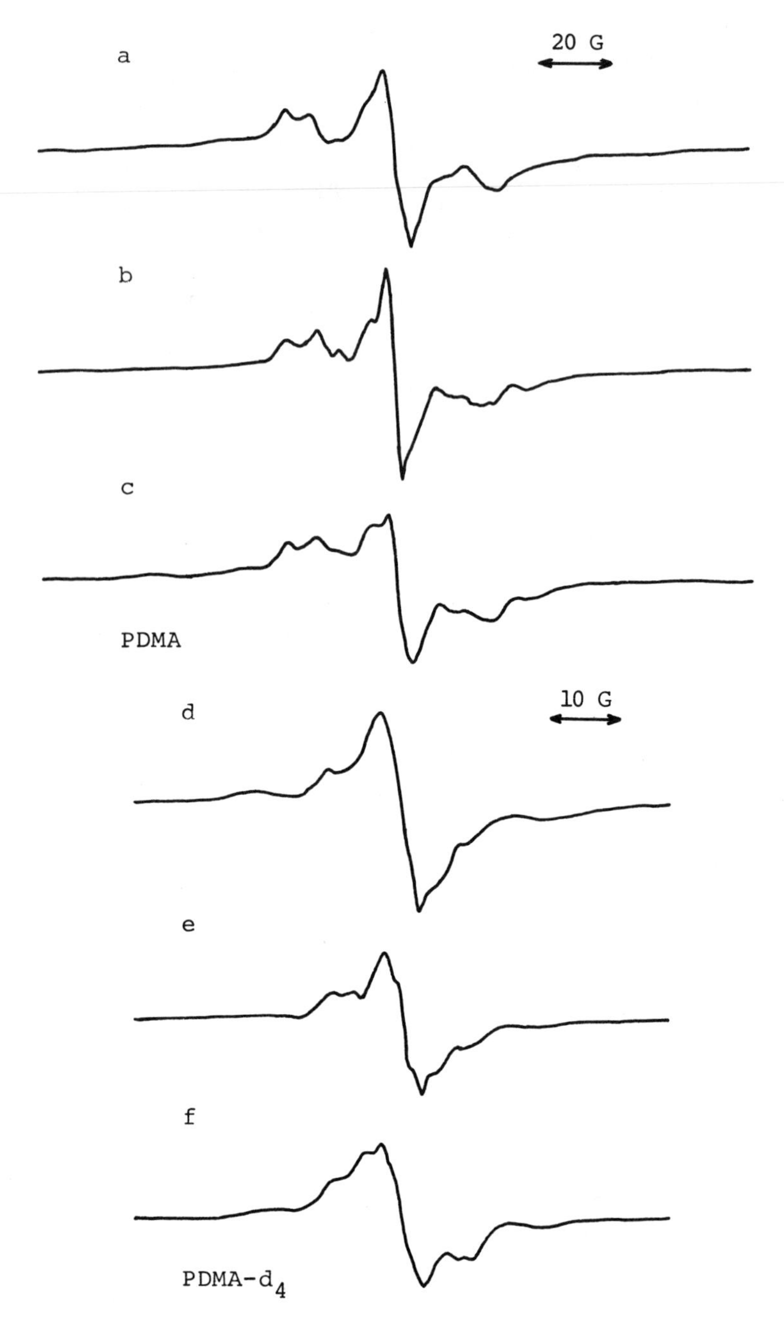

Figure 6. ESR spectra of glassy PDMA and PDMA-d_4 *bleached by UV-light after the irradiation at 77 K to 5.34 Mrad, and recorded at 77 K (a,d) and 163 K (b,e). Spectra c and f were observed at 103 K after the measurement at 163 K.*

the methylene protons of methacryl groups are responsible for the spectra in Figures 6a and 6b and support their assignment to 5-membered cyclic radical(III).

BDMA irradiated at 77 K in the solid state contains some unknown species, judging from the spectrum shown in Figure 7a. However, it disappears on increasing the temperature of the system and spectrum shown in Figure 7b was observed at 173 K. This spectrum changes to the pattern given in Figure 7c on measurement at 103 K. This spectral change between Figures 7b and 7c is reversible, depending on temperature. ESR spectra of glassy BDMA are depicted in Figure 8. They are taken after photo-bleach at 77 K because an anion radical of methacryl group was detected. ESR spectra of BDMA in crystalline and glassy state and their temperature dependence just correspond to those observed in crystalline and glassy PDMA, respectively. These facts permit us to consider that 5-membered cyclic radical(III) is formed in both glassy and crystalline states of BDMA irradiated at 77 K.

Conclusion

All these results show that cyclization reaction occurs at 77 K in the three substituents studied. Crystal data of MDMA tell us that the formation of 5-membered ring is the most favorable reaction in the crystalline state of MDMA(17). Based upon this result PDMA and BDMA are considered to have favorable conformation for the cyclization to form 5-membered cyclic radical(III) in both the crystalline and glassy states. To detect the initiation radical (II) irradiation and ESR measurements have to be carried out at lower temperature than 77 K.

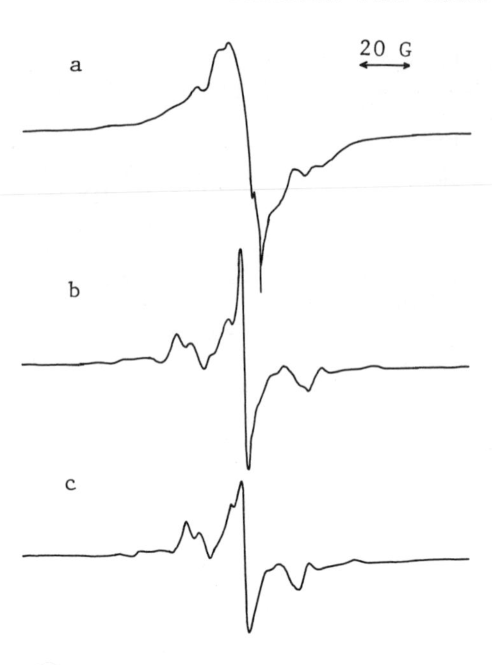

Figure 7. ESR spectra of polycrystalline BDMA irradiated at 77 K to 5.34 Mrad, and recorded at 77 K (a) and 173 K (b). Spectra c was observed at 103 K after the measurement at 173 K.

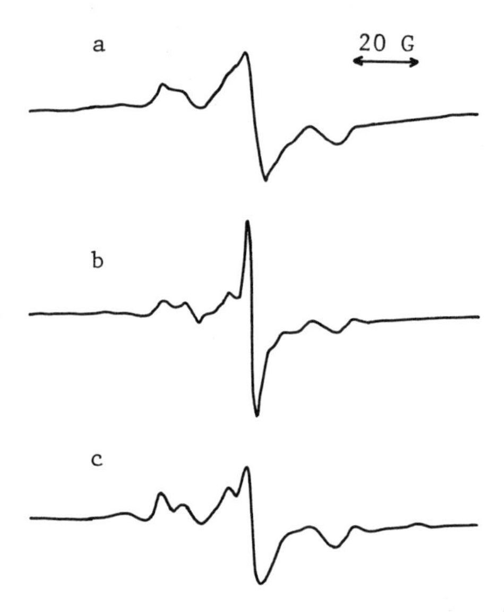

Figure 8. ESR spectra of glassy BDMA bleached by UV-light after the irradiation at 77 K to 5.34 Mrad, and recorded at 77 K (a) and 193 K (b). Spectra c was observed at 103 K after the measurement at 193 K.

Acknowlegements

The Grant-in-Aid for individual research(365322) from the Ministry of Education, Japan, is acknowledged. A part of this work was done under the Visiting Researchers' Program of Kyoto University Research Reactor Institute.

Literature Cited

1. Sokolova, T. A.; Rudkovskaya, G. D. Vysokomol. Soedin. 1961, 3, 706.
2. Götzen, F.; Schröder, G. Makromol. Chem. 1965, 88, 133.
3. Azori. M.; Plate, N. A.; Rudkovskaya, G. D.; Sokolova, T. A.; Kargin, V. A. Vysokomol. Soedin. 1966, 8, 759.
4. Butler, G. B.; Myers, G. R. J. Macromol. Sci. Chem. 1970, A5, 135
5. Kodaira, T.; Aoyama, F. J. Polym. Sci. Polym. Chem. Ed. 1974, 12, 897.
6. Kodaira, T.; Aoyama, F.; Morishita, K.; Tsuchida, M.; Nogi, S. Kobunshi Ronbunshu, 1974, 31, 682.
7. Kodaira, T.; Niimoto, M.; Aoyama, F.; Yamaoka, H. Makromol. Chem. 1978, 179, 1791.
8. Kodaira, T.; Sakai, M.; Yamazaki, K. J. Polym. Sci. Polym Lett. Ed. 1975, 13, 521.
9. Kodaira, T.; Ishikawa, M.; Murata, O. J. Polym. Sci. Polym. Chem. Ed. 1976, 14, 1107.
10. Kodaira, T.; Yamazaki, K.; Kitoh, T. Polym. J. 1979, 11, 377.
11. Kodaira, K.; Murata, O.; Edo, Y. J. Polym. Sci. Polym. Lett. Ed. 1980, 18, 737.
12. Aso, C.; Tagami, S.; Kunitake, T. J. Polym. Sci. Part A-1, 1969, 7, 497.
13. Corfield, G. C.; Crawshaw, A. J. Polym. Sci. Part A-1, 1969, 7, 1179.
14. Casorati, E.; Martina, A.; Guaita, M. Makromol. Chem. 1977, 178, 765.
15. Kodaira, T.; Morishita, K.; Yamaoka, H.; Aida, H. J. Polym. Sci. Polym. Lett. Ed. 1973, 11, 347.
16. O'Donnell, J. H.; McGarvey, B.; Moravetz, H. J. Amer. Chem. Soc. 1964, 86, 2322.
17. Higuchi, T. Private communication.
18. Rudkovskaya, G. D.; Sokolova, T. A. Zh. Org. Khim. 1966, 2, 1220.
19. Mannich, C.; Ritsert, K. Chem. Ber. 1924, 57, 1116.
20. Kurita, Y. Nippon Kagaku Zasshi, 1964, 85, 833.
21. Kodaira, T. Unpublished results.

RECEIVED February 1, 1982.

10

Mössbauer Studies of Polymers from 1,1′-Divinylferrocene

G. C. CORFIELD, J. S. BROOKS, and S. PLIMLEY

Sheffield City Polytechnic, Sheffield S1 1WB, England

Linear, saturated, polymers obtained by radical initiation of 1,1'-divinylferrocene have the Mossbauer parameters (δ, 0.23 mm s^{-1}; ΔE_Q, 2.29 mm s^{-1}) expected for cyclopolymers with three-carbon bridged ferrocene units in the main chain. However, cationic initiation yields polymers which exhibit differences from the radical polymers in Mossbauer (δ, 0.27 mm s^{-1}; ΔE_Q, 2.40 mm s^{-1}) and other spectroscopic studies. The cationic polymers may contain a bicyclic unit or a ladder structure in the chain.

The use of modern spectroscopic methods can reveal information on the microstructures of cyclopolymers which previously went undetected (1). Here we report the application of Mossbauer spectroscopy to a study of the polymers of 1,1'-divinylferrocene (DVF).

The polymerisation of DVF was first investigated by two independent research groups (2-7) who reported that soluble polymers could readily be obtained using free radical and cationic initiators. These authors proposed that cyclopolymerisation had occurred with the formation of linear polymers having three-carbon bridged ferrocene units (I), and some acyclic units (II), in the chain. Evidence for structure (I) as the predominant unit in the polymer chain was provided by the low level of vinyl unsaturation detectable by NMR or infrared spectroscopy and the observation that bands attributable to a bridged ferrocene (8) were to be found in the infrared spectra of these polymers.

Recent investigations (9, 10) of bridged ferrocenes by Mossbauer spectroscopy have demonstrated that a short three-carbon bridge causes changes in the iron-ring geometry which leads to differences in Mossbauer parameters between three-carbon bridged ferrocenes and compounds with four- or five-carbon bridges. We have re-investigated the polymerisation of

0097-6156/82/0195-0123$06.00/0
© 1982 American Chemical Society

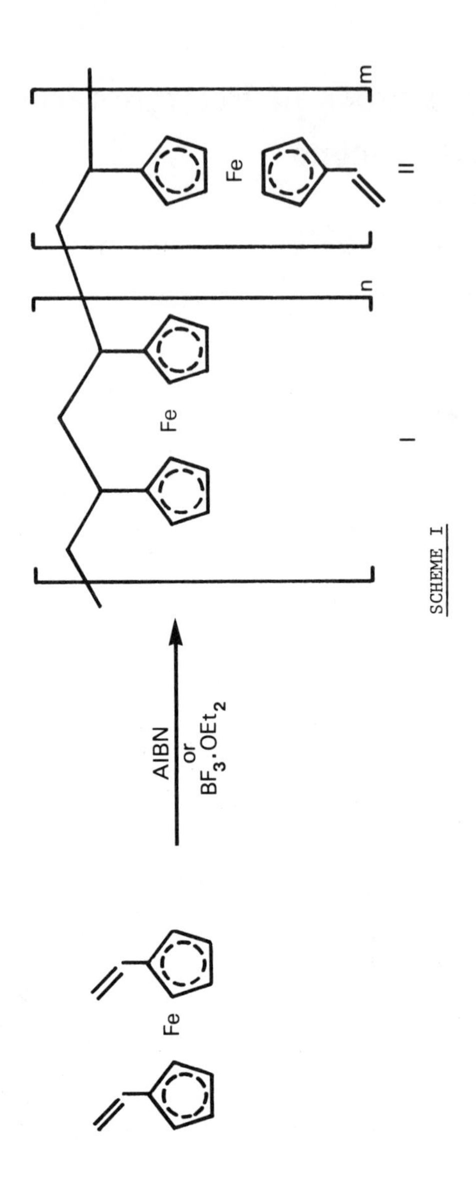

SCHEME I

DVF and studied the polymers by Mossbauer spectroscopy, using the isotope Fe-57, to seek supporting evidence for the three-carbon bridged unit (I) as the predominant unit in the chain.

Experimental

DVF was prepared from ferrocene by established procedures (2, 3, 4, 11, 12, 13), recrystallised from methanol (m.p. 39.5-41.0 °C), and shown to be pure by thin layer chromatography (silica gel with toluene as eluent).

Polymerisations using free radical and cationic initiators were carried out essentially as described by Kunitake *et al* (2, 3, 4).

^{57}Fe Mossbauer spectra were obtained using a digital constant acceleration spectrometer having a symmetrical triangular velocity drive waveform. A 10 m Ci ^{57}Fe in Pd source was used and all experiments were carried out at room temperature. The spectrometer was calibrated between runs using the magnetic splitting of enriched ^{57}Fe absorber foil. The data were folded to determine the zero velocity position and the folded data were fitted with Lorentzian functions by a non-linear least squares fitting program (14).

Results and Discussion

Polymerisation. DVF has been polymerised using radical and cationic initiators, and typical results are given in Table I. Radical initiation in dilute solutions gave products which were predominantly soluble in benzene. Cationic initiation resulted in higher conversions, again mainly to soluble polymers. These results, which are similar to those obtained by Kunitake *et al* (2, 3, 4), indicate that linear polymers, possibly cyclopolymers, have been obtained.

TABLE I POLYMERISATION OF 1,1'-DIVINYLFERROCENE

Initiator/ mol.dm^{-3}	*Monomer mol.dm^{-3}*	*Solvent*	*Temp °C*	*Conversion*[a] *% Soluble,*	*% Insoluble*[b]
AIBN/ 1.56x10^{-3}	0.10	C_6H_6	64	50	3
$BF_3.OEt_2$/ 0.01	0.20	CH_2Cl_2	0	48	16

[a]Precipitated into methanol

[b]Solubility in benzene

Infrared Spectroscopy. The infrared spectra of the polymers (Figure 1) were in accord with previous work (2, 3, 4). Of particular interest in the spectrum of the polymers obtained by radical initiation [PDVF(radical)] are the distinct bands at 810 cm^{-1} and 850 cm^{-1}, which have been observed in a large number of heterobridged ferrocene compounds (8). This pair of characteristic bands is not clearly seen in the polymers obtained by cationic initiation [PDVF(cationic)] where a broad absorption, centred at approximately 830 cm^{-1}, is observed. Other differences between the spectra can be seen, particularly in the *fingerprint* region of the spectrum. These differences have been interpreted (2, 3, 4) as due to differences in the extent of cyclisation in the polymers. However, an alternative explanation is that this indicates that the polymers have significantly different structures.

NMR Spectroscopy. The ^{1}H NMR spectra of the polymers (Figure 2) were also similar to those previously published. The spectrum of PDVF (radical) contained no detectable vinyl unsaturation, whereas that of PDVF (cationic) showed a significant amount of unsaturation, but considerably less than that expected if one vinyl group remained unreacted on each ferrocene unit. However, it is of interest that the signal ascribed by Kunitake *et al* (2, 3, 4) to vinyl unsaturation is a single broad peak at δ5.90 ppm rather than the pair of signals expected at approximately δ4.9 and 6.1 ppm for a vinyl substituent on a ferrocene ring. Also, the aliphatic protons patterns are different for PDVF (radical) and PDVF (cationic). Kunitake *et al* observed this and attributed it to a difference in polymer conformations, influenced by the degree of cyclisation. Again, an alternative explanation is that the polymers are structurally different.

^{13}C NMR spectra of these polymers (Figure 3) also exhibit clear differences, which support our suggestion that the polymers are structurally different. A detailed analysis and discussion of the ^{13}C NMR spectra will appear in a subsequent publication.

Mössbauer Spectroscopy. Mössbauer spectroscopic studies of the polymers are reported here, and these provide the evidence to substantiate the structural differences between the polymers.

The Mossbauer effect can be used to compare nuclear transition energies in two materials with high precision. This is useful in obtaining chemical information as the nucleus is sensitive to changes in its electronic environment. Two hyperfine interactions which give rise to information about the electronic environment are isomer shift (IS) and quadrupole splitting (QS). Figure 4 shows a Mossbauer spectrum, typical of that obtained from ferrocene compounds, with these two parameters distinguished. IS results from the electrostatic interaction between the charge distribution of the nucleus and those electrons which

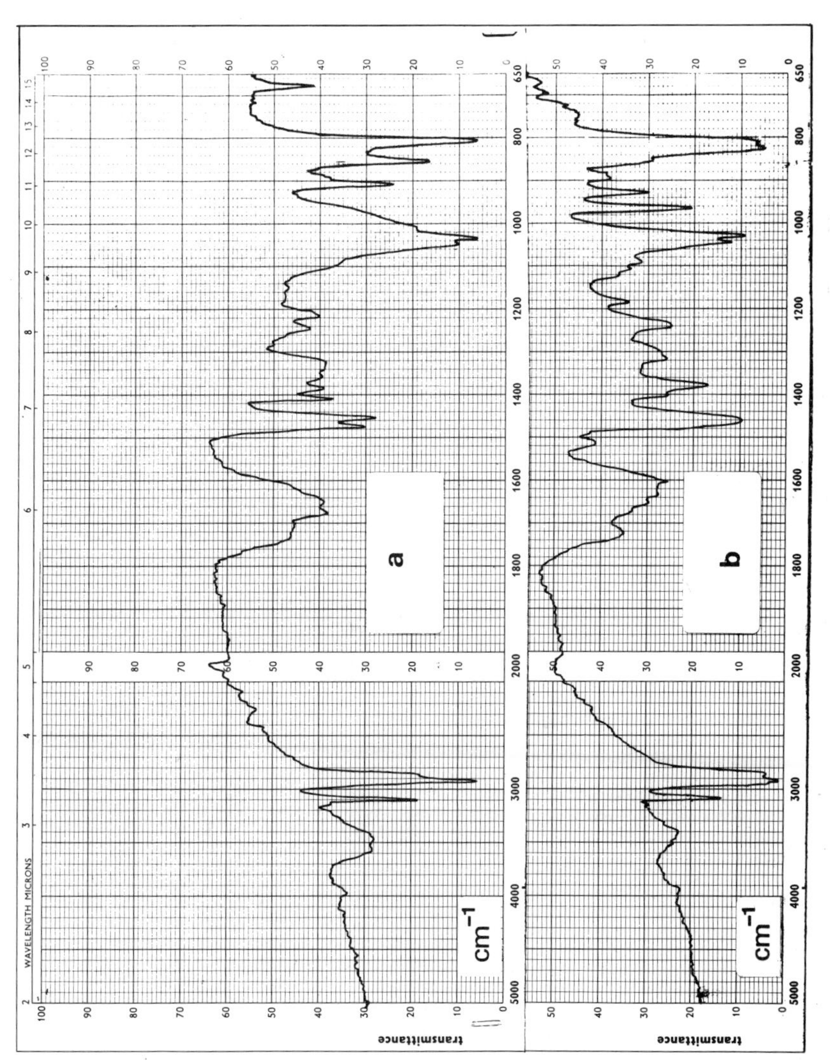

Figure 1. IR spectra of PDVF(radical) (a) and PDVF(cationic) (b).

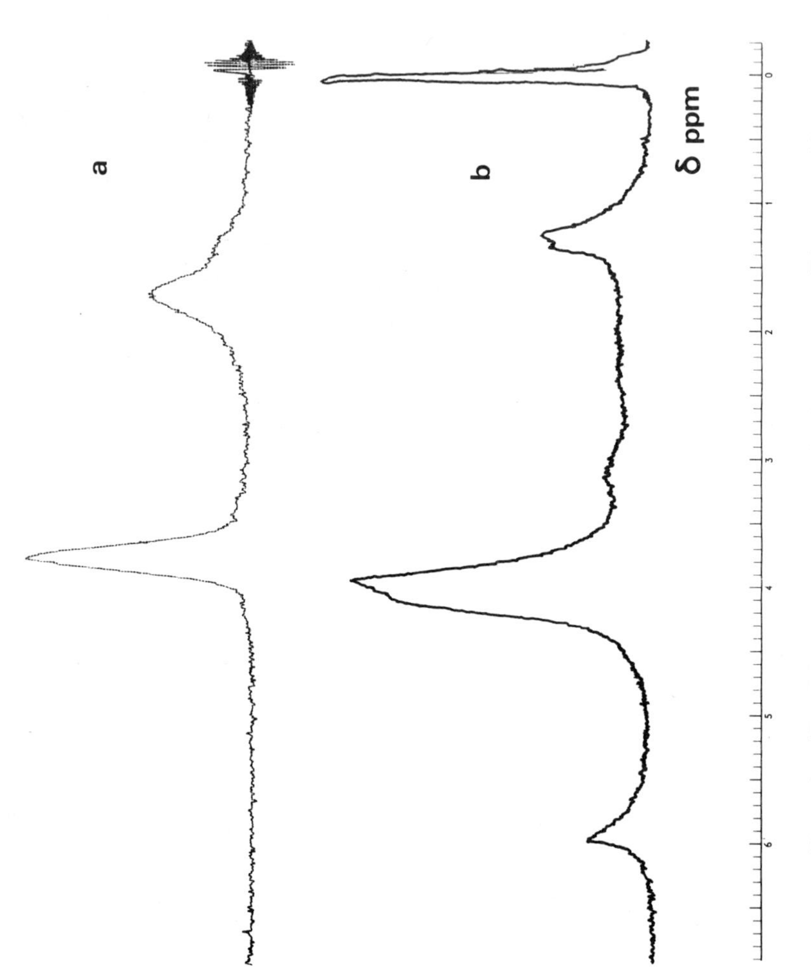

Figure 2. ^{1}H NMR spectra of PDVF(radical) (a) and PDVF(cationic) (b).

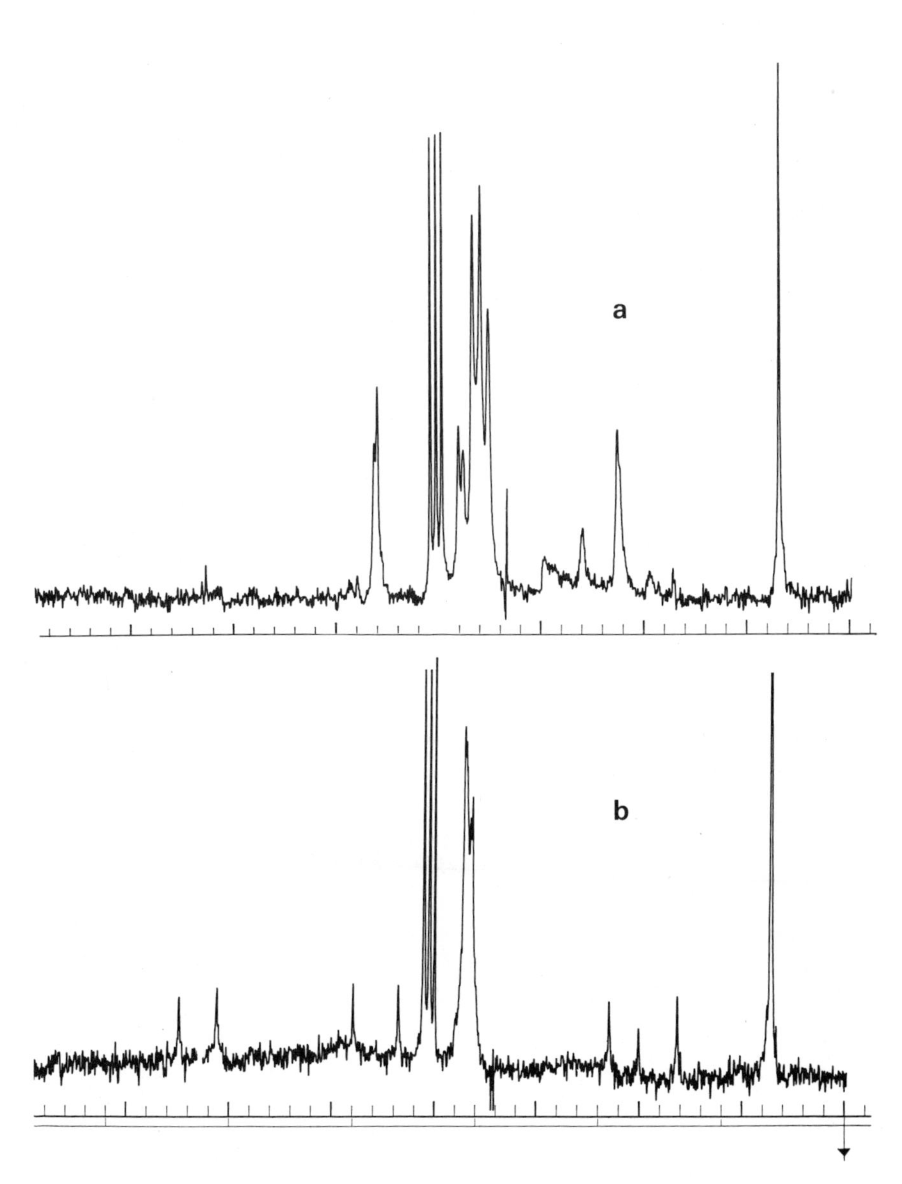

Figure 3. ^{13}C *NMR spectra of PDVF(radical) (a) and PDVF(cationic) (b).*

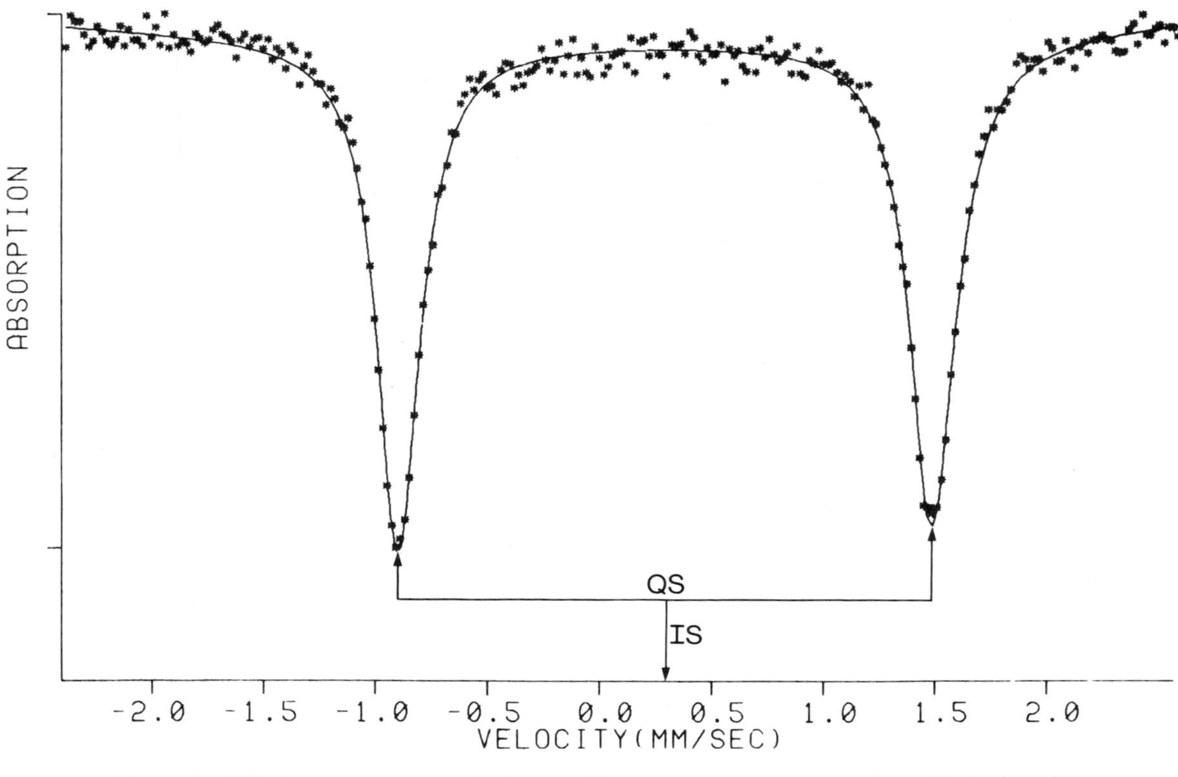

Figure 4. *Mössbauer spectrum typically found for ferrocene compounds. Key: IS, isomer shift; and QS, quadrupole splitting.*

have a finite probability of being found in the region of the nucleus. Thus, IS is influenced by the s-electron density at the nucleus; these being the only orbitals which have a finite probability of interacting with the nuclear charge density. The s-electron density, and hence IS, of the iron atom is related therefore to its bonding and valency, since other orbitals can affect the s-electron nuclear density by shielding effects. QS results from the interaction between the nuclear quadrupole moment and the electric field gradient of the nucleus. Variations in the distribution of electrons within the d-orbitals of the iron atom, due to variations in the substituents attached to the atom, cause changes in the electric field gradient tensor and hence give rise to differences in QS.

IS leads to a shift of the centre of gravity of the Mossbauer spectrum away from zero Doppler velocity. QS results in partial removal of the degeneracy of the nuclear excited state resulting, for Fe-57, in a doublet in the Mossbauer spectrum. The Mossbauer spectra are computer fitted using IS (δ/mm s^{-1}) and QS (ΔE_Q/mm s^{-1}) as position parameters for each pair of Lorentzian lines, with the *full width at half height* (Γ/mm s^{-1}) and the relative depth as line shape parameters. The computer program varies the fitting parameters to minimise χ^2, and the solid line drawn through the data points represents the *best fit* to the data.

Table II gives Mossbauer parameters for the compounds which comprise the synthetic route from ferrocene to DVF. It can be seen that for these compounds, which are all in the same oxidation state, IS remains fairly constant. Ring substitution produces negligible changes in IS since the molecular orbitals of ferrocene are not formed with orbitals of a single ring atom but with the ring as a whole. The variations in QS are more significant. The electric field gradient at the nucleus arises largely from the π-bonded ligands and little change is observed when a σ-bonded substituent is introduced to the ferrocene ring, as in 1,1'-bis(1-hydroxyethyl)ferrocene. However, where π orbital overlap can occur, there is some redistribution of the electrons and variations are observed, as with 1,1'-diacetylferrocene and DVF.

Investigation of PDVF by Mossbauer spectroscopy shows that PDVF (radical) and PDVF (cationic) have significantly different parameters (Table III and Figure 5). Thus the repeating units in these polymers must have different structures, which influence the ferrocene ring system in different ways. Mossbauer measurements on a variety of bridged ferrocene compounds (9, 10) have demonstrated that for compounds with only three carbons in a heteroannular bridge (for example, III) the strain causes a significant decrease in the values of IS and QS (Table III) due to a decrease in the iron-ring distances, caused by the tilting of the ring system (Figure 6). With less-strained bridges comprising four (IV) or five (V) carbons the IS and QS values are

Table II

Mössbauer Parameters of Monomer and Intermediates

Compound	*Isomer Shift*,[a] δ/mm s^{-1}	*Quadrupole Splitting*, ΔE_Q/mm s^{-1}	*Line Width* Γ/mm s^{-1}
Ferrocene	0.27(2)	2.39(2), 2.397[b]	0.12(1)
1,1'-Diacetylferrocene	0.26(2)	2.19(2)	0.13(1)
1,1'-Bis(1-hydroxyethyl)ferrocene	0.27(2)	2.39(2)	0.12(1)
1,1'-Divinylferrocene	0.27(2)	2.27(2)	0.12(1)

[a] Relative to Fe in Pd

[b] Theoretical QS, Trautwein *et al* (15)

Table III

Mössbauer Parameters of Polymers and Model Compounds

Compound	*Isomer Shift*,[a] δ/mm s^{-1}	*Quadrupole Splitting*, ΔE_Q/mm s^{-1}	*Line Width* Γ/mm s^{-1}
PDVF (radical)	0.23(2)	2.29(2)	0.13(1)
PDVF (cationic)	0.27(2)	2.40(2)	0.12(1)
PVF (radical)	0.28(2)	2.40(2)	0.12(1)
PVF (cationic)	0.27(2)	2.39(2)	0.13(1)
Methylferrocene[b]		2.39(5)	
3-carbon ring model (III)[c]	0.231(2)	2.256(3), 2.30[b], 2.285[d]	
4-carbon ring model (IV)[c]	0.244(2)	2.351(5)	
5-carbon ring model (V)[c]	0.253(4)	2.344(5)	

[a]Relative to Fe in Pd
[b]Lesikar (18)
[c]Nagy *et al* (9, 10)
[d]Theoretical Q.S., Trautwein *et al* (15)

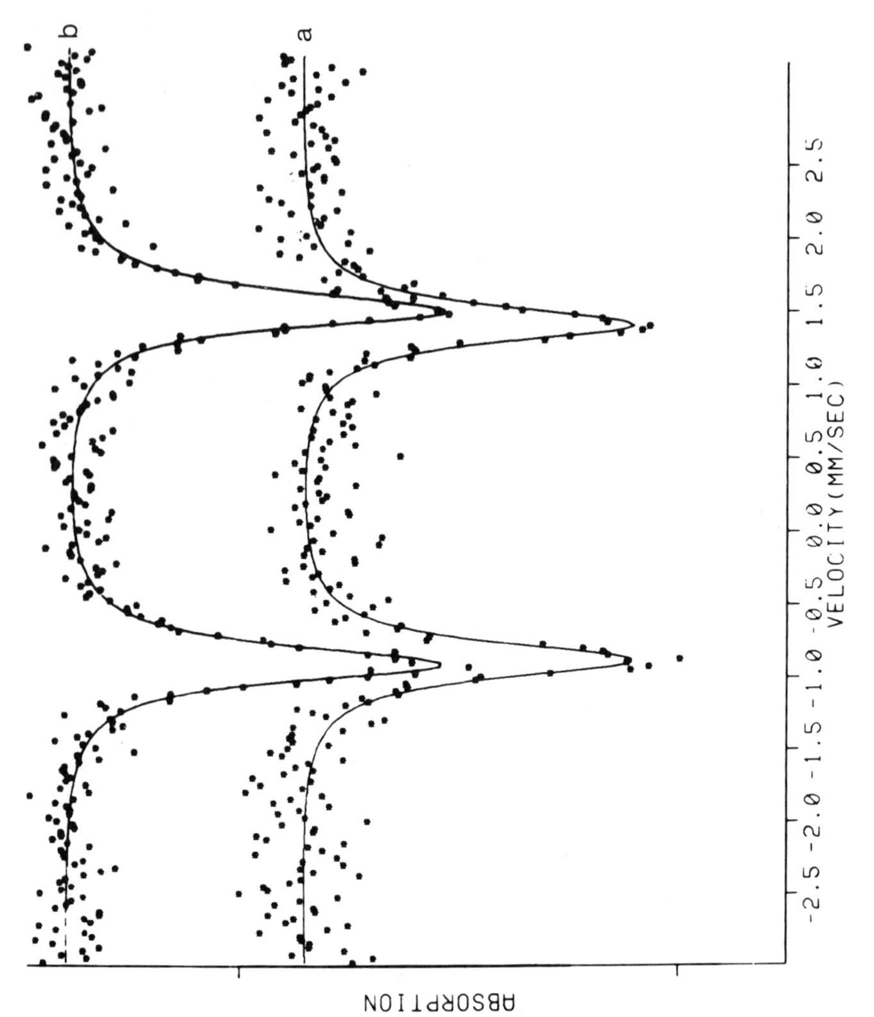

Figure 5. Mössbauer spectra of PDVF(radical) (a) and PDVF(cationic) (b).

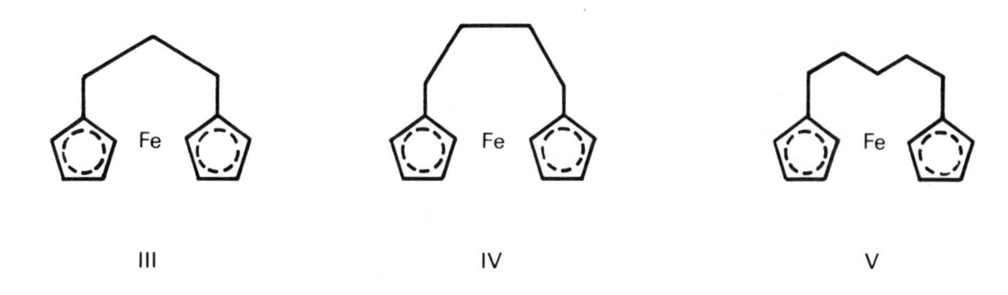

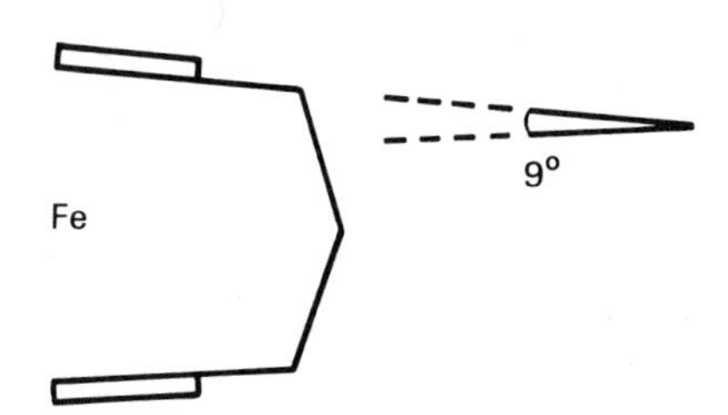

Figure 6. Three-carbon bridged ferrocene compound having a tilted ring system with a dihedral angle of 9°.

Fe

Fe

Fe

Fe

VI

VII

not as greatly affected. PDVF(radical) exhibits IS and QS values which are similar to those observed and predicted for the model compound having three carbon atoms bridging the ferrocene rings. This provides strong evidence that this is a cyclopolymer with a three-carbon bridge (I) as the repeating unit. However, PDVF(cationic) has IS and QS values similar to those that have been obtained for the unstrained ferrocene units in polyvinylferrocene [PVF(radical) and PVF(cationic), Table III]. It has been observed (16, 17) that the bonding of the iron atom to the rings is not affected by alkyl substituents in the cyclopentadiene rings, and methylferrocene exhibits a similar QS (18). The experimental data was computer fitted as a single quadrupole doublet. The co-existence of strained (ΔE_Q = 2.29 mm s^{-1}) and unstrained (ΔE_Q = 2.40 mm s^{-1}) units would result in a Mossbauer spectrum having two overlapping quadrupole doublets. Clear evidence for this would be given by corresponding changes in line widths and asymmetry in line intensities. No significant line broadening (Tables II and III) or intensity asymmetry (Figure 5) was observed for the polymers studied.

Conclusions

These results therefore indicate that the soluble, saturated polymers produced by radical initiation of DVF are cyclopolymers containing a three-carbon bridged ferrocene unit (I). However, the Mossbauer spectra obtained for polymers produced by cationic initiation do not show evidence for the three-membered bridge. Certainly PDVF(cationic) contains some acyclic unsaturated units, not necessarily pendant vinyl groups (II), but the low level of unsaturation suggests that cyclic units exist. Thus, for example, a five-carbon bridged bicyclic unit (VI) or a ladder structure (VII) may be found in these polymers, and would be consistent with the Mossbauer parameters obtained. Further polymerisation studies and Mossbauer and other spectroscopic studies of these polymers are in progress.

Acknowledgements

This work is supported by the Ministry of Defence and is carried out in collaboration with PERME (Waltham Abbey). We thank Drs. P. Golding and G.F. Hayes for helpful discussions during the course of the work. Mossbauer equipment was purchased with a grant from the Science Research Council.

Literature Cited

1. Corfield, G.C.; Butler, G.B. "Cyclopolymerisation and Cyclocopolymerisation"; Chapter 1 in "Developments in Polymerisation Vol. 3"; Haward, R.N., Ed.; Applied Science Publishers: London, 1982.

2. Kunitake, T.; Nakashima, T.; Aso, C. J. Polym. Sci., Part A-1 1970, 8, 2853.
3. Kunitake, T.; Nakashima, T.; Aso, C. Makromol. Chem. 1971, 146, 79.
4. Aso, C.; Kunitake, T.; Tagami, S. "Progress in Polymer Science, Japan Vol. 1"; Imoto, M.; Onogi, S., Eds.; Halsted Press: Tokyo, 1971; p 170.
5. Sosin, S.L.; Jashi, L.V.; Antipova, B.A.; Korshak, V.V. Vysokomol. Soedin. Ser. B 1970, 12, 699.
6. Sosin, S.L.; Jashi, L.V.; Antipova, B.A.; Korshak, V.V. Vysokomol. Soedin. Ser. B 1974, 16, 347.
7. Korshak, V.V.; Sosin, S.L. "Organometallic Polymers"; Carreher, C.E.; Sheats, J.E.; Pittman, C.U., Eds; Academic Press: New York, 1977; p 26.
8. Neuse, E.; Crossland, R.K.; Koda, K. J. Org. Chem. 1966, 31, 2409.
9. Nagy, A.G.; Dezsi, I.; Hillman, M. J. Organometal. Chem. 1976, 117, 55.
10. Hillman, M.; Nagy, A.G. J. Organometal. Chem. 1980, 184, 433.
11. Rosenblum, M.; Woodward, R.B. J. Amer. Chem. Soc. 1958, 80, 5443.
12. Yamakawa, K.; Ochi, H.; Arakawa, K. Chem. Pharm. Bull. (Japan) 1963, 11, 905.
13. Rausch, M.D.; Siegel, A. J. Organometal. Chem. 1968, 11, 317.
14. Lang, G.; Dale, B. Nucl. Inst. Meths. 1974, 116, 567.
15. Trautwein, A.; Reschke, R.; Dezsi, I.; Harris, F.E. Journal de Physique 1976, 12, C6-463.
16. Good, M.L.; Buttone, J.; Foyt, D. Ann. N.Y. Acad. Sci. 1974, 239, 193.
17. Korecz, L.; Abou, H.; Ortaggi, G.; Graziani, M.; Belucco, U.; Burger, K. Inorg. Chim. Acta 1974, 9, 209.
18. Lesikar, A.V. J. Chem. Phys. 1964, 40, 2746.

RECEIVED February 1, 1982.

11

New Phase Transfer Catalysts Containing Oxyethylene Oligomers

L. J. MATHIAS and J. B. CANTERBERRY

University of Southern Mississippi, Department of Polymer Science, Hattiesburg, MS 39401

A new difunctional monomer (the divinylether of tetraethylene glycol) has been synthesized from the inexpensive and readily available tetraethylene glycol and acetylene. The monovinyl ether, however, is obtained with the literature procedure as the major product. While attempted polymerization of the latter with $BF_3 \cdot OEt_2$ gave mainly an oligomeric acetal, the divinyl ether gave an insoluble gel which is presumed to possess mainly a cyclized structure along with random cross-links. This polymer displayed phase transfer catalytic activity comparable to that of 18-crown-6 with NaCN in the nitrile substitution of hexylbromide, while with KCN it was less active than the crown ether. The oligomeric acetal did not catalyze this reaction significantly.

Applications of phase transfer catalysis to organic synthesis are widespread and growing at an enormous rate. While a large number of phase transfer catalysts (PTC's) have been investigated (1), the most commonly used are the various onium ion salts, the crown ethers and cryptands, and the oligomers and polymers of oxyethylene. The most generally attractive of these are the crown ethers because of high activity, selectivity in binding, stability, and the ability to interact strongly with primary ammonium ions (2). Crown ethers, however, are also some of the most expensive PTC's available. One of the ways in which the expense of the crown ethers, as well as many other catalytic groups, can be reduced is by incorporating them into insoluble polymeric matrices which allow their ready recovery by filtration and subsequent reuse. This has in fact been accomplished for the crown ethers (3), several onium salts (4), and various oligooxyethylenes (5). Catalytic activity of many of these polymer-bound species has been

0097-6156/82/0195-0139$06.00/0
© 1982 American Chemical Society

found to be comparable to or slightly below that of the corresponding low molecular weight compounds (6). In addition, the polymeric catalysts sometimes display properties which result from the unique characteristics of the polymer matrix such as substrate size selectivity and cooperative cation binding (7).

Despite the availability of polymeric PTC's and their demonstrated activity and recyclability, the need exists for developing new materials. The synthetic effort and initial expense of the polymeric catalysts is high, a fact reflected in the cost of some commercially available compounds (8). In addition, polystyrene is generally used as the polymer matrix and the inherent hydrophobicity of this backbone must affect the ability of the pendant catalytic group to promote interaction between the two immiscible phases. We have been devoting a large effort to the synthesis and characterization of new crown ether-containing polymers (9). and polyoxyethylene analogs (10) which have potential as PTC's. We would now like to report results on the synthesis of new vinyl derivatives of oxyethylene oligomers which readily undergo polymerization to give soluble or insoluble materials, depending on reaction conditions. Most importantly, these polymers are prepared in high yield from inexpensive starting materials and possess good phase transfer catalytic activity.

Experimental

Commercial tetraethylene glycol, acetylene, solvents and additional reagents were used as obtained. Monomer and polymer characterization included ^{1}H NMR (Varian EM-390), ^{13}C NMR (Varian CFT-20) and IR (Perkin-Elmer 580).

C_2H_2, KOH, 180° → 1 + 3

Monomer Synthesis. Tetraethylene glycol (100 g, 0.52 m) was placed in a gas washing flask along with potassium hydroxide (15 g, 0.27 m). The gas dispersion tube was inserted and the flask placed in an oil bath. Nitrogen gas was bubbled through the solution as the bath temperature was gradually raised to 175-185°. Acetylene was then bubbled in (in place of the nitrogen) for 8-10 hours while the temperature was maintained between 175° and 185°. Acetylene addition was then terminated. The flask was removed from the oil bath and allowed to cool to room temperature. The viscous, dark brown reaction mixture was mixed with 100 ml each of water and carbon tetrachloride (CCl_4) in a seporatory funnel. After vigorous shaking, the two layers were separated.

The aqueous layer was then extracted twice with equal volumes of CCl_4. The combined organic layers were evaporated under reduced pressure to give a viscous brown oil. This material was dried in vacuo overnight and then distilled at 90-120° and 0.5 mm Hg. The distillate, which consists of both 1 and 3, was then dissolved in an equal volume of CCl_4 and extracted once with an equal volume of water. Both the organic and aqueous layers were then evaporated under reduced pressure to give essentially pure 1 and 3, respectively. Final drying was carried out under a vacuum of 0.1 mm Hg for several days. Alternatively, the separated compounds could be vacuum distillated at, respectively, 110-113° with 0.5 mm Hg and 93-94° with 0.3 mm Hg. Typical yields of 1 and 3 were 13% and 40%, respectively.

Polymer Synthesis. A typical conversion of 1 to insoluble polymer 2 involved flame-dried glass-ware and 1,2-dichloroethane (DCE) which had been distilled from P_2O_5 and stored over 4A sieves. The monomer (2g, 1.12×10^{-2} m) was dissolved in 15 ml of DCE in a reaction flask equipped with rubber septa. This mixture was carried through three vacuum freeze-thaw cycles. A fresh initiator solution was made by dissolving BF_3 etherate (0.36g, 2.53×10^{-3} m) in 10 ml of DCE. Iniation involved injecting 20 µl of this solution into the reaction flask to give a mole ratio of monomer to initiator of 2.2×10^3. Polymerization was very rapid at room temperature and caused almost immediate gelling of the reaction mixture. The light-brown polymer could be isolated by breaking up the gel, stirring with ether and filtering. Vacuum drying gave a white or light-brown powder in essentially quantitative yield.

Use of benzene as the polymerization solvent, but with a molar ratio of monomer to initiator of 3.2×10^2, led to a much darker colored product. However, on treatment with water, filtering and drying, nearly white product could be obtained in high yield.

Attempts to convert monomer 3 to vinyl polymer involved essentially the same conditions as used for 1 although no gelation was observed. Polymer did not precipitate on mixing with ether and product isolation required solvent evaporation under reduced pressure. The viscous, light-yellow oil thus obtained did not solidify even with extensive drying.

Phase Transfer Comparisons. An inhomogeneous mixture of 1-bromohexane (1.5 g, 1.43×10^{-2} m) and an equal volume of a saturated solution of KCN or NaCN in water was heated at 85° in the presence of 8 mole-% of catalyst (based on molecular weight of repeat unit). No stirring was employed. The insoluble polymeric catalyst was suspended at the interface between the two immiscible layers. The reaction was followed with ^1H NMR using the hydrogens adjacent to the bromide and nitrile. Relative rates of reaction were evaluated by comparing reactions carried out simultaneously under the same reaction conditions.

Results and Discussion

The rationale for the synthesis of monomer 1 (the divinyl ether of tetraethylene glycol) was based on three considerations. First, 1 would be available in a single step from inexpensive and readily available chemicals, thus overcoming one of the major drawbacks of the crown ether phase-transfer catalysts. Second, 1 was expected to undergo a cationic cyclopolymerization process to give mainly repeat units of structure 2 capable of cation complexation. Third, those monomer units which did not cyclize were expected to provide cross-linking to render the polymeric catalyst insoluble and capable of recovery by filtration for reuse. Indeed, even were cyclization found to be a minor process, it was hoped that the oxyethylene matrix would nonetheless possess adequate catalytic ability to function as a PTC.

BF_3, $ClCH_2CH_2Cl$, 25°

1 2

Monomer Synthesis. The synthesis of 1 was based on the published procedure of Reppe for the divinyl ether of triethylene glycol (11). The reaction conditions given in the Experimental were those found to give the best yields of 1 with this method. Some variations in reaction time and temperature as well as catalyst concentration were examined in attempts to improve the yield. Milder conditions reduced the rate of reaction without increasing the final yield, while harsher conditions led to extensive decomposition. Several additional mono- and divinyl ethers of oligooxyethylenes have been synthesized with this procedure and in general, yields were low.

The outlined procedure unfortunately gave a mixture of 1 and 3 with the latter predominating. Surprisingly, two extractions combined with one or two vacuum distillations were sufficient to separate and purify these two compounds. The materials thus obtained were either clear and colorless or possessed a clear, light green or blue color for which we are unable to account. This color apparently did not inhibit conversion to polymer. Figure 1 gives the ^{13}C NMR spectra of starting material and monomer 1.

Polymerizations. A variety of polymerization conditions were examined for both the monovinyl and divinyl compounds. It was hoped that cationic polymerization of monomer 3 would give a vinyl polymer possessing tetraethylene glycol pendant groups.

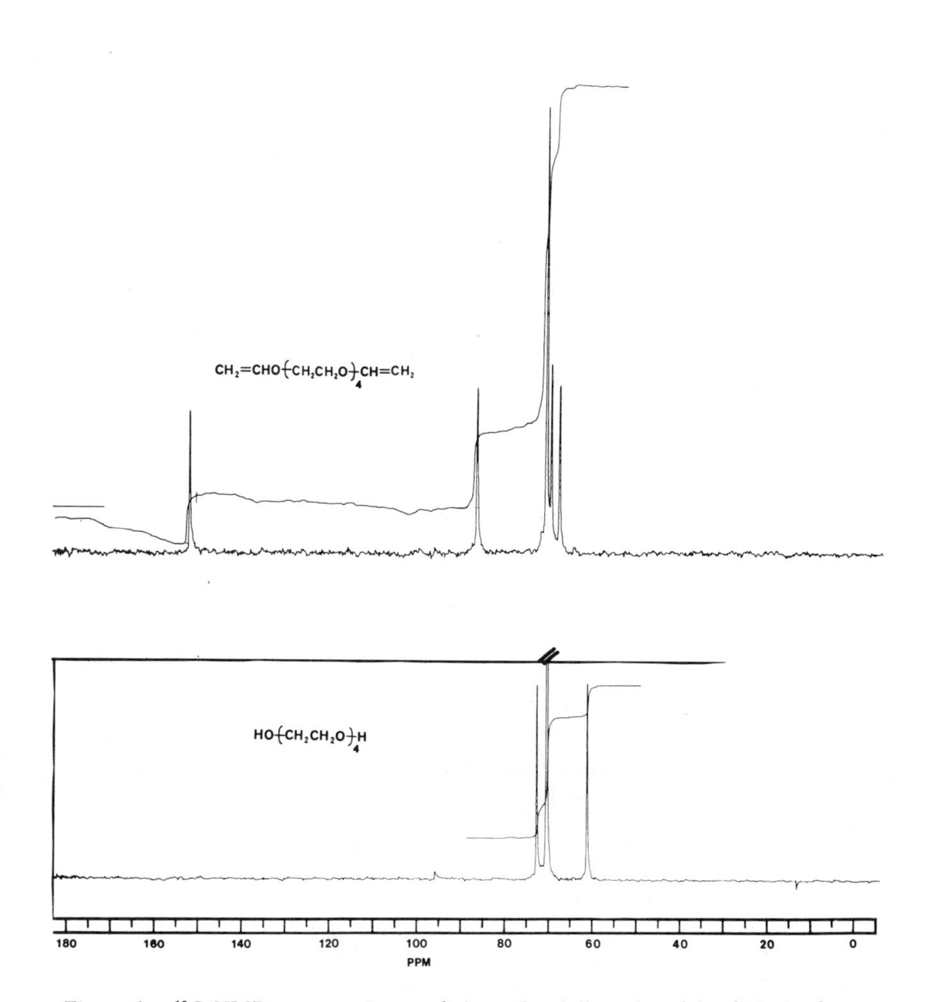

Figure 1. ^{13}C *NMR spectra of tetraethylene glycol (lower) and its divinyl ether (upper) taken on neat samples.*

Unfortunately, this was not the case and apparently only acetal formation occurred.

The polymerization of the divinyl monomer 1 to insoluble product was carried out under a variety of conditions with the most reproducible given in the Experimental section. While cross-linking was obvious, a number of points provide indirect evidence for a low degree of such cross-linking and therefore a high degree of cyclopolymerization. First, the products obtained were soft, flexible gels which on drying and reswelling with halogenated hydrocarbons underwent ca. 10-fold changes in volume. Such behavior would be unlikely if cross-linking were high. Second, polymer swollen with $CDCl_3$ gave excellent ^{13}C NMR spectra such as that given in Figure 2. No residual vinyl groups were observed in the ^{13}C NMR or the IR spectra. The sharpness of the peaks, comparable to those of poly(ethyl vinyl ether), indicates similar molecular mobility on the NMR timescale. Again, this is inconsistent with a high degree of cross-linking. Third, while the spectrum of 2 has two sharp peaks assignable to terminal hydroxyethyl groups which may have arisen from side-reactions involving loss of one of the monomer vinyl groups, the relative ratio of these peaks to the polymer peaks is low. In fact, in most samples these peaks are altogether absent. All other peaks in the spectra of 2 can be accounted for by the proposed cyclic repeat units. These three points seem most consistent with a cyclopolymerization process.

The question arises of whether a cyclopolymerization process to give such macrocycles in high yield is feasible or not. It has been demonstrated for diallyl phthallate that 10- and 11-membered rings are formed in a 1:6 ratio with >98% cyclization under free radical conditions (12). Furthermore, while α,ω difunctional polymethylene monomers cyclize to a very limited extent when 11- to 17- membered rings are formed, heteroatom-containing difunctional monomers cyclize 60-90% for comparable ring sizes (13). Butler and Lien have presented evidence for a high degree of 13- membered ring formation during the cationic and free radical polymerization of 1,2-bis(2-ethenyloxyethoxy)-benzene (14), an aryl-containing homolog of 1. It is possible that for divinyl ethers such as 2 preferential association between the cationic initiator and the two vinyl groups or between the terminal vinyl ether and the propagating species generated by initial intermolecular attack favors the subsequent intramolecular attack over a competing second intermolecular addition. Further work is underway on this system to hopefully clarify this process.

The effect of temperature on the conversion of 1 to polymer was also examined with the expectation that lower temperatures would facilitate the cationic polyaddition. It was found that the rate of polymerization sharply decreased down to a temperature of ca. -30° where conversion essentially stopped. Samples initiated in the range -78° to -30° displayed no reaction at these temperatures even after several hours, although when allowed to warm above -30° they underwent rapid polymerization. This unex-

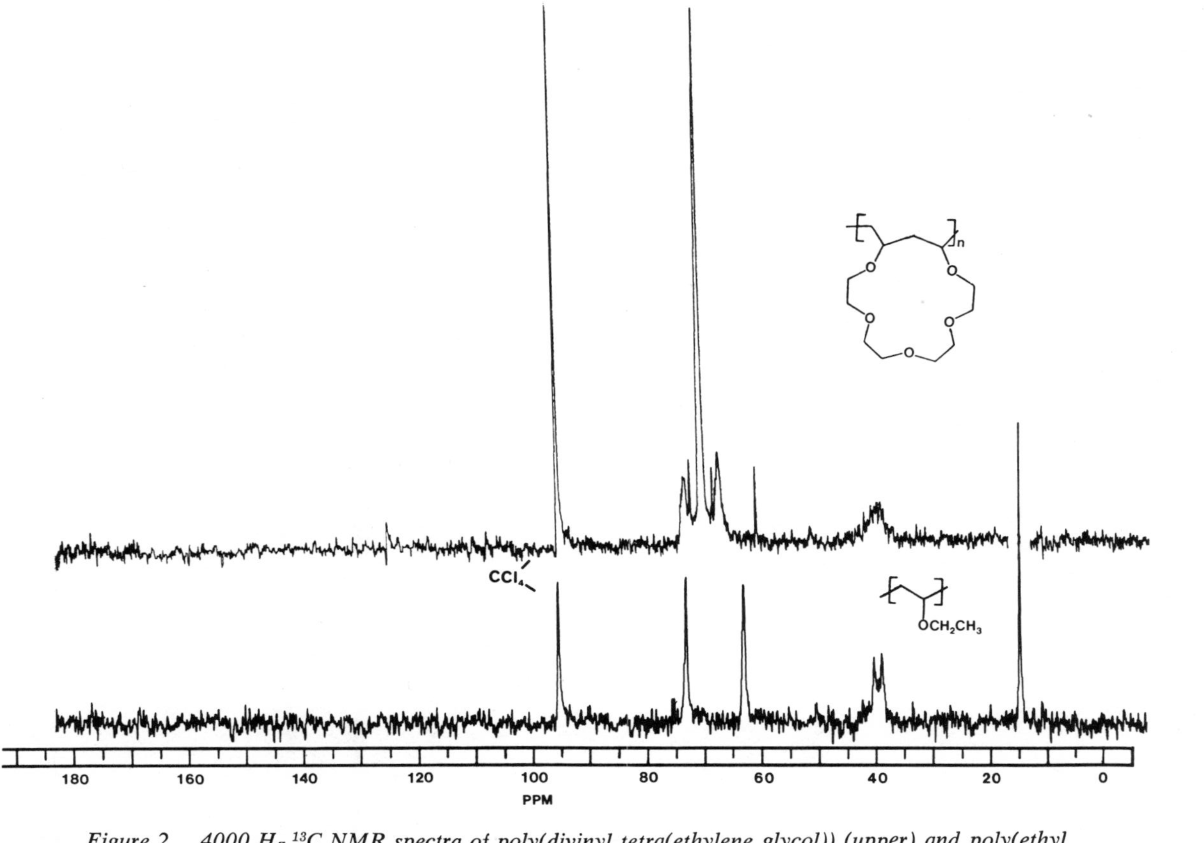

Figure 2. 4000 H_Z ^{13}C NMR spectra of poly(divinyl tetra(ethylene glycol)) (upper) and poly(ethyl vinyl ether) (lower).

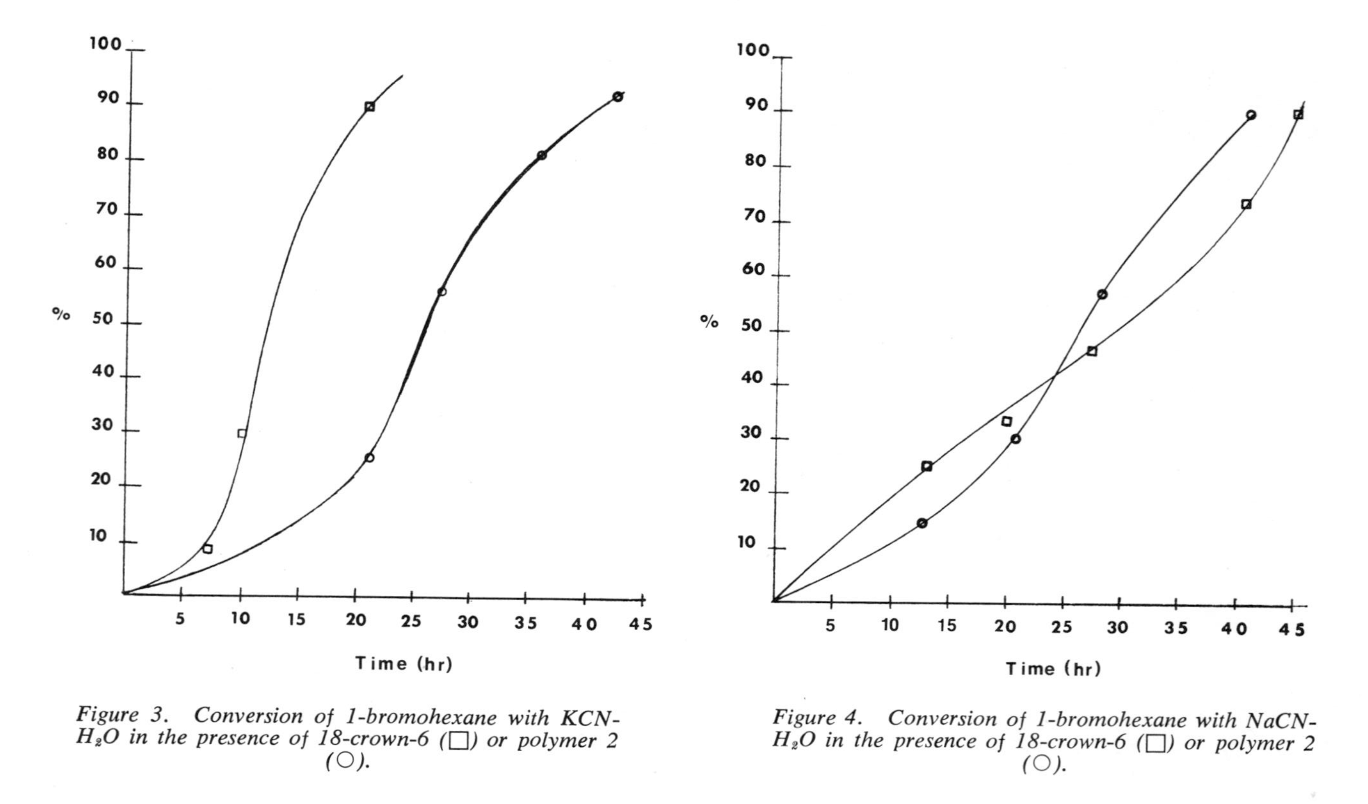

Figure 3. Conversion of 1-bromohexane with KCN-H_2O in the presence of 18-crown-6 (□) or polymer 2 (○).

Figure 4. Conversion of 1-bromohexane with NaCN-H_2O in the presence of 18-crown-6 (□) or polymer 2 (○).

pected behavior is also under further investigation for monomer 1 and several homologs and monovinyl ether analogs.

Phase Transfer Catalysis. Initial evaluations of the ability of polymer 2 to function as an insoluble or triphase catalyst involved examination of relative rates compared to 18-crown-6. The reaction employed was the substitution of the bromine of 1-bromohexane by nitrile anion. The alkyl bromide composed one phase and a concentrated aqueous solution of either KCN or NaCN made up the other phase. The crown ethers functioned as soluble PTC's while polymer 2 suspended at the interface served as an insoluble catalyst. A reaction temperature of 85° was employed to give reasonable conversion times. The reactions were followed with ^{1}H NMR by monitoring the integrated intensities of the hydrogens adjacent to the halide and nitrile groups. The two triplets associated with these peaks are well separated at 3.23δ and 2.15δ allowing direct comparison of their relative ratios in the reaction mixtures. While this method has limited accuracy, it does allow rapid initial evaluations. Figure 3 shows the relative rates of conversion of 1-bromohexane to the nitrile with KCN. With 18-crown-6 and polymer 2, 90% conversion was achieved after 20-25 h for the former and 42 h for the latter.

Figure 4 gives similar data using NaCN as the nitrile source. Comparison of Figures 3 and 4 reveals two important points. First, the polymer is a better catalyst with NaCN than is the soluble crown ether under these conditions, a result which is somewhat surprising but completely reproducible. Second, the rate of reaction with the polymeric catalyst is essentially the same for both salts with 90% conversion occurring after about 40 h. Several possible explanations for this lack of effect of cation on the rate of reaction suggest themselves. It is possible that polymer 2 does not possess the cyclic repeat unit structure postulated. However, the spectral and physical properties of 2 are consistent with macrocycle formation during polymerization, and the catalytic activity of 2 per repeat unit is higher than analagous polymers with pendant oligooxyethylene groups and polyoxyethylene itself. An alternative possibility is that the 16-crown-5 structure proposed for 2 is able to accommodate both K^{+} and Na^{+} with comparable facility, especially when the potential for cooperative cation binding involving adjacent or non-adjacent crown units along the polymer is included. In this case, the importance of the size of the crown cavity might be less crucial than a high degree of polymer flexibility which would allow ready cooperative interaction. Further effort is required with analogs of 2 and copolymers of various homologs of 1 with monovinyl ethers. This work is presently underway.

Acknowledgement. Grateful acknowledgement is made to Auburn University for a Grant-in-Aid that allowed the initiation of this work.

Literature Cited

1. Starks, C. Chemtech 1980, 110.
2. Cram, D.J.; Cram, J.M. Accts. Chem. Res. 1978, 11, 8.
3. Smid, J.; Shah, S.C.; Sinta, R.; Varma, A.J.; and Wong, L. Pure Appl. Chem. 1979, 51, 111.
4. Molinari, H.; Montanari, F.; Quici, S.; and Tundo, P. J. Amer. Chem. Soc. 1979, 101, 3920.
5. Manecke, G. and Reuter, P. Pure Appl. Chem. 1979, 51, 2313.
6. Regen, S.L.; Heh, J.C.K. and McLick, J. J. Org. Chem. 1979, 44, 1961.
7. Regen, S. and Lee, D.P. Israel J. Chem. 1978, 17, 284.
8. Parish Chemical Company
9. Mathias, L.J. and Al-Jumah, K.B. J. Poly. Sci. Polym. Chem. Ed. 1980, 18, 2911.
10. Mathias, L.J. and Canterberry, J.B. Macromolecules 1980, 13, 1723.
11. Reppe, W.J. Liebigs Ann. Chem. 1956, 601, 81.
12. Matsumoto, A.; Iwanimi, K.; Kitamura, T.; Oiwa, M.; Butler, G.B.; Polym. Preprints 1981, 22, 36.
13. Schulz, R.C.; Stenner, R.; Makromol. Chem. 1966, 91, 10.
14. Butler, G.B.; Lien, Q.S.; Polym. Preprints 1981, 22, 54.

RECEIVED March 15, 1982.

12

Synthesis of Macrocyclic Ring-Containing Polymers Via Cyclopolymerization and Cyclocopolymerization

GEORGE B. BUTLER and QCHENG S. LIEN[1]

University of Florida, Department of Chemistry and Center for Macromolecular Science, Gainesville, FL 32611

The principle of cyclopolymerization has been applied to the synthesis of macrocyclic ether-containing polymers which may simulate the properties of crown ethers. 1,2-Bis(ethenyloxy)benzene (a 1,7-diene) and 1,2-bis(2-ethenyloxyethoxy)benzene (a 1,13-diene) are typical of the monomers synthesized. Homopolymerization of the 1,7-diene *via* radical and cationic initiation led to cyclopolymers of different ring sizes; homopolymerization of the 1,13-diene led to cyclic polymer only *via* cationic initiation. Both monomer types were copolymerized with maleic anhydride to yield predominantly alternating copolymers having macrocyclic ether-containing rings in the polymer backbone.

Cyclopolymerization has been applied extensively to synthesis of monomers containing stable ring structures, particularly five- and six-membered rings (1). However, little attention has been given to synthesis of larger ring-containing polymers *via* this mechanism. Barnett, Butler and Crawshaw (2) designed monomers to produce seven-membered rings, and Marvel and Garrison (3) investigated the formation of ring sizes larger than seven *via* Ziegler-Natta catalysts on suitable monomers. Eight-membered ring systems expected from cyclopolymerization of o-allylphenyl acrylate (4) have been studied as well as ten-membered ring systems derived from o-(β-acryloyloxyethoxy)styrene (5). The latter monomer was postulated to lead to extensive cyclization as the result of extensive conformational control in the monomer which favored cyclization. Subsequent to completion of this work, it has been shown (6) that divinyloxy monomers such as 1,4-divinyloxybutane and diethylene glycol divinyl ether undergo cyclopolymerization initiated by iodine.

[1] Current address: Loctite Corporation, Newington, CT 06111.

0097-6156/82/0195-0149$06.00/0
© 1982 American Chemical Society

It was the purpose of this investigation to synthesize suitable monomers which could conceivably undergo cyclopolymerization or cyclocopolymerization to lead to large ring-containing polymers. A more specific purpose was to synthesize polymers which would contain ring structures in the polymer backbone which may simulate the properties of the crown ethers (7). Thus a variety of monomers, including 1,2-bis(ethenyloxy)benzene, 1,2-bis(ethenyloxy)-4-methylbenzene, 1,2-bis(2-ethenyloxyethoxy)benzene and 1,t-butyl-3,4-bis(ethenyloxyethoxy)benzene, as well as model compounds 2,3-dimethyl-1,4-benzodioxane and cis- and trans-2,4-dimethyl-1,5-benzodioxepane, were synthesized and studied. This paper deals with cyclopolymerization studies of the four monomers.

Monomer Synthesis

Monomers 1 and 2 were synthesized according to the reaction schemes shown:

OH, OH (catechol, R) + 2NaOH + $2ClCH_2CH_2OH$ —(n-BuOH, Reflux)→ 1,2-bis(2-hydroxyethoxy)benzene (R)

R = H or CH_3

+ $SOCl_2$ —($CHCl_3$, Pyridine, Reflux)→ 1,2-bis(2-chloroethoxy)benzene (R) (1)

+ $CH_3CH_2C(CH_3)_2OK$ —(t-Amyl Alcohol, Reflux)→ 1,2-bis(ethenyloxy)benzene (R)

1, R = H; 2, R = CH_3

The Synthesis of 1,7-Dienes

The preparation of 1,2-bis(ethenyloxy)benzene [1] and 1,2-bis(ethenyloxy)-4-methyl benzene [2] is outlined in Eq. 1. Catechol was converted to 1,2-bis(2-hydroxyethoxy)benzene, which was reacted with thionyl chloride to give 1,2-bis(2-chloroethoxy)-benzene.

The elimination reaction of this chloride was carried out with potassium 2-methyl-2-butoxide to give the desired monomer. Monomer 2 was prepared in the same way using 4-methyl catechol. The structures were confirmed by infrared (IR), proton magnetic resonance (PMR), 13-C nuclear magnetic resonance (CMR) and elemental analysis. Model compounds 3 - 6 were synthesized as aids in establishing the structure of polymers derived from 1.

The Synthesis and Structural Determination of Model Compounds

The preparation of six-membered ring model compounds is outlined in Eq. 2.

HO OH + 2 TsCl ⟶ (TsO, OTs ditosylate)

⟵ Catechol, NaOH, DMF (2)

cis- 3
trans- 4

Racemic 2,3-butanediol was converted to the racemic ditosylate, which was then treated with catechol under the influence of NaOH in dimethylformamide (DMF) to yield the desired compounds, cis- and trans-2,3-dimethyl-1,4-benzodioxane [3] and [4]. The solid trans isomer [4] was obtained by repeated recrystallization from methanol. Column chromatography of the mother liquid using Al_2O_3 as adsorbent and petroleum ether and diethyl ether in a 1:1 ratio as eluent, gave pure liquid cis isomer [3]. By comparison with the PMR spectra (8) of pure trans- and cis-2,3-dimethyldioxane, the solid isomer was identified as trans, and the liquid isomer as cis. The compounds were further characterized by IR, NMR, CMR and elemental analyses. The CMR spectral data for 3 and 4 are summarized in Table I.

The methine proton absorbs at δ 3.81 in cis-isomer and δ 4.18 in trans-isomer. The IR spectra of both isomers are similar in general, showing aromatic ether at 1265 cm^{-1}, phenyl at 750, 850 and 1600 cm^{-1}.

Preparation of the seven-membered cyclic model compounds is outlined in Eq. 3. Racemic 2,4-pentanediol was first converted to the racemic ditosylate, which was then treated with catechol

TABLE I
CMR Chemical Shifts for 2,3-Dimethyl-1,4-benzodioxane

Carbon	cis	trans
1	121.58	121.50
2	117.57	117.24
3	143.06	144.24
4,5	71.98	74.36
6	14.41	16.94

3 cis
4 trans

under the influence of NaOH in DMF to yield the desired compounds, cis- and trans-2,4-dimethyl-1,5-benzodioxepane [5] and [6]. The pure solid cis-isomer was obtained by repeated fractional recrystallization from methanol. Column chromatography of the mother liquid, using Al_2O_3 as adsorbent and a solution of petroleum ether and diethyl ether in a 1:1 ratio as eluent, gave pure liquid trans-isomer.

= 2 TsCl → (OTs, OTs); NaOH, Catechol / DMF

cis - [5]
trans- [6]

(3)

The methylene protons in the trans-isomer are identical, but are different in the cis-isomer. Decoupling at the methine proton absorption gave a singlet of methine protons in trans-isomer and an AB quartet in the cis-isomer. The methine protons absorb at δ 3.9 in the cis-isomer, but at δ 4.55 in the trans-isomer. The IR spectra of both isomers are similar. The CMR spectral data for 5 and 6 are summarized in Table II.

TABLE II
CMR Chemical Shifts for 2,4-Dimethyl-1,5-benzodioxepane

Carbon	cis	trans
1	124.07	122.60
2	123.01	121.06
3	152.06	150.16
4,6	77.44	75.43
5	46.75	45.47
7	22.60	21.54

Cyclopolymerization of 1,7-Dienes

Monomers 1 and 2 were polymerized via cationic (Eq. 4) and radical (Eq. 5) initiators to give soluble linear cyclopolymers. The results are summarized in Table III and Eqs. 4 and 5.

All the polymers obtained were soluble in common organic solvents such as THF, benzene, pyridine, chloroform and acetone. The IR spectra of the polymers showed only slightly detectable absorption of the vinyloxy group at 965 cm^{-1} or 1640 cm^{-1}. The PMR spectra of the polymers also showed only slightly detectable vinylic proton absorption at $\delta = 6.4$, suggesting that the polymers are linear cyclopolymers, as shown in Eqs. 4 and 5. More detailed structural information was obtained by comparison of the spectra of the polymers with those of the model compounds, 3, 4, 5 and 6.

TABLE III
Cyclopolymerization of 1,7-Dienes

Monomer	Conc.(M)/ Solvent	Initiator	Polym. Temp.	% Conv.	η_{int} [a]	Mn [b]	T_s [c] °C
1	1.0/CH_2Cl_2	$BF_2O(Et)_2$ 2%	Ambient	60	0.09	3700	
1	0.5/benzene	Benzoyl peroxide 4%	70°C	30	0.086	2100	130-145
1	0.5/benzene	$BF_3O(Et)_2$ 2%	Ambient	70		2800	200-215
1	0.5/CH_2Cl_2	ϕ_3CSbF_6 1%	Ambient	67.2		6700	190-215
2	0.5/benzene	$BF_3O(Et)_2$ 2%	Ambient	50		2270	180-200

[a] Intrinsic viscosity (dl/g,benzene)

[b] Number-average molecular weight

[c] The temperature of softening points

The CMR spectral data of the polymers of 1 are summarized in Table IV.

TABLE IV
CMR Spectral Data on Cyclopolymers
of 1,2-Bis(ethenyloxy)benzene [1]

Polymer	Carbon	Cis	Unresolvable	Trans
7	1,2		123.08	
	3	151.88		150.02
	4,6		78.57	
	5		43.58	
	7		22.84	
9	1		121.50	
	2		117.24	
	3		143.39	
	4,5	74.55		76.46
	6		26.90	

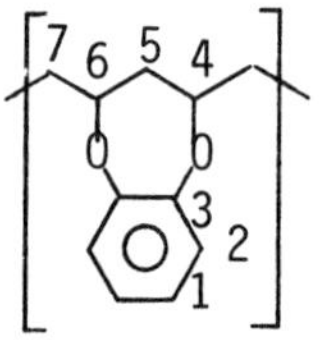

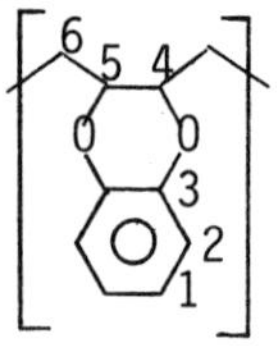

Polymers 7 and 9 match with model compounds 5-6 and 3-4, respectively, in PMR, CMR and IR spectral data. Almost all the carbons of the polymers could be directly correlated with the carbons of the model compounds. Especially important were the methylene carbon (C_5) resonances in the 7-membered ring compounds 5 and 6 which occurred at 46.75 and 46.47 ppm, and were found in CMR of polymer 7 at 43.58 ppm but were completely absent in that of polymer 9.

Synthesis of 1,13-Dienes

The 1,13-dienes, 1,2-bis(2-ethenyloxyethoxy)benzene [11] and 1,2-bis(2-ethenyloxyethoxy)-4-t-butylbenzene [12] were prepared by treating 2-chloroethyl vinyl ether with the corresponding catechol under the influence of NaOH in 1-butanol (n-BuOH) as shown in Eq. 6.

OH OH R

R = H, t-Butyl

$$+ \ 2\ ClCH_2CH_2OCH = CH_2 + 2\ NaOH \quad \xrightarrow{\text{n-BuOH}} \qquad (6)$$

11, R = H

12, R = t-Butyl

O O O O

Cyclopolymerization of 1,13-Dienes

Cyclopolymers having 13-membered rings as repeating units were obtained by cationic initiation of 1,2-bis(2-ethenyloxyethoxy)benzene under the conditions shown in Eq. 7 and summarized in Table V. The polymer was soluble in essentially all common organic solvents but insoluble in water and petroleum ether. The absorptions of the vinyloxy group were completely absent in the IR spectrum of polymer 13. The CMR showed complete absence of vinyloxy carbons, but saturated aliphatic backbone carbons appeared. The solubility and spectroscopic evidence is consistent with the proposed structure 13.

TABLE V
Cyclopolymerization of 1,13-Dienes

Monomers	Init. %	Conc./ Solvent	Conv.	Time	[η] dl/g[a]	M_n	T_s(°C)
11	$(C_6H_5\text{-C-O})$ 0 4%	0.45M/ benzene	oily residue	4 days	-	-	-
11	$BF_3O(Et)_2$ 2%	0.2M/ benzene	90%	12 hrs.	0.25	13500	135-160
11	$BF_3O(Et)_2$ 2%	0.08M/ benzene	75%	12 hrs.	0.14	2520	120-140
12	$BF_3O(Et)_2$ 2%	0.33M/ n-hexane	60%	10 min.	cross-linked	-	-
12	$BF_3O(Et)_2$ 2%	0.033M/ benzene	viscous oil	24 hrs.	-	-	-

[a] In a 1:1 mixture of THF and nitrobenzene.

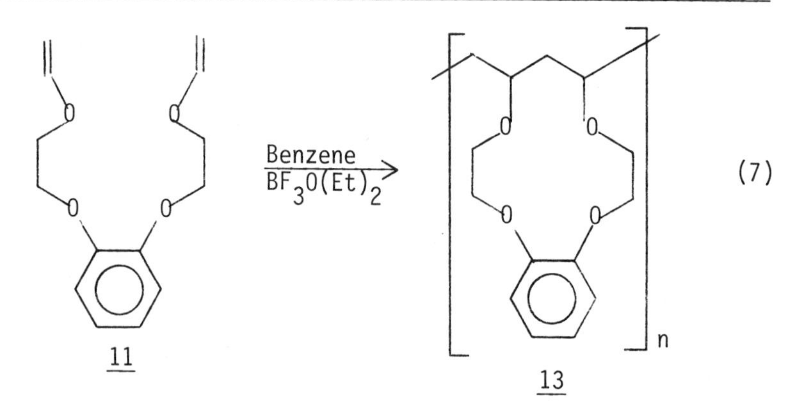

Monomer 12 led only to a viscous oil *via* cationic initiation in dilute solution, and to a crosslinked polymer [14] at higher concentration. Neither monomer 11 nor 12 gave polymers *via* radical initiation.

Cyclocopolymerization of 1,7-Dienes with Maleic Anhydride

The copolymerization of monomers 1 and 2 with maleic anhydride (MA), respectively, was shown to produce soluble, linear copolymers of low molecular weight. The results are summarized in Eq. 8 and Table VI. The polymers [15] and [16] obtained were soluble in common organic solvents such as acetone, chloroform, pyridine and DMF. All spectral evidence, as well as solubility of the polymers, support the structure proposed in Eq. 8.

R = H [1]
R = CH_3[2]

+ 2n

AIBN

(8)

R = H [15]
R = CH_3 [16]

TABLE VI
Cyclocopolymerization of 1,7-Dienes with MA

Monomer	Conc./ Solvent	Init./%	Temp./Time	Conv.	M_n	T_s(°C)
1	0.5M/CH_2Cl_2	AIBN/2%	60°/30 hrs.	35.34%	8710	240-245
1	0.45M/ cyclohexanone	AIBN/2%	70°/23 hrs.	33.3%	1900	235-245
2	0.5M/benzene	AIBN/2%	70°/30 hrs.	27.2%	2250	240-250

The MA content of copolymers obtained at constant total concentration but varying ratios of comonomers were determined (9) and are summarized in Table VII. The copolymers obtained had a composition in an approximate comonomer ratio of 1 to MA of 1:2.

TABLE VII
Influence of Comonomer Ratio on Conversion

D/A[a]	1/3	1/2	1	2/1	3/1
Conversion[b]	35.34%	54.73%	42.30%	22.01%	24.50%
MA%	63	64	63	60	50

[a] D: 1,2-bis(ethenyloxy)benzene [1]; A: MA total concentration 0.5M in CH_2Cl_2, 60°C, 30 hrs.
[b] Based on the total weight of comonomers.

The maximum conversion was at the feed ratio of 1:2 which indicates some regular placement of the two comonomers in the copolymer. Since the copolymer had a molar composition of approximately 2:1 of MA to 1, the major repeating unit of the copolymer is predicted to be that shown by 15-m. The slightly low MA content of the copolymer can thus be justified by the repeating unit 15-n.

When two comonomers vary widely in their electron affinity, a charge-transfer (CT) complex may be formed between them, and this complex may participate in copolymerization. The existence of a CT complex could be detected by many methods such as PMR, CMR (10) and ultraviolet (UV). The interaction in a CT complex is considered to involve an electron transfer from the highest occupied molecular orbital (HOMO) of an electron donating comonomer to the lowest occupied molecular orbital (LUMO) of an electron accepting comonomer (11). This electron transfer results in a new absorption in UV, a characteristic of the CT complex.

MA has a strong electron affinity, while monomer 1 has a weak electron affinity. A CT complex between [1] and MA was identified having a γ_{max} = 355 nm (UV). Since the CT band of MA and DVE was at γ_{max} = 278 nm (12), the absorption at γ_{max} = 355 was considered to be mainly from the interaction between MA and the benzene moiety of monomer 1. The composition of the CT complex, as determined by the continuous variation method, was a 1:1 complex. The optical density of the CT band at different wavelengths was plotted against the various mole fractions of monomer 3 in a solution of constant total concentration, each plot showing maximum absorbance at molar ratio of 1:1.

The equilibrium constant of the CT complex was also measured by UV using the inverse Hanna-Ashbaug-Foster-Fyfe (HAFF) (10,11) equation, and was found to be 0.243.

It has been suggested that CT complexes may actually participate in the polymerization. For example, Butler and Campus (12) presented evidence for participation of the CT complex of MA and divinyl ether (DVE) in a terpolymerization study using fumaronitrile, DVE and MA. Recently, a computer program (13) was published for evaluating the CT copolymerization model when operating in competition with the terminal model.

To what extent the CT complex of MA and 1 participate in the copolymerization is not certain. However, this possibility cannot be ruled out at present.

Cyclocopolymerization of a 1,13-Diene with MA

Monomer 11, a 1,13-diene, undergoes cyclocopolymerization with MA to yield a linear, soluble cyclocopolymer [17] of reasonably high molecular weight, 13,100 (see Table VIII). This cyclocopolymer represents the first example of a structure which includes both monomers in a ring having greater than nine members. All spectral evidence supports a 15-membered ring structure as illustrated in Eq. 9. The polymerization conditions used are summarized in Table VIII.

TABLE VIII
Cyclocopolymerization of Monomer 11 with MA

Conc./Solvent	Init./%	Temp.	Conv.	M_n	T_s(°C)
0.4M/CH_2Cl_2	AIBN/1%	60°C	40%	13100	>275
0.4M/Cyclohexanone	AIBN/2%	70°C	88.91%	6100	Decomposed 235-300°C

2 MA + 11 —(AIBN, Cyclohexanone)→ [17]n (9)

The MA content of polymer 17 was determined and found to be 65.71%, very close to the theoretical 67% for a composition of 1:2 molar ratio of monomer 11 to MA. A CT complex between MA and monomer 11 was observed. This was supported by a CT band at γ_{max} = 363 nm in the UV spectrum. The composition of the CT complex was determined to be 1:1 by the continuous variation method. The equilibrium constant was K_{eq} = 0.28 from the inverse HAFF equation.

Experimental

Monomer Synthesis - 1,3-Bis(ethenyloxy)benzene [1]. A flask, equipped with nitrogen inlet, mechanical stirrer and condenser, was flame dried under N_2. After cooling, 1.5 l of t-amyl alcohol was placed in the flask, and 92 g (2.35 mol) potassium was added slowly under N_2. The N_2 inlet was replaced with a pressure-equilizing addition funnel, and the condenser protected with a drying tube. A solution of 240 g (1.02 mole) 1,2-bis-(2-chloroethoxy)benzene in 250 ml hot t-amyl alcohol was dropped into the flask and refluxed for 36 hours.

The KCl formed was filtered and washed with 500 ml benzene. The combined filtrate and extract was washed twice with 100 ml H_2O, dried over anhydrous M_gSO_4 and condensed in an evaporator. Frac. dist. (vac) gave 107 g (0.66 mole) product. b.p. 49°C/ 0.75 mm. Yield: 64%. Anal. Calcd. for $C_{10}H_{10}O_2$: C, 74.05; H, 6.20. Found: C, 74.48; H, 6.34. IR (neat): 965, 1640 ($-OCH_2=CH_2$), 750, 1600 (AR), and 1200 (C-O-C) cm^{-1}. PMR ($CDCl_3$): peaks at δ 6.9 (S, 4H). $-O(H_a)C=C(H_b)(H_c)$

H_a at δ 6.4 (Q, J_{tr} = 14 Hz, J_{cis} = 6.5 Hz, 2H), H_b at δ = 4.45 (Q, J_{gem} = 2 Hz, J_{tr} = 14 Hz, 2H); H_c at δ 4.15 (Q, j_{gem} = 2 Hz, J_{cis} = 6.5 Hz). ^{13}CMR in d_6-benzene from TMS C_1, 125.41 ppm; C_2 = 119.1; C_3 = 147.38; C_4 , 93.87; C_5, 149.51. Mass spectrum (m/e, relative intensity): peaks at 162 (M^+, 11.6), 134 (100), 121 (86.1), 81 (20.3), 65 (20.3), 28 (41.9).

1,2-Bis(ethenyloxy)-4-methylbenzene [2]. This monomer was prepared by the same procedure as monomer 3. The physical data of monomer 2 are recorded as follows: b.p. 55-56°C/0.75 mm. Yield: 58%. Anal. Calcd. for $C_{11}H_{12}O_2$: C, 74.12; H, 7.91. Found: C, 75.14; H, 7.07, IR 1630, 950, ($-OCH=CH_2$), 700 and 1590 cm^{-1} (Ar). PMR ($CDCl_3$): δ 6.75 (d, 3H), 2.25 (S, 3H); p-vinoxyl, Ha, δ 6.36 (Q, J_{cis} = 6 Hz, J_{trans} = 14 Hz), Hc, δ = 4.09 (J_{cis} = 6 Hz, J_{gem} = 2 Hz), Hb, δ = 4.38 (J_{trans} = 14 Hz, J_{gem} = 2 Hz);

m-vinoxyl, Ha, 6.36 (Q, J_{cis} = 6 Hz, J_{trans} = 14 Hz), Hc, 4.27 (J_{gem} = 2 Hz, J_{cis} = 6 Hz), Hb, 4.42 (J_{trans} = 14 Hz, J_{gem}= 2 Hz). CMR in d_6-benzene from TMS, C_1, 149.63 ppm; C_2, 120.47; C_4, 128.00; CH_3 at C_4, 20.60; C_5, 124.88; C_6, 119.94; $C_{\alpha(vinyl)}$, 147.07; $pC_{\beta(vinyl)}$, 93.69; $mC_{\beta(vinyl)}$, 93.11. Mass spectrum: m/e (relative intensity) 176 (M^+, 22.3), 148 (100), 135 (77.3), 77 (30.9), 28 (58.3).

1,2-Bis(2-ethenyloxyethoxy)benzene [11]. In a flask, equipped with argon inlet, mechanical stirrer and condenser with drying tube, was placed 60 g (0.55 mole) 99% pure catechol and 800 ml t-BuOH. Potassium (42.5 g, 1.088 mole) was added slowly to the refluxing solution under argon. The argon inlet was replaced with a pressure-equilizing addition funnel through which 1.5 mole of 2-chloroethyl vinyl ether was added to the solution, and the solution was refluxed for five days. The salt was filtered and washed three times with 1.5 l petroleum ether. The combined filtrate and extract was condensed in an evaporator. Fract. dist. (vac) gave 50 g product 11. b.p. 138-141°C/10^{-4} mm Hg. Yield: 40%, m.p. 46-47°C. Anal. Calcd. for $C_{14}H_{18}O_4$: C, 67.18; H, 7.24. Found: C, 67.20; H, 7.20. IR (KBr) 975, 1610 ($-OCH=CH_2$); 1190 ($-CH_2OCH_2-$); 815, 720 (Ar) cm^{-1}. PMR δ 6.85 (S, 4H), 4.05 (m, 2H);

$$\begin{matrix} {}_{\alpha}H & & H_{\beta} \\ & C=C & \\ -O & & H_{\beta} \end{matrix}$$

H_{α} at δ 6.48 (q, J_{cis} = 7 Hz, J_{trans} = 14 Hz), H_{β} and other protons 4.3-3.85 (m, 12H). Mass spectrum: m/e (relative intensity), 250 (M^+, 11.8), 136 (86.6), 121 (47.9), 80 (22.2), 71 (25.1), 45 (100), 43 (78.2*), 41 (23.3), 28 (77.8*).

1-t-Butyl-3,4-bis(ethenyloxyethoxy)benzene [12]. In a flask, equipped with nitrogen inlet, mechanical stirrer and condenser with a drying tube, was placed 32.2 g (0.2 mole) 4-t-butylcatechol, 18.5 g (0.45 mole) NaOH and 600 ml n-butanol. After refluxing 20 minutes to dissolve all NaOH, 53 g (0.5 mole) 2-chloroethyl vinyl ether was dropped through a pressure-equalizing addition funnel under N_2. After 48 hours of refluxing, n-butanol was removed in a rotary evaporator and the residue dissolved in 300 ml H_2O and extracted three times with 500 ml petroleum ether. The combined extract was dried over $MgSO_4$ and condensed in a rotary evaporator. Fract. dist. (vac) gave 20 g of the desired product. Yield: 33%; b.p. 164-168°C/10^{-3} mm Hg. Anal. Calcd. for $C_{18}H_{26}O_4$: C, 70.59; H, 8.50. Found: C, 70.56;

H, 8.55. IR (neat) 985, 1652 ($-OCH-CH_2$); 815 (Ar); 1200 (C-O-C) cm^{-1}. PMR ($CDCl_3$), δ 6.85 (q, 3H), 6.45 (q, 2H, J_{cis} = 7 Hz, J_{trans} = 14 Hz); 4.0 (m, 12H), 1.25 (S, 9H). Mass spectrum: m/e (relative intensity), 306 (M^+, 29.3), 177 (100), 71 (17.5), 45 (43.5), 43 (48.9*), 41.25 (1).

Synthesis of Model Compounds

2,3-Dimethyl-1,4-benzodioxane [3] [4]. Into a stirred solution of 62 g (1.1 mole) KOH, 55 g (0.5 mole) catechol and 1.5 l DMF in a flask, equipped with mechanical stirrer, pressure-equalizing addition funnel and condenser with drying tube, was added a solution of 2,3-butanediol ditosylate in 500 ml DMF. After stirring at room temperature for 24 hours and refluxing 24 hours, it was mixed with twice the volume of water and extracted with twice the volume of diethyl ether. The extract was dried over anhydrous $MgSO_4$ and condensed in a rotary evaporator. Fract. dist. (vac) gave 24 g mixture of cis- [3] and trans-2,3-dimethyl-1,4-benzodioxane [4]. Yield: 29.23%. b.p. 96-100°C/6 mm. Anal. Calcd. for $C_{20}H_{12}O_2$: C, 73.12; H, 7.36. Found: C, 73.27; H, 7.40.

Repeated recrystallization from methanol gave 14 g of trans-2,3-dimethyl-1,4-benzodioxane [4]. m.p. 73-74°C. Column chromatography of the mother liquid, using Al_2O_3 as adsorbent and a 1:1 mixture of petroleum ether and diethyl ether as eluent, and then fractional distillation under vacuum gave 9 g of pure cis-2,3-dimethyl-1,3-benzodioxane [3]. b.p. 96°C/3.5 mm.

PMR cis-isomer [3]: δ 1.3 (d, J = 6 Hz, 6H), 3.81 (M, 2H), 6.80 (S, 4H). trans-Isomer [4]: δ 1.20 (d, J = 7 Hz, 6H), 4.18 (m, 2H), 6.72 (S, 4H). IR: (No major difference between 3 and 4) 1600 cm^{-1}, 750 cm^{-1} (Ar), 940 cm^{-1}, 1085 cm^{-1}, 1265 cm^{-1} ($\overline{Ar}$-O-C). CMR data: see Table I. Mass spectrum: m/e (relative intensity), [4], 164 (M^+, 90.8), 135 (56.3), 110 (100), 80 (32.0), 55 (36.4), 28 (31.3).

cis- and trans-2,4-Dimethyl-1,5-benzodioxepane [5] [6]. These compounds were prepared by similar methods as used for 3 and 4 using 2,4-pentanediol as starting material.

Anal. Calcd. for $C_{11}H_{14}O_2$: C, 74.12; H, 7.91. Found: C, 74.31, H, 8.05.

(2S, 4R)-Dimethyl-1,5-benzodioxepane [5]. Yield: 6%. m.p. 114-116°C, a white flake crystal. PMR: δ 1.38 (d, J = 6.5 Hz, 6H): 1.92 (m, 2H), 3.9 (m, 2H), 6.95 (S, 4H). CMR: See Table II. Mass spectrum: m/e (relative intensity), 178 (M^+, 45.7), 121 (76.2), 110 (100), 69 (54.1), 41 (61.5).

(2S, 4S)-Dimethyl-1,5-benzodioxepane [6]. Yield: 3%. b.p. 110°C/5 mm. PMR: δ 1.35 (d, J = 6.5 Hz, 6H), 1.95 (q, J_{trans} =

9 Hz, J_{cis} = 7 Hz, 2H), 4.55 (m, 2H), 6.82 (S, 4H). CMR: See Table II. IR: (No major difference between 5 and 6), 1580 cm^{-1}, 750 (Ar), 930 cm^{-1}, 1255 cm^{-1} (Ar-O-C).

Polymerization Studies

One representative example of each polymerization is given to illustrate the general procedures.

Radical Polymerization of 1,7-Dienes 1 and 2. Charged to a 300 ml polymer thick-wall glass tube were 24 g (0.148 M) monomer 1, 250 ml benzene (0.5 M) and 1.9 g (7.84 x 10^{-3} M, 4%) benzoyl peroxide. The polymer tube was connected to a high vacuum line, frozen in a dry ice-acetone slurry, and then evacuated. The solution was subjected to three cycles of freeze-thaw processes before the polymer tube was sealed under vacuum. The sealed tube was placed in an oil bath at 60°C for one week. The solution remained clear but became viscous. The polymer [9] was precipitated as a white powder by pouring the solution into methanol, filtered and redissolved in benzene and reprecipitated from methanol three times before it was dried in a vacuum pistol at 60°C for three days. Conversion was 31%. $\overline{M}_n$ = 2.1 x 10^3 g/mole. Anal. Calcd. for $C_{10}H_{10}O_2$: C, 74.05; H, 6.20. Found: C, 73.90; H, 6.17.

Cationic Polymerization of 1,7-Dienes. A 500 ml three-necked, round-bottomed flask, equipped with a gas inlet, mechanical stirring bar and septum, was flamed and dried under argon. In the flask were placed 5 g (0.0308 M) monomer 1 and 480 ml CH_2Cl_2 (0.064 M) and then 0.1477 (1%) ϕ_3CSbF_6 under argon. The red brown solution was stirred at room temperature for 38 hours. At the end of the polymerization, a few ml of ammonia-methanol was added to terminate the polymerization. The polymer [7] was precipitated by pouring the solution into methanol. It was redissolved in acetone and reprecipitated from petroleum ether, then dried in a vacuum pistol at 60°C for three days. Conversion, 67%. $\overline{M}_n$ = 6.07 x 10^3 g/mole.

The benzene skeleton stretching of compounds 5 and 6, seven-membered ring model compounds, occurred at 1580 cm^{-1} in IR, while that of six-membered ring model compounds 3 and 4 absorbed at 1600 cm^{-1} in IR. The IR of polymer 7 from cationic polymerization had a peak at 1580 cm^{-1} but none at 1600 cm^{-1}, while that of polymer 9 from radical polymerization had a peak at 1600 cm^{-1} but none at 1580 cm^{-1}.

Only the general similarity between the PMR spectra of polymers and that of model compounds was observed because the peaks were broad and poorly defined. However, CMR did give detailed and clear-cut information about the polymer structures. (Compare the data of Tables I-IV.)

Cationic Polymerization of 1,13-Dienes 11 and 12. By a procedure analogous to that used for 1,7-dienes, the 1,13-dienes were polymerized via cationic initiation. Conversion to 13: 90%. Anal. Calcd. for $C_{14}H_{18}O_4$: C, 67.18, H, 7.24. Found: C, 66.94; H, 7.29.

The IR absorptions of the vinyloxy group at 1610 cm^{-1} and 975 cm^{-1} were completely absent in the spectrum of polymer 13. Instead, peaks at 740 cm^{-1} and 1595 cm^{-1} from benzene ring, 920, 1120 and 1260 cm^{-1} from C-O-C were observed. PMR spectra showed three peaks, aromatic protons at δ 6.9, protons of carbon adjacent to oxygen at δ 4.0 and saturated aliphatic protons at δ 1.80 but no vinylic protons at δ 6.2. The integration results agreed well with the numbers of different protons. CMR showed complete absence of vinyloxy carbons at 86.82 ppm, but the saturated aliphatic backbone carbons appeared at 40.91 ppm.

Copolymerization of MA and 1,7-Diene 1. Charged to a 300 ml thick glass wall polymer tube were 4 g (0.025 mole) monomer 1, 4.8 g (0.049 mole) MA, 0.080 g (2%) AIBN and 200 ml (0.45 M) cyclohexanone. After it was subjected to three cycles of freeze-thaw degassing processes and sealed under high vacuum, it was placed in an oil bath at 70°C for 23 hours. The solution remained clear but became viscous. At the end of the polymerization time, the polymer was precipitated by pouring the solution into n-hexane, washed thoroughly with diethyl ether and dried at 90°C under vacuum for 48 hours. Conversion, 15: 34%. $\overline{M}_n$: 1900 g. Anal. Calcd. for $C_{18}H_{14}O_8$ (a perfect 1:2 alternating structure): C, 60.34; H, 3.94. Found: C, 62.14; H, 4.43. It was soluble in acetone, cyclohexanone, chloroform and DMF.

The IR spectra of the polymers [15 and 16] showed absorptions for the cyclic anhydride unit at 1780 and 1850 cm^{-1}, the benzene ring at 760 and 1590 cm^{-1} and ether linkage at 1230 and 920 cm^{-1}, but no absorptions of the vinyloxy group at 960 and 1640 cm^{-1}. PMR spectra also showed the aromatic protons at δ 7.0, the protons of carbons next to oxygen at δ 4.0 and the saturated aliphatic protons at δ 1.2.

Copolymerization of MA and 1,13-Diene 11. Charged to a 100 ml thick glass wall polymer tube were 2 g (0.008 M) monomer 11, 1.824 g (0.016 M) MA, 0.0263 g (2%) AIBN and 60 ml cyclohexanone. The remainder of the procedure was analogous to that used for copolymerization of 1 with MA. Conversion to 17: 89%. $\overline{M}_n$: 6060 g/mole. Anal. Calcd. for $C_{22}H_{22}O_{10}$: C, 59.19; H, 4.96. Found: C, 58.99; H, 5.09. It was soluble in acetone, cyclohexanone, DMF and DMSO.

The IR spectrum of polymer 17 when polymerized in cyclohexanone, showed the absorptions of cyclic anhydride at 1770 and 1860 cm^{-1}, ether linkage at 1250 cm^{-1}, benzene ring at 740 and 1590 cm^{-1}, and only a slight absorption of ethenyloxy at 1615 cm^{-1}. PMR of the polymer showed aromatic protons at δ 6.75 and the protons of carbons adjacent to an oxygen at δ 3.9. No absorption of vinylic protons could be distinguished from the background noise.

The MA content of polymer 17 was determined to be 65.71%, very close to the theoretical 67% for a composition of a 1:2 regularly alternating cyclocopolymer of monomer 11 and MA.

Literature Cited

1. Butler, G.B.; Corfield, C.G.; Aso, C. in "Progress in Polymer Science"; Vol. 4, Jenkins, A.D., Ed., Pergamon Press: New York, 1974; Ch. 3.
2. Barnett, M.D.; Crawshaw, A.; Butler, G.B. J. Am. Chem. Soc. 1959, 81, 5946.
3. Marvel, C.S.; Garrison, W.E. J. Am. Chem. Soc. 1958, 80, 1740.
4. Yokota, K.; Kakuchi, T.; Takada, Y. Polym. J. 1976, 8(6), 495.
5. Chu, S.C.; Butler, G.B. Polym. Ltrs. 1977, 15, 277.
6. Seung, S.L.N.; Young, R.N. J. Polym. Sci., Polym. Ltrs. Ed. 1978, 16, 367.
7. Pederson, C.J. J. Am. Chem. Soc. 1967, 89, 7017.
8. Gatti, G.; Segre, A.L.; Morandi, C. Tetrahedron 1967, 23, 4385.
9. Fritz, J.S.; Lisichi, N.M. Anal. Chem. 1951, 23, 589.
10. Griffith, R.C.; Grant, C.M.; Roberts, J.D. J. Org. Chem. 1975, 40 (25), 3726.
11. Foster, R. "Organic Charge-Transfer Complexes"; Academic Press: New York, 1969; p. 20.
12. Butler, G.B.; Campus, A.F. J. Polym. Sci. 1970, A-1, 8, 545.
13. Pittman, C.U. Jr.; Rounsefell, T.D. Macromolecules 1975, 8 (1), 46.

RECEIVED February 9, 1982.

Radiation-Induced Loss of Unsaturation in 1,2-Polybutadiene

MORTON A. GOLUB and ROBERT D. CORMIA[1]

Ames Research Center, NASA, Moffett Field, CA 94035

An IR study was made of the radiation-induced loss of unsaturation and methyl production in 1,2-polybutadiene (VB). G(-1,2), which depends on the initial vinyl content, decreased from ∿550 for VB with 98.5% 1,2 initially, to ∿270 for VB with 85% 1,2 initially. The latter G(-1,2) compares favorably with the value of 196 (von Raven and Heusinger, 1974) for VB with a stated 91.5% 1,2 double bonds (but 85% estimated in this work). G(-*trans*-1,4), which depends on the initial vinylene content, ranged from ∿21 for VB with 14% *trans*-1,4 to near-zero for VB with <1% *trans*-1,4 initially. Methyl production amounted to one methyl group formed for every 4-5 vinyl units consumed in the radiation-cyclized VB, in contrast to one methyl formed for every two vinyls reacted to give monocyclic structures in cationic cyclization. The IR spectra of γ-irradiated VB were very similar to those of UV-irradiated or thermally-treated VB at the same residual vinyl contents. Both the carbonium ion chain mechanism (von Raven and Heusinger) and the radical chain mechanism (Okamoto and Iwai, 1976), for the radiation-induced cyclization of VB were dismissed in favor of a nonionic, nonradical "energy chain" mechanism. The latter mechanism apparently holds for the cyclization of VB whether induced by γ-rays, UV radiation or heat.

The radiation-induced cyclization and crosslinking of 1,2-polybutadiene (VB) were reported by von Raven and Heusinger (1) and by Okamoto and Iwai (2) to occur with very high G values (∿20-200 and ∿7-14, respectively, the values increasing with

[1] Current address: Surface Science Laboratories, Inc., Mountain View, CA 94043.

0097-6156/82/0195-0167$06.00/0
© 1982 American Chemical Society

molecular weight). The cyclization and crosslinking in VB were assumed to involve intra- and intermolecular reactions of vinyl groups, respectively, proceeding either by a carbonium ion (1) or radical (2) chain mechanism. Both mechanisms picture the cyclization as yielding fused cyclohexane rings (Reaction 1), and the

+/or

(1)

concomitant crosslinking as involving analogous intermolecular reactions (typified by Reaction 2). An alternative view was advanced that the radiation-induced cyclization of VB, like the

+/or

(2)

corresponding photo- and thermal cyclizations, results in bicycloheptane structures (Reaction 3) instead of fused cyclohexane rings (3). Reaction 3 was invoked because all attempts to

Δ

$h\nu$

+/or

(3)

achieve a free-radical postpolymerization of VB to a ladder-like polymer were unsuccessful, and because there were precedents for Reaction 3 in certain diolefins (3,4). Moreover, the close similarity of the IR spectra of thermally, photochemically and radiation chemically cyclized VB argued against a cationic or carbonium ion mechanism for the radiation-induced cyclization (3). This argument is reinforced by the fact that the cationically cyclized VB consists of predominantly monocyclic (or isolated six-membered) rings, with one methyl group formed for every two vinyl units reacted (1,5), whereas the highly cyclized VB, obtained by UV irradiation or by heating under nonpyrolytic conditions, and apparently also by γ-irradiation, contains about one methyl group for every four vinyl units consumed (3). Although ions and radicals are doubtless formed in γ-irradiated VB (1,2), they are probably less important than excited double bonds (3) in promoting the radiation-induced loss of 1,2 unsaturation culminating in extensive cyclization and some crosslinking.

Since there was only one IR spectrum of γ-irradiated VB in the literature (1) suitable for quantitative comparison with the spectra of UV-irradiated or thermally-treated VB (3), we obtained new IR data on γ-irradiated VB. Our aim was to determine if the growth of methyl groups with progressive loss of vinyl unsaturation in the radiation cyclization of VB indeed conforms to that of the corresponding photo- and thermal cyclizations, or if it follows, instead, that of the acid-catalyzed or cationic cyclization of VB. Such information was expected to lead to a better understanding of the mechanisms of the various cyclizations of VB and of the resulting cyclized structures.

Experimental

Three samples of VB were used in this study: Polymer A (98.5% 1,2, 0.7% *cis*-1,4 and 0.8% *trans*-1,4 double bonds; $\bar{M}_n$ = 219,000; $\bar{M}_w/\bar{M}_n$ = 1.08) and Polymer B (90.1% 1,2, 3.9% *cis*-1,4 and 6.0% *trans*-1,4 double bonds; $\bar{M}_n$ = 270,000; $\bar{M}_w/\bar{M}_n$ = 1.1), both kindly supplied by Dr. A. F. Halasa and Mr. J. E. Hall, The Firestone Tire & Rubber Company, Akron, OH; and Polymer C (86.0% 1,2, ∿14% *trans*-1,4 double bonds; $\bar{M}_n$ ∿200,000; $\bar{M}_w/\bar{M}_n$ ∿2.0, with structure and molecular weight similar to those of von Raven and Heusinger's Polymer 6 (1)), obtained from Institut Français du Pétrole, Solaize, France. Most of the work was performed with Polymer A, while Polymers B and C served to bridge the gap between Polymer A and Polymer 6. The VB samples were purified by reprecipitation from benzene solution with methanol as precipitant. Thin films of the purified VB samples, cast from benzene stock solutions onto NaCl disks, were sealed in vacuum in Pyrex tubes and irradiated in a Gammacell 220 ^{60}Co source. IR spectra of the VB films before and after γ-irradiation were obtained with a Perkin-Elmer Model 180 spectrophotometer. Because excessive scatter was encountered in determining residual 1,2 unsaturation of the γ-irradiated films by means of the characteristic 11.0-μm band, an alternative method was followed: First, a calibration plot of $A_{7.1}/A_{6.9}$ *vs* percent 1,2 unsaturation (Figure 1) was constructed from IR data on thermally-treated VB films, where the residual vinyl content was obtained as $100(A_{11.0})_h/(A_{11.0})_o$, the subscripts h and o denoting absorbances of heated and unheated films, respectively. The suitability of the calibration plot is shown by the fact that $A_{7.1}/A_{6.9}$ values for various polybutadienes with different vinyl contents (Polymers EB, FI and DI of a prior paper (6), with 48, 30 and 10% 1,2 units, respectively, and Polymers B and C of this paper, all depicted by squares in Figure 1) fit well on that plot. Second, the value of $A_{7.1}/A_{6.9}$ was obtained for each γ-irradiated film, and the percent 1,2 unsaturation determined from Figure 1. Changes in the *trans*-1,4 content were estimated by means of the characteristic 10.3-μm peak. The methyl production was estimated by the $A_{7.3}/A_{6.9}$ ratio as before (3,6). To compare the growth of methyl with loss of 1,2 unsatura-

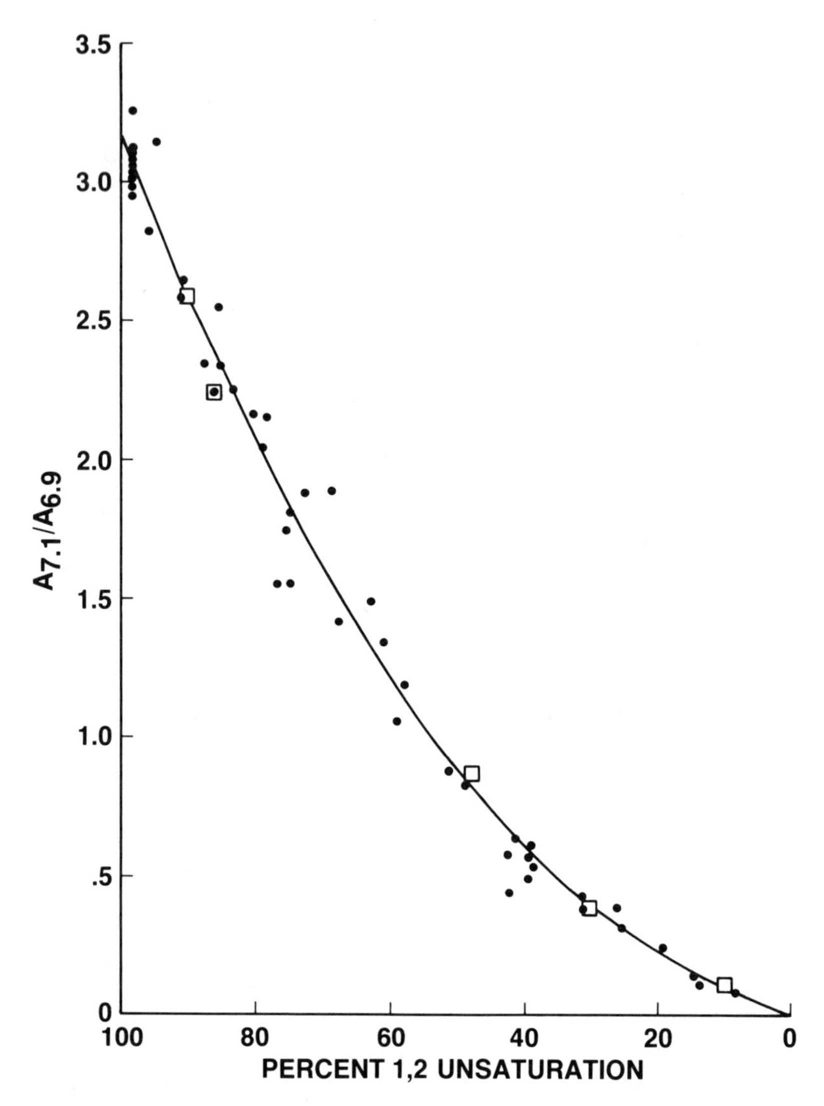

Figure 1. Calibration plot for residual vinyl content in cyclized 1,2-polybutadiene. Values calculated for Polymers B, C, EB, FI, and DI with different initial vinyl contents are represented by □.

tion for the different cyclizations, new photocyclizations of VB were also carried out as described (3).

Results and Discussion

Figure 2 shows that VB Polymer A undergoes pronounced radiation-induced loss of 1,2 unsaturation (indicated by decreased absorption at 11.0, 10.1, 7.1 and 6.1 μm, and increased absorption at 6.9 μm) accompanied by formation of methyl groups (new band at 7.3 μm) and negligible change in *trans*-1,4 unsaturation (10.3 μm). The dashed lines are baselines used to measure absorbances of the pertinent bands. The spectra of Polymers B and C were similar to those shown in Figure 2 except for displaying more intense 10.3-μm peaks initially (reflective of their higher *trans*-1,4 contents) which experienced some diminution on γ-irradiation. The changes in Figure 2B, besides being qualitatively the same as (or similar to) those observed in the IR spectra of photocyclized (or thermally cyclized) VB (3), are qualitatively identical to those observed in the IR spectrum of von Raven and Heusinger's γ-irradiated VB Polymer 6 (see Figure 4 in Reference 1), except for reduction of the *trans*-1,4 content in Polymer 6 (and seen also in Polymers B and C). For an $A_{7.1}/A_{6.9}$ value of 0.247 in Figure 2B, and using the calibration plot of Figure 1, we calculated that the vinyl content in VB Polymer A decreased from an initial 98.5% to ∿20.8% 1,2 after a dose of 274 Mrad. Using the same $A_{7.1}/A_{6.9}$ approach, we calculated that the vinyl content in Polymer 6 decreased from an initial 85% 1,2 (by our estimate, or 91.5% as stated by von Raven and Heusinger (1)) to ∿31% 1,2 after a dose of 150 Mrad; we estimated also that the *trans*-1,4 content correspondingly decreased from ∿15% to ∿6.7%. By way of comparison, the unsaturation of our Polymer C decreased from 86.0% 1,2 and 14.0% *trans*-1,4 initially to 42.8% 1,2 and 10.7% *trans*-1,4 after a dose of 43.3 Mrad.

Figure 3 presents a superposition of kinetic data for the radiation-induced loss of 1,2 unsaturation in VB, the data having been normalized for the different initial vinyl contents in the various polymers. The residual unsaturation-dose data for Polymer A were plotted directly, yielding the solid line shown; data for Polymers B, C and 6 were arbitrarily plotted at doses greater than the actual doses by amounts equal to shift factors of 2.0, 2.7 and 3.0 Mrad, respectively. These shift factors correspond to the doses required to decrease the 1,2 unsaturation of Polymer A from 98.5% initially to 90.1, 86.0 and 85.0% (the initial vinyl contents of Polymers B, C and 6), respectively. By this normalization procedure, the data for the three VB polymers used in this work (A, B and C) and the single data point available from the literature (Polymer 6) could be fitted to a common kinetic plot. From the *initial* slope to the solid line in Figure 3, $G_0(-1,2)$ was found to have the rather high value of ∿550. Taking the tangent to the line at 85% 1,2, we obtain $G_{3.0}(-1,2)$ ∿270, which compares

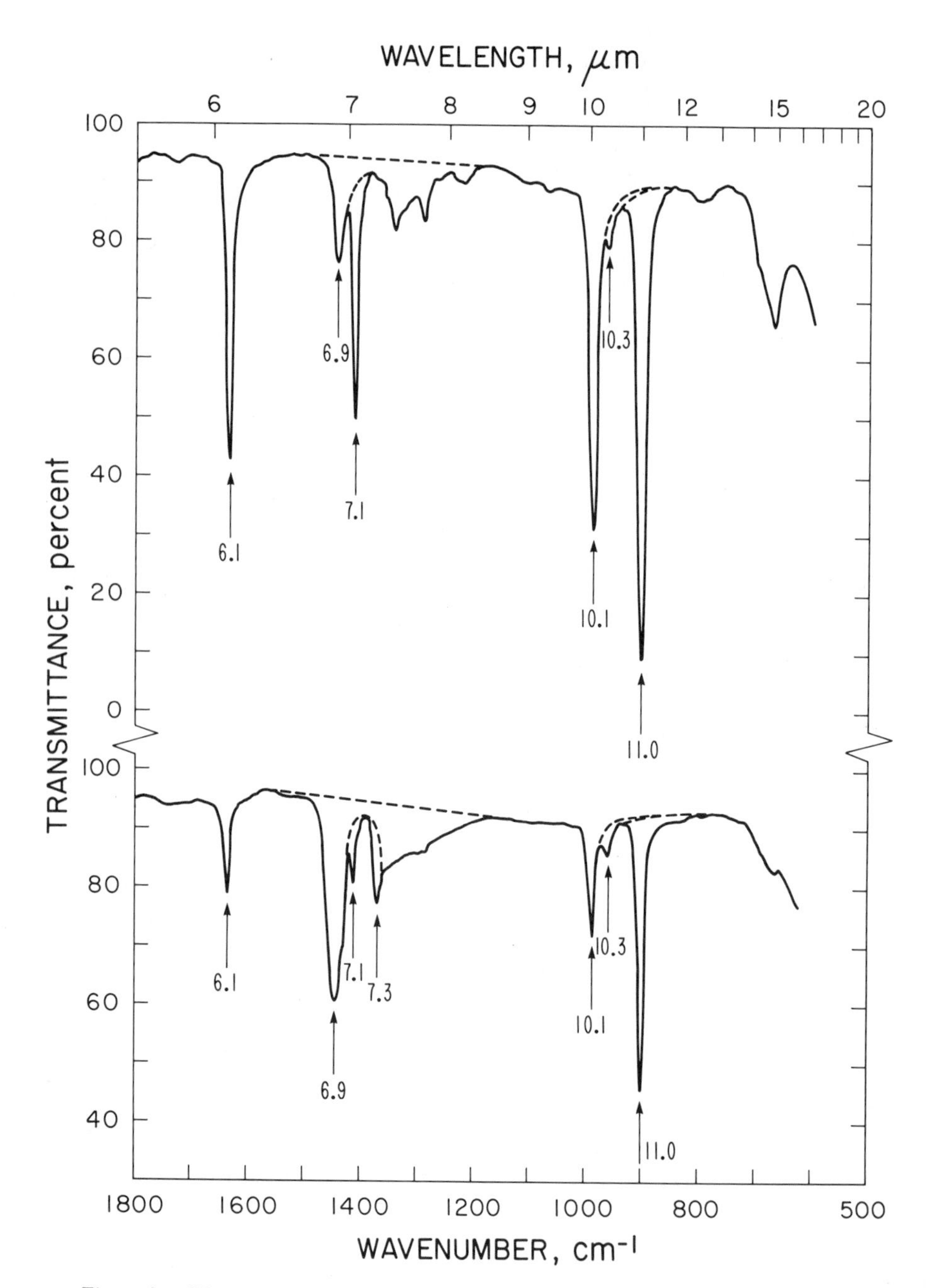

Figure 2. IR spectra of 1,2-polybutadiene film, Polymer A, before (top) and after (bottom) γ-irradiation in vacuum to a dose of 274 Mrad.

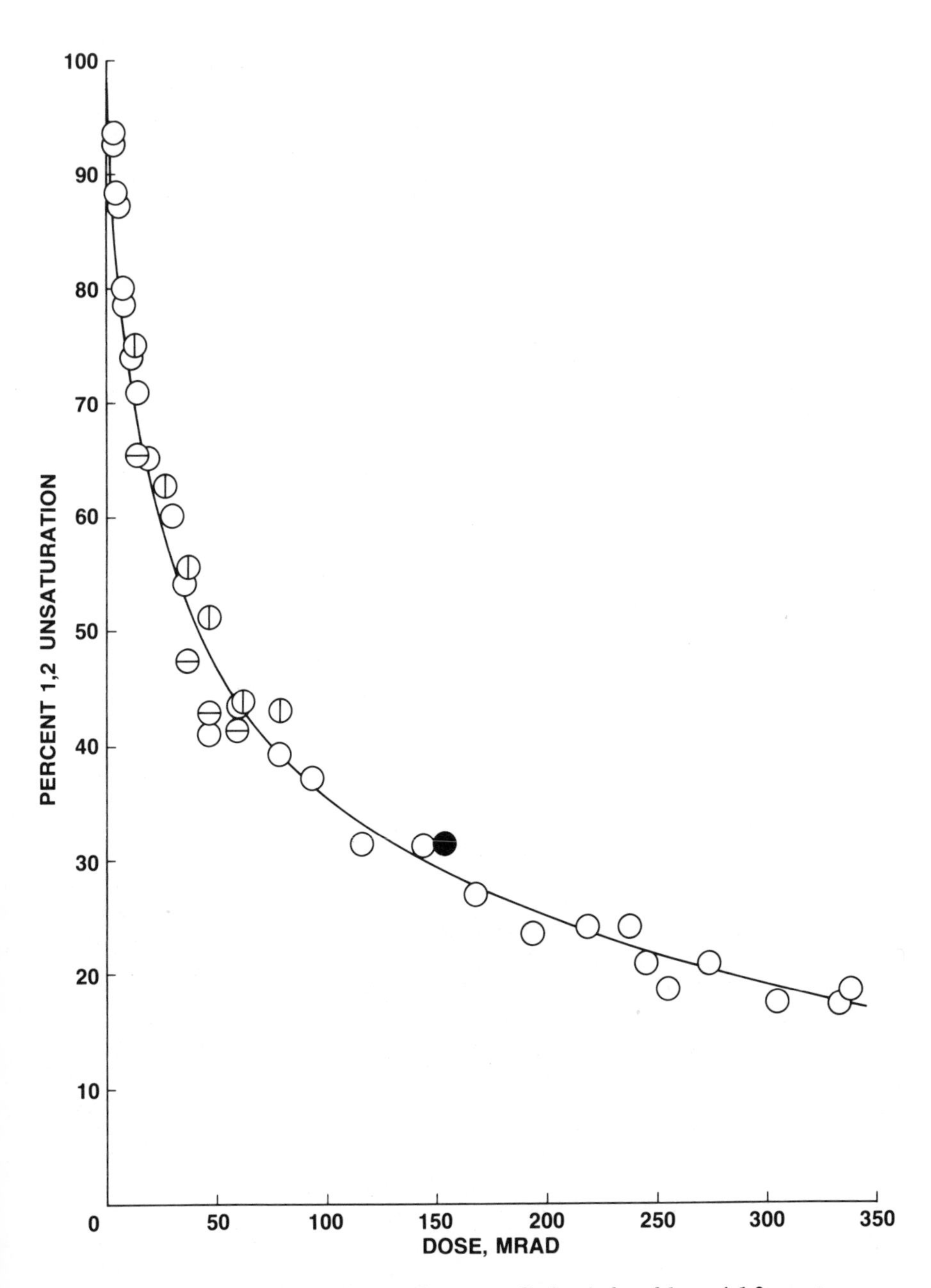

Figure 3. Superposition of kinetic data for radiation-induced loss of 1,2-unsaturation in polybutadiene with different initial vinyl content. Key: ○, 98.5% unsaturation; ⦶, 90.1% unsaturation; ⊖, 86.0% unsaturation; and ●, 85.0% unsaturation.

favorably with a G(-1,2) = 196 reported for Polymer 6 (1). The discrepancy between the latter values is no doubt due to the fact that G(-1,2) decreases very rapidly with loss of 1,2 units, especially in the early part of the cyclization, thereby preventing a precise determination of the radiation chemical yield. In any case, the large G(-1,2) values observed for loss of vinyl double bonds in VB imply a chain reaction. However, instead of accepting the proposed carbonium ion (1) or radical (2) chain mechanisms, we prefer the nonionic, nonradical "energy chain" mechanism postulated previously (3) for the thermal, photo- and radiation-induced cyclizations of VB.

We estimated G(-*trans*-1,4) to be ∿21 for Polymer C. Von Raven and Heusinger (1) reported G(-*trans*-1,4) = 107 for their very similar Polymer 6, but this is obviously a computational error and should be of the order of 11-17 (based on the initial slopes of curves for loss of 1,2 and *trans*-1,4 in their Figure 5 and their value of G(-1,2) = 196). In any event, the G(-*trans*-1,4) for a VB containing ∿14% *trans*-1,4 is considerably smaller than G(-1,2) for such a polymer, although somewhat larger than G(-1,4) ∿8 for a high-1,4 polybutadiene (7). On the other hand, G(-*trans*-1,4) drops practically to zero (as it is in Polymer A) with decrease in the initial *trans*-1,4 content of VB. The question of the relative extents of radiation-induced loss of 1,2 and 1,4 units in polybutadienes containing different vinyl/vinylene ratios merits further study.

The formation of methyl groups accompanying the radiation-induced loss of 1,2 unsaturation in VB is shown in Figure 4. The data point for γ-irradiated Polymer 6 (filled circle, calculated from Figure 4 in Reference 1) falls in line with the data for γ-irradiation of Polymers A, B and C (other circles) and does *not* fit the dotted line representing methyl formation in the acid-catalyzed or cationic cyclization of VB. To avoid clutter, data points for the new photo- and thermal cyclizations are not plotted in Figure 4, but are represented by the solid and dashed lines, respectively. Evidently, thermal cyclization of VB produces somewhat more methyl groups for a given loss of vinyl units than does photochemical cyclization, while radiation cyclization gives methyl results which overlap those for thermal and photocyclizations. Since the cationically cyclized VB is regarded as predominantly monocyclic, with one methyl group formed for every two vinyl double bonds consumed (1,5), we infer (3) from Figure 4 that radiation cyclized VB contains about one methyl for every 4-5 vinyl double bonds consumed.

Since the IR spectra of photo- and radiation-cyclized VB are very similar, as are the corresponding rates of methyl formation, the mechanism for cyclization is probably essentially the same for these two cases, if not also for the thermal case. The common denominator is presumably *initial* excitation of the vinyl double

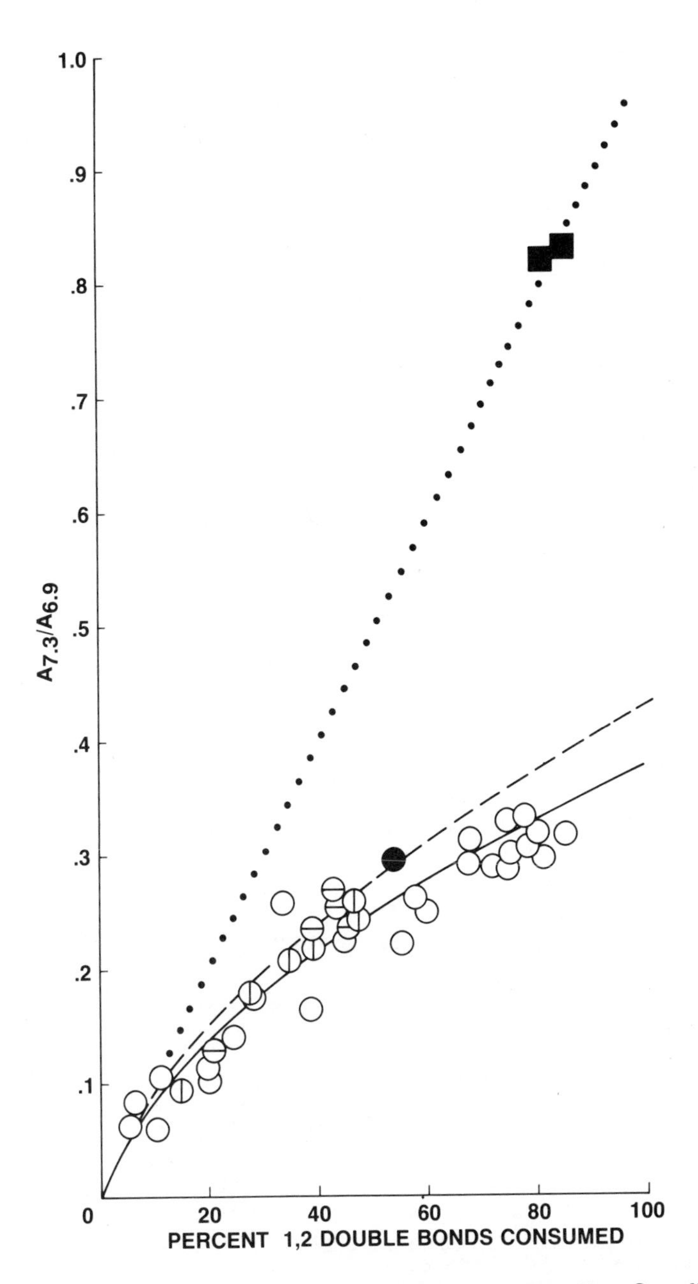

Figure 4. Methyl formation accompanying cyclization of VB. Key: ○, radiation; – – –, thermal; ——, photochemical; and · · ·, cationic cyclization.

The different circles have the same key as in Figure 3; the right and left squares were calculated from IR spectra in References 1 and 8, respectively.

bonds (ensuant on absorption of photons or interaction with γ-rays or simple heat input) leading to their disappearance by cycloaddition (Reaction 3) with, at most, a minor involvement of Reactions 1 and 2 in the radiation case. Since ring closure in Reaction 3 is exothermic (∿50-80 kcal/mole, depending on strain energy in the resulting bicycloheptanes), the energy released to the surroundings could serve to excite another vinyl double bond to undergo cycloaddition (the activation energy for thermal cyclization being ∿34 kcal/mole), thereby propagating an "energy chain" cyclization (3). In any event, there is no need to invoke the cationic chain mechanism of von Raven and Heusinger for the radiation-induced loss of unsaturation in VB inasmuch as the end-result is indistinguishable from that of the photochemical analog. The rather high yield of methyl groups in the photo-, radiation- or thermally-induced cyclization of VB may arise from the following hydrogen transfer process (3, 6):

H transfer

(4)

occurring at the same time as Reaction 3. Finally, the cross-linking encountered in the UV- or γ-irradiated or thermally-treated VB could result from *inter*molecular cycloaddition of vinyl double bonds (analogous to the *intra*molecular Reaction 3) along with a small contribution from reactions between polymeric radicals generated in Reaction 4.

Acknowledgment

The authors thank Mrs. Linda K. Yurash for experimental assistance.

Literature Cited

1. Von Raven, A.; Heusinger, H. *J. Polym. Sci. Polym. Chem. Ed*. 1974, 12, 2255
2. Okamoto, H.; Iwai, T. *Kobunshi Ronbonshu, Eng. Ed*. 1976, 5, 292.
3. Golub, M. A.; Rosenberg, M. L. *J. Polym. Sci. Polym. Chem. Ed*. 1980, 18, 2543.
4. Golub, M. A. *Rubber Chem. Technol*. 1978, 51, 677; and references cited therein.
5. Carbonaro, A.; Greco, A. *Chim. Ind. (Milan)* 1966, 48, 363.
6. Golub, M. A. *J. Polym. Sci. Polym. Chem. Ed*. 1981, 19, 1073.
7. Golub, M. A. *J. Phys. Chem*. 1965, 69, 2639.
8. Binder, J. L. *J. Polym. Sci*. 1966, B4, 19.

RECEIVED March 1, 1982.

C3 Cyclopolymerization: Cationic Cyclopolymerization of 1,3-Bis(*p*-vinylphenyl)propane and Its Derivatives

JUN NISHIMURA and SHINZO YAMASHITA

Kyoto Institute of Technology, Department of Chemistry, Faculty of Polytechnic Science, Sakyoku, Kyoto 606, Japan

The cyclopolymerization of the title compound (St-C_3-St), C3 cyclopolymerization, is reviewed. The polymerization has the similar behavior to the fluorescence emission of 1,3-diphenylpropane and its derivatives (n=3 rule or C3 rule). The monomer and its derivatives were prepared by the convenient method from the corresponding α-phenethylalcohol derivatives, using dimethyl sulfoxide-$ZnCl_2$-CCl_3COOH system. St-C_3-St gave a cyclopolymer only by cationic initiators, and the presence of cyclized units in the main chain was elucidated by several spectroscopic analyses and also by the isolation of cyclocodimers obtained from the reaction of the monomer with styrene in the presence of the catalytic amount of CF_3SO_3H.

The cyclopolymerization is revealed to be extremely sensitive to the structure of the monomer. Among the monomers investigated, St-C_3-St, 1,4-bis(p-vinylphenyl)butane (St-C_4-St), 1,3-bis(p-vinylphenyl)butane, 1,3-bis(p-vinylphenyl)-2-methylpropane, and 1,3-bis-(4-vinylnaphthyl)propane (VN-C_3-VN) were cyclopolymerized.

The intra- and intermolecular interaction between cationic center and π-electron system have attracted many chemists and much knowledge on the interaction has been accumulated (1). In 1975, we intended to apply such a kind of interaction to cyclopolymerization. Thus, it is expected that the transition state leading to a strained cyclic unit in the polymerization could be stabilized by the intramolecular attractive interaction between cationic growing-end and π-system: There had been few examples of cyclopolymerizations giving strained units. A paracyclophane unit was employed as a strained cyclic unit in the polymer main chain. In Table I are summarized some cyclophanes with their strain energies.

According to the calculation reported by Boyd (4), most of the strain of [3.3]paracyclophane (I) is due to the repulsion of

0097-6156/82/0195-0177$06.00/0
© 1982 American Chemical Society

Table I Strain Energies of Some Cyclophanes[a]

Cyclophane	Strain Energy kcal/mol(kJ/mol)	Cyclophane	Strain Energy kcal/mol(kJ/mol)
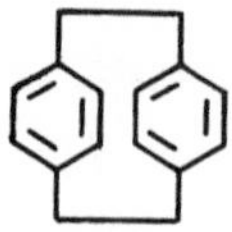	34 (142)	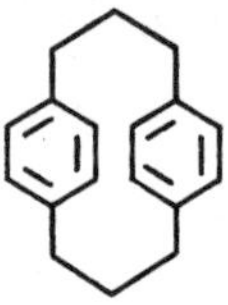	12 (50)
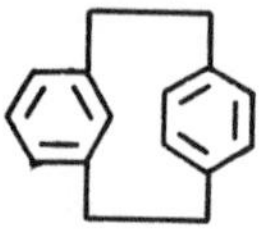	25 (105)	C_6 / C_6	2 (8)
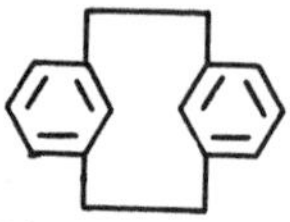	14 (59)	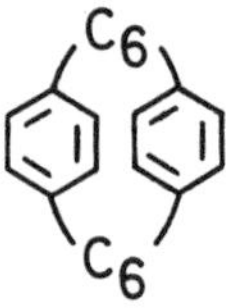	

[a] See references 2, 3 and 4. The values in parentheses were given by the present authors.

the face-to-face oriented π-clouds of benzene rings. When the strain energy could be reduced or compensated by the donor-acceptor interaction between two benzene rings, the strained paracyclophane unit would be readily introduced in the main chain by a polymerization. In fact, the title compound, 1,3-bis(p-vinylphenyl)propane (St-C_3-St), was polymerized by some cationic initiators to afford the polymer containing [3.3]paracyclophane unit in the main chain (5).

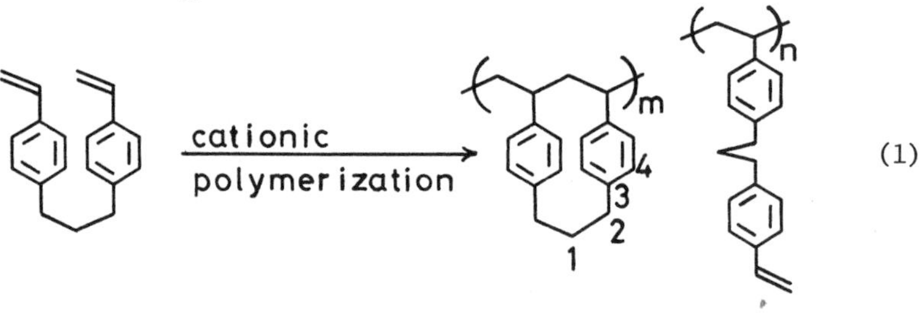

(1)

In this paper, we would like to present the results on the cyclopolymerization as a review focusing on the structure of the monomers and their cyclopolymerizabilities.

Preparation of St-C_3-St and its Derivatives

Since monomers used for the cyclopolymerization are rather unusual, but are considered to have the usage as crosslinking agents, a brief discussion of their preparation seems to be beneficial.

In 1963, St-C_3-St and its derivatives were prepared for the first time by Wiley and Mayberry (6) for the study of their copolymerizations with styrene. They used the pyrolysis of α-phenethylalcohol derivatives over Al_2O_3. Recently, Schwachula and Lukas (7) modified the dehydration of α-phenethylalcohol in dimethyl sulfoxide (DMSO), which was originally discovered by Traynelis et al. (8), and carried out the preparation of St-C_1-St in DMSO in the presence of a catalytic amount of H_2SO_4. We also independently modified the dehydration in DMSO with $ZnCl_2$-CCl_3COOH (9).

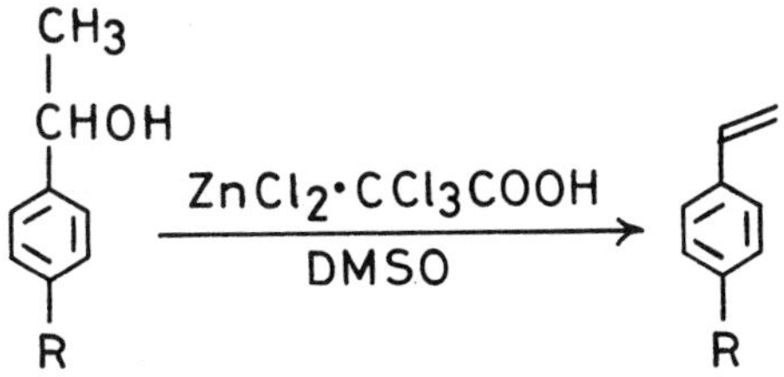

$$\text{R-C}_6\text{H}_4\text{-CH(OH)CH}_3 \xrightarrow[\text{DMSO}]{\text{ZnCl}_2\cdot\text{CCl}_3\text{COOH}} \text{R-C}_6\text{H}_4\text{-CH=CH}_2 \qquad (2)$$

Our method is efficient and gives several distyryl compounds related to St-C_3-St in reasonable yields. Results of the preparation are listed in Table II. These monomers now can be easily prepared.

Table II Preparation of Monomers[a]

Monomer	Yield (%)	Monomer	Yield (%)
St-C_1-St	52	[structure]	43[b]
St-C_2-St	68		
St-C_3-St	87	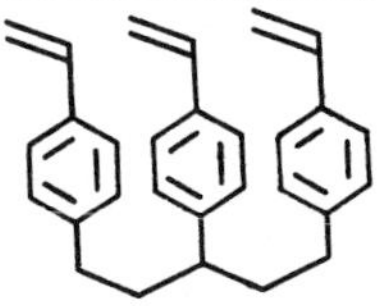	
St-C_4-St	88		
St-C_5-St	73		
VN-C_3-VN	54		
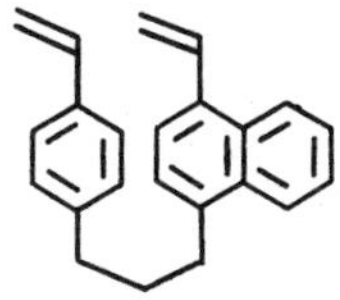 [structure]	59[b]	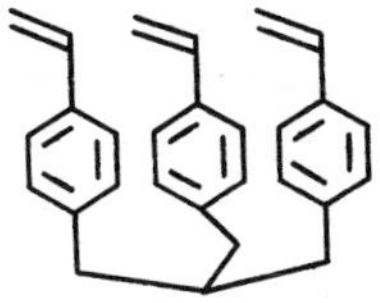 [structure]	55[b]

a Reference 9. b Unpublished data.

Structure of Cyclopolymer from St-C_3-St or VN-C_3-VN

It is generally difficult to confirm the presence of cyclized unit in a polymer. We can see one of the complete, beautiful structural proofs in the studies of Butler and his associates (10, 11). Such a kind of proof, however, is not done usually. The polymers, which are made of divinyl monomers and soluble despite of low residual unsaturation, are apt to be regarded as cyclopolymers. The criterion is considered to be valid if the cyclic unit is five- or six-membered, and the degree of polymerization is high enough. But more evidences other than the solubility should be necessary when the cyclic unit has a peculiar structure like a cyclophane.

The chemistry of cyclophanes has been studied extensively by Cram and others (12), and their following spectroscopic properties are revealed: Characteristic absorption around 11 μ in IR spectroscopy (13), high field shift of aromatic protons in ^{1}NMR spectroscopy due to the shielding effect of the opposite aromatic nucleus (14), abnormal absorption at ca. 240 nm and red-shifted, broad B-band in UV spectroscopy (15), red-shifted CT-band with TCNE in visible spectroscopy (VS) (14, 16), and characteristic properties in fluorescence spectroscopy (FS) (17, 18) and ^{13}CNMR spectroscopy (19), which are discussed below.

Table III Spectral Data

	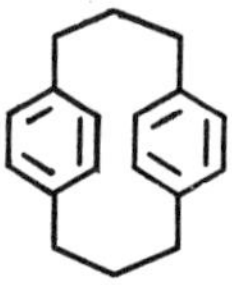 [3.3]paracyclophane (structure)	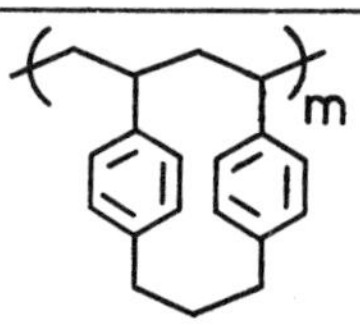 poly(St-C$_3$-St) cyclopolymer unit, $(\ )_m$ (structure)	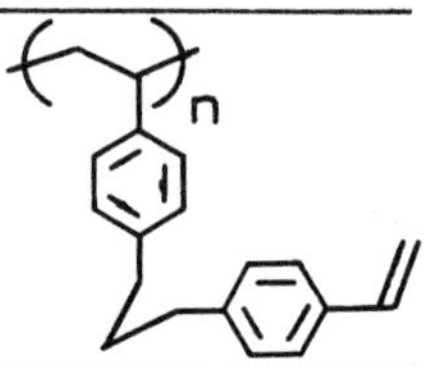 linear polymer unit, $(\ )_n$ (structure)
IR (cm^{-1})	919	928	904, 988
UV	Absorption at ca. 240 nm. Characteristic B-band.	Absorption at ca. 240 nm. Characteristic B-band.	Minimum at ca. 240 nm.
VS (CT-complex with TCNE, nm)	599	608	520
^{1}NMR (Aromatic protons, δ)	6.4	6.5	7.1

Poly(St-C$_3$-St) obtained by cationic initiators has the same characteristic spectroscopic properties as [3.3]paracyclophane. The spectral data of the cyclopolymer and the linear polymer are shown in Table III, Figures 1 and 2, together with the cyclophane.

It is known that [3.3]paracyclophane, which has the almost highest transannular interaction of the less distorted benzenes (12), has the fluorescence emission at longer wavelength (356 nm) (18) than the excimer of 1,3-diphenylpropane (332 nm). The fluorescence spectrum of the cyclopolymer, poly(St-C$_3$-St), recorded under the same conditions as for [3.3]paracyclophane is illustrated in Figure 1 (20). Both have the fluorescence at the same wavelength, and therefore the polymer is supported to contain [3.3]paracyclophane units as sequence units. The fluorescence emission at 312 nm is ascribed to the residual styryl groups.

The structure of poly(St-C_3-St) can be also clarified by ^{13}C-NMR spectroscopy. Although the polymer gave a rather complicated spectrum, carbons in the linkage and a part of aromatic carbons, numbered from 1 to 4 in the structure illustrated in equation (1), were able to be assigned easily. The chemical shifts of these carbons are almost the same as those of [3,3]paracyclophane.

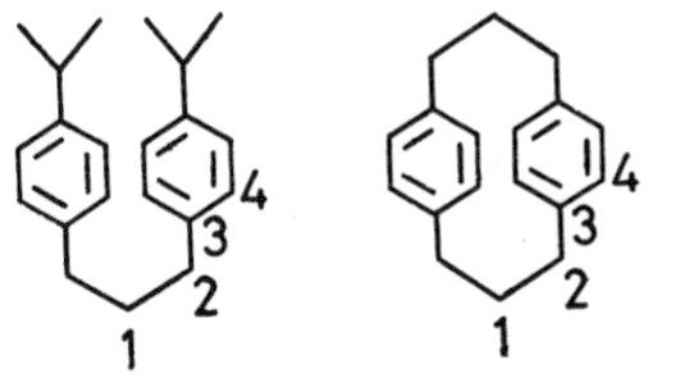

(I)

The comparison of the chemical shifts with those of the model compounds is shown in Figure 2 (20). The carbons at position 1 and 3 are affected significantly by the strain.

Recently we isolated the cyclodimers shown in equation (3), under a cationic condition (21), so that the structure of the cyclopolymer, or the cyclophane-containing polymer is now firmly established.

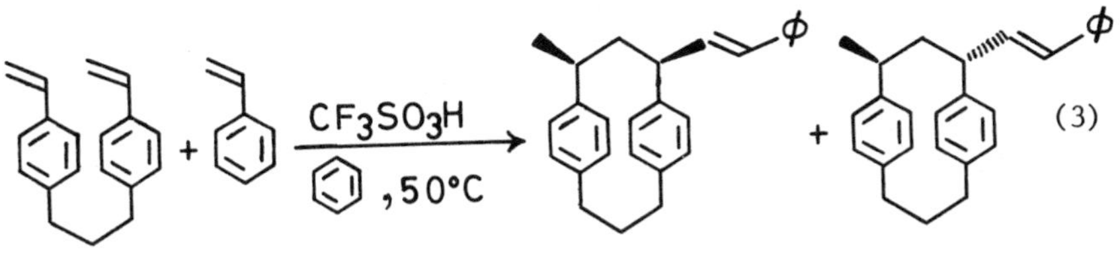

CF_3SO_3H, 50°C (3)

The next interesting subject on the structure would be the elucidation of the microstructure of the polymer, using the spectroscopic data of model compounds, cyclocodimers and their derivatives.

The structure of the naphthalenophane-containing cyclopolymer was also clarified in the same manner as for poly(St-C_3-St) (22). All spectroscopic results shown in Table IV are consistent with the proposed structure for the polymer. Again the recent cyclocodimerization finally concluded the presence of naphthalenophane units in the polymer clearly (21).

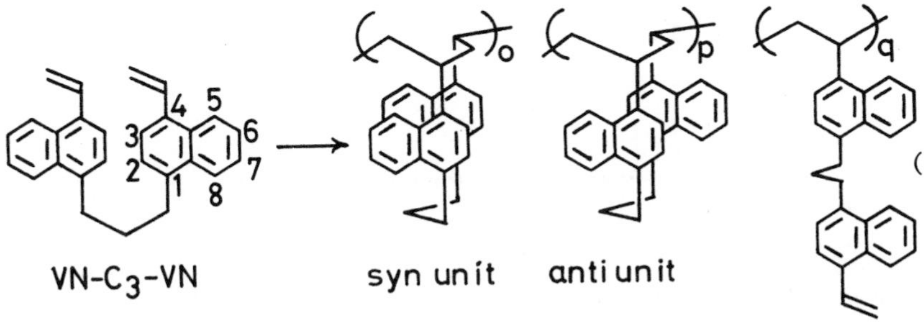

VN-C_3-VN → syn unit anti unit (4)

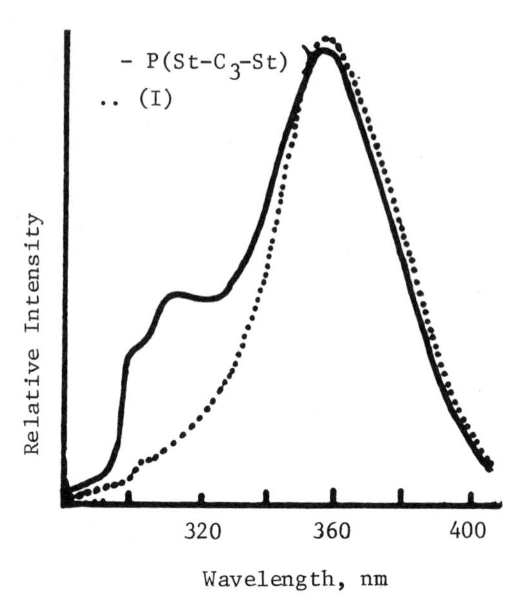

*Figure 1. Fluorescence spectra of the cyclopolymer and linear polymer of bis(*p*-vinylphenyl)propane.*

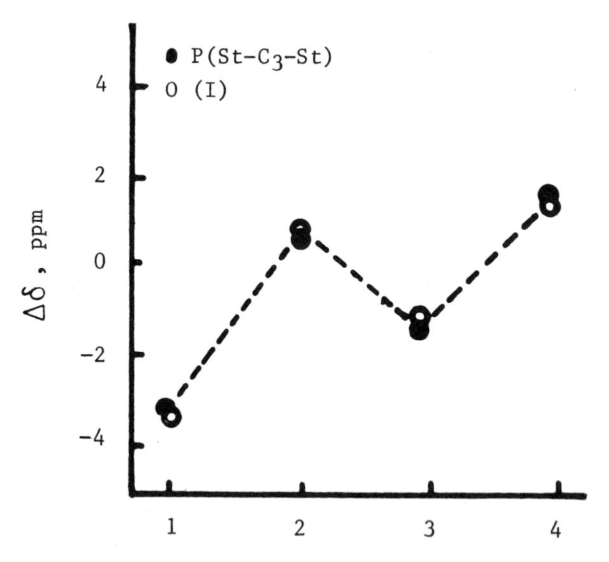

*Figure 2. Chemical shift differences in the cyclopolymer and linear polymer of bis(*p*-vinylphenyl)propane. The standard was 1,3-bis(*p*-isopropylphenyl)propane.*

As shown in Figure 3, poly(VN-C_3-VN) have a maximum fluorescence emission at 356 nm due to residual vinylnaphthyl groups,

Table IV Spectroscopic Data of Poly(VN-C_3-VN) and Related Compounds

	Poly-(VN-C_3-VN)	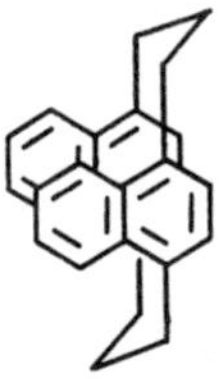	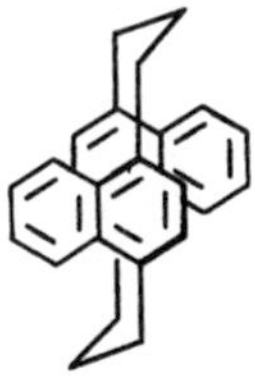	VN-C_3-VN
VS (with TCNE, nm)	802	760	745	640
FS (nm)	356 465	- 465	complex[a]	358 -
^{1}HNMR (aromatic protons, δ)	8.0 (br m) 7.3 (br m) 6.8 (br m)	7.27 (m,8H)[b] 6.90 (s,4H)[b]	7.77 (m,8H)[b] 5.98 (s,4H)[b]	7.95 (m,4H) 7.30 (m,10H)[c]

a See Figure 3.
b Reference 23.
c Including vinyl protons.

and another at 465 nm due to syn-[3.3](1,4)naphthalenophane unit, but exhibited little emissions from the excimer of open-chain dinaphthylpropane units and anti-[3.3](1,4)naphthalenophane units. This evidence and that from NMR spectroscopy, which showed almost no resonance at δ7.77 for anti units, clearly led the conclusion that poly(VN-C_3-VN) contained syn units predominantly among two possible isomeric cyclized units.

In conclusion, differing from polymers having ordinary cyclic units, the structure of these cyclophane-containing polymers can be elucidated readily and firmly by several spectroscopic methods.

Several Effects on the Cyclopolymerization

Can any kind of initiator produce the cyclopolymer from St-C_3-St? Since the monomer is a kind of styrene derivative, so it could be polymerized by a variety of initiators; cationic, radical, anionic, and coordination catalysts, and the question could be answered by the results of the polymerization.

When common rings are incorporated in cyclopolymers as sequence units, radical and cationic initiators are known generally to give highly cyclized polymers (24). However, anionic initiators give a little or almost no cyclized units in the main

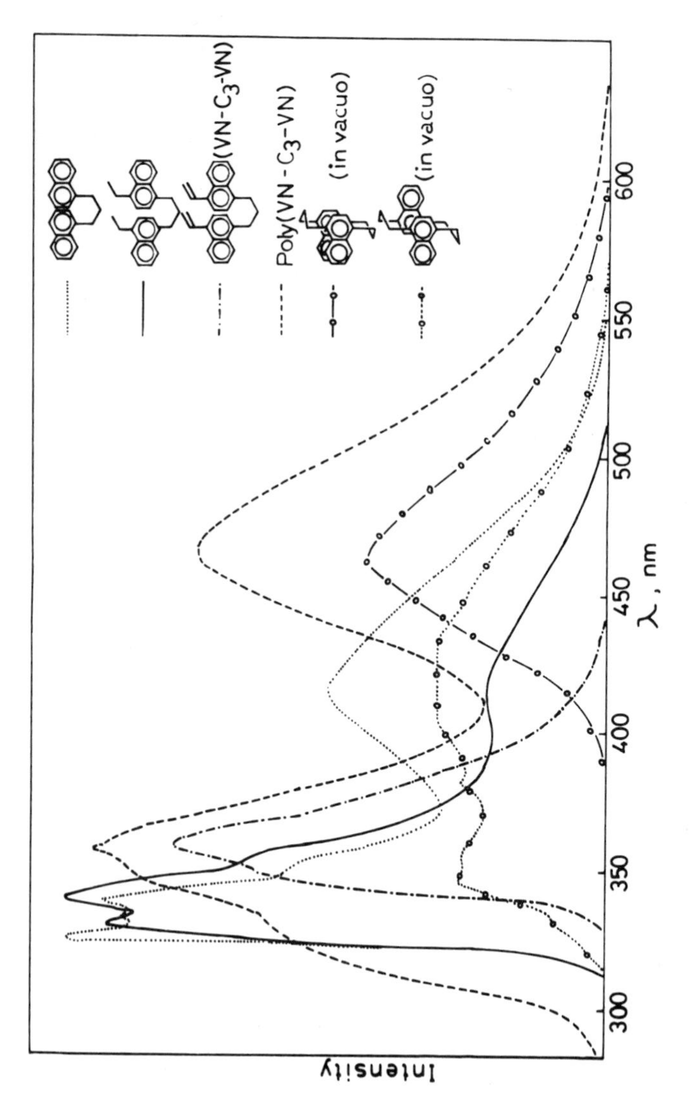

Figure 3. Fluorescence spectra. (*Reproduced, with permission, from Ref. 22.*)

chain. McCormick and Butler suggested that anionic species must form the unstable 4nπ-system with a double bond intramolecularly to introduce cyclized units in the main chain, so that the cyclization does not take place smoothly (25).

We found that radical, anionic, and coordination polymerization gave benzene-insoluble polymer or soluble polymer as shown in Table V (26). These polymers, however, did not exhibit the characteristic spectra for [3.3]paracyclophane units above mentioned. Therefore they were concluded not to be cyclopolymer. Only cationic initiators could induce cyclopolymerization of St-C_3-St (27, 28).

Before the discussion on the fact that the radical initiators failed to cyclopolymerize St-C_3-St, we need the close examination of examples for the synthesis of the [3.3]paracyclophane skeleton. Because, for the syntheses of strained cyclic compounds like [3.3]-paracyclophane, the properties of intermediates formed during reactions seem to affect the cyclization of open-chain starting materials.

Cram and his associates (15) used the following acyloin reaction for the cyclization of the starting material to [3.3]paracyclophane skeleton, because the reaction of the diphenylpropane derivative (II) gave very little amount of the desired cyclized product, even if the high dilution technique was employed.

(5)

(II)

Recently Yoshino and his associates (29) reported the anionic method to make [3.3]paracyclophane skeleton with malonate synthesis. Using the high dilution technique, the yield was 0.05%.

(6)

Table V Polymerization of St-C_3-St

Entry	Catalyst (mM)		Solv.[a]	$[M]_o$ (M)	Temp. (°C)	Time	Yield (%)		CPH[b] (%)	DB[c] (%)	$\overline{Mn}$
							sol	gel			
1	BPO	3	B	0.06	80	12 h	47	0	0	75	2900
2	AIBN	3	B	0.06	80	3.8 h	38	0	0	68	2900
3	n-BuLi	2.5	H	0.015	-78	d	0	100	-	-	-
4	NaNapht	1.5	T	0.030	-78	300 s	0	89.5	-	-	-
5	PhMgBr	33	M	0.16	30	120 s	17.4	69.1	0	71	5400
6	Ni°	235	D	1.06	50	10 min	23.0	45.9	0	46	-
7[e]	$AcClO_4$	1.5	C	0.06	30	30 min	91.7	0	91	0	-

a B, H, T, M, D, and C mean benzene, n-hexane, THF, mixed solvent of HMPA, benzene and ether (5: 2: 0.5), o-dichlorobenzene, and methylene chloride, respectively.

b Cyclophane unit content.

c Residual double bond content.

d The system was gelled as soon as the catalyst was added.

e Cationic polymerization.

On the other hand, Nishida and his associates (30, 31) made a [3.3]paracyclophane derivative from the reaction of dispiro-[2.2.2.2]deca-4,9-diene and TCNQ without the high dilution technique and moreover in high yield. They showed the presence of a zwitter ionic intermediate. The reaction suggests us that if such an electronic attractive force is existed in an intermediate, even the strained [3.3]paracyclophane skeleton can be rather easily prepared.

(7)

The reason why only the cationic polymerization of St-C_3-St produced the cyclopolymer containing [3.3]paracyclophane units would be due to the formation of $(4n + 2)\pi$-system with fulfilling the necessity of orbital symmetry (second order conservation rule of orbital symmetry) and the intramolecular attractive interaction of the styryl group (a donor) and the cation (an acceptor) (5).

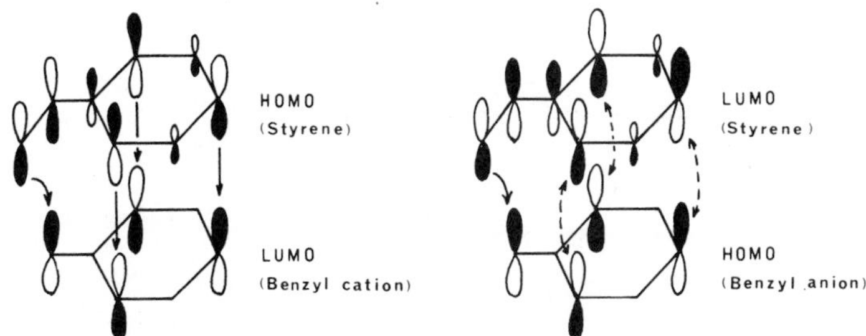

Figure 4. Models for polymer growing ends.

Even radical initiators failed to produce the cyclopolymer, probably because there may be rather strong repulsive π-π interaction between benzene rings, as seen in the examples of [3.3]paracyclophane synthesis, when the open-chain intermediate is going to cyclize.

Hereafter we would like to discuss the cyclopolymerizability influenced by the modification of the monomer structure from the following three respects; linkages between two styryl groups, substituents on the vinyl groups, and aromatic nuclei bearing vinyl groups.

The addition polymerization of St-C_n-St formally gives [3.n]-paracyclophane unit-containing cyclopolymer. As seen in Table I, [3.3]paracyclophane has the strain energy of 12 kcal/mol (50 kJ/mol), [3.2]paracyclophane of more than 12 kcal/mol but less than

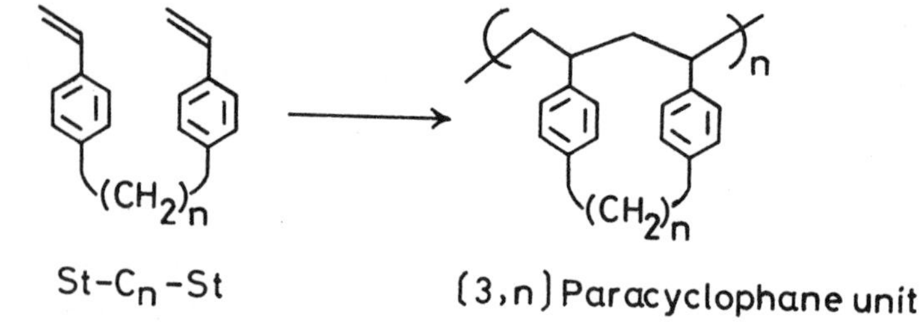

(8)

34 kcal/mol (142 kJ/mol), and [3.1]paracyclophane has not been prepared because of the strain. It is very interesting to know how small cyclophane units can be introduced by the addition polymerization of St-C_n-St. The cationic polymerization of St-C_1-St and St-C_2-St gave polymers which became easily insoluble after drying, even if the initial monomer concentration was less than 0.04 M. Therefore, their cyclopolymerizations hardly occurred. In the system where the transfer reaction occurs actively, these monomers gave soluble polymers. However, they are not cyclopolymers, but linear unsaturated polymers as shown below (27).

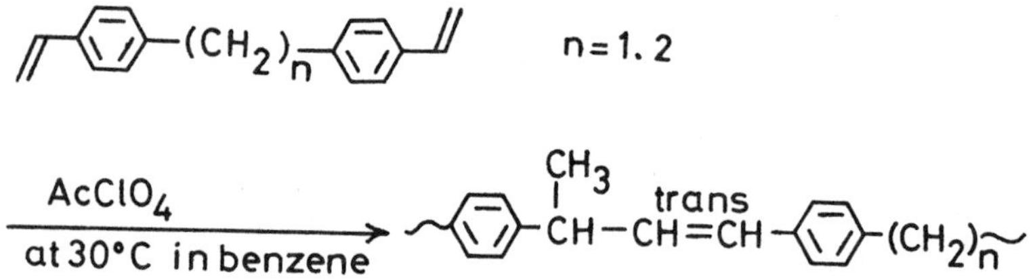

(9)

St-C_4-St gave cyclopolymer by cationic polymerization, although k_c/k_{11} obtained by $1/f_c$ vs $[M]_o$ plot was one third of that of St-C_3-St at 0°C in $(CH_2Cl)_2$ by BF_3OEt_2 as shown in Figure 5 (32).

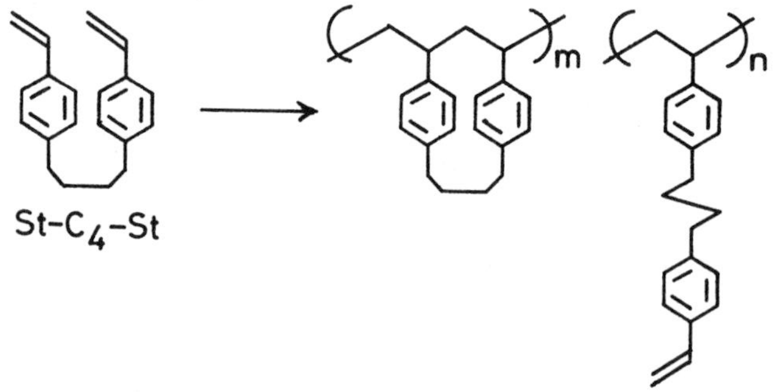

(10)

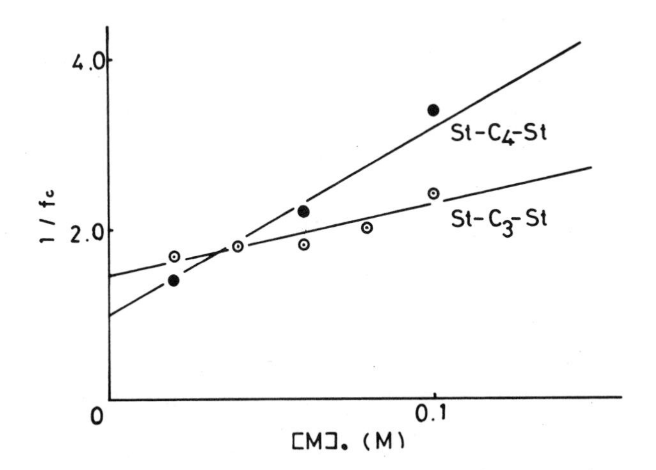

Figure 5. Correlation between $[M]_0$ and $1/f_c$ for bis(p-vinylphenyl)propane (⊙), and 1,4-bis(p-vinylphenyl)butane (●).

It means that St-C_4-St is cyclopolymerized less readily than St-C_3-St by BF_3OEt_2. It is because a C-C bond in methylene linkage of St-C_4-St should be arranged in eclipsed conformation at the transition state (IV).

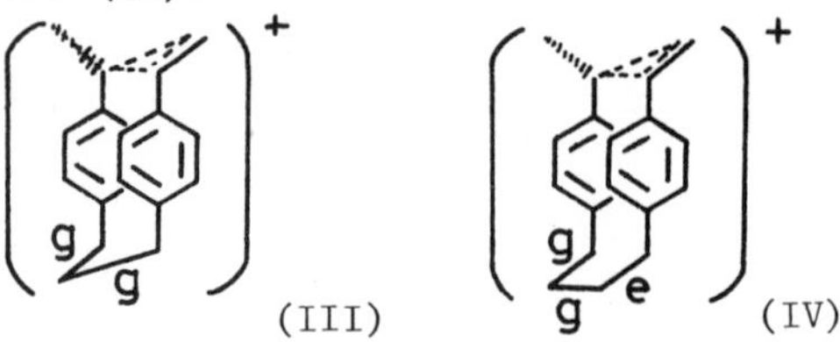

(III) (IV)

The introduction of one methyl group on the methylene linkage did not affect the cyclopolymerization significantly (33). The effect of the conformation of the three carbon (C3) linkage with two or more methyl groups, however, is considered to be recognized in the polymerization, from the analogy to the results of the excimer fluorescence of 1,3-diphenylpropane derivatives (34, 35).

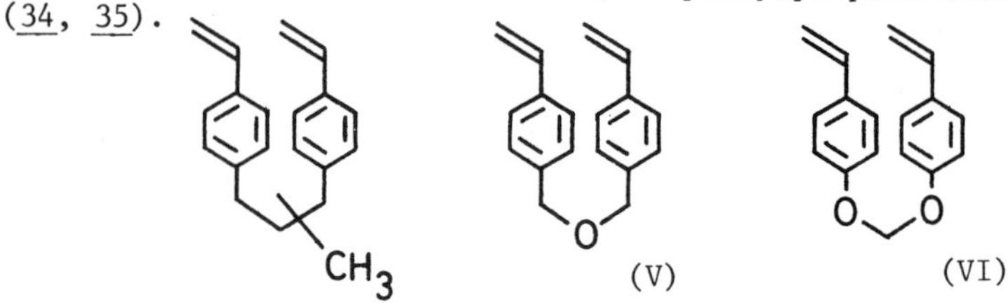

(V) (VI)

Monomers (V, VI) with one and two oxygen atoms in the linkage were prepared and polymerized by BF_3OEt_2 in CH_2Cl_2 at 0 °C with 0.06 M of initial monomer concentration (33). The polymer yielded was insoluble in ordinal solvents after drying. Although the detailed analytical data of the polymer were not obtained, it is quite likely that the cyclopolymerization occurred hardly in these cases. It is ascribed to little formation of the sandwich type transition state like (III), since the trans conformation of C-O bond is preferable.

The cyclopolymerization is affected very significantly by the

$SnCl_4$, 0 °C → $-(CH_2)_3-$ etc. (11)

MS-C_3-MS

substituents at α-position of vinyl groups. The α-methylstyrene derivative, MS-C_3-MS was not cyclopolymerized by $SnCl_4$ in $(CH_2Cl)_2$ at 0 °C, but gave polymer containing indan rings to some extent (36). It may be due to too strong steric repulsion between methyl groups and/or too significant stabilization of carbocationic site with additional methyl group to form the sandwich-type transition state (VII). As shown in Table VI, for an example of the steric effect appearing in the sandwich-type orientation of two benzene rings, we observed the considerable steric effect of p,p'-dimethyl groups on the formation of excimer (VIII) (37): i.e., intensity ratio ratio (I_D/I_M) between the excimer and monomer fluorescence became about one third of that of the non-substituted compound (X = H), after the substitution of methyl groups at p-positions.

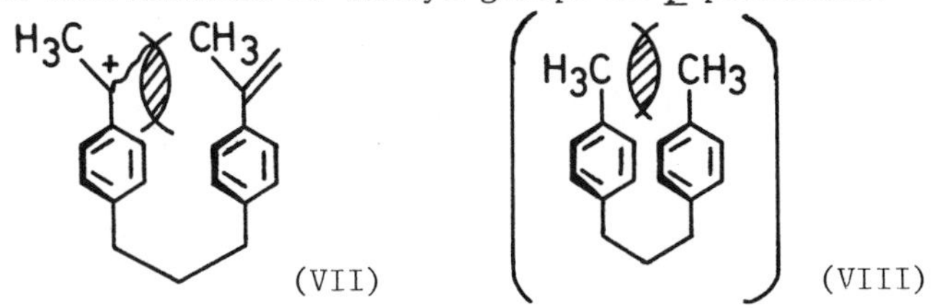

Table VI Fluorescence Maxima and Lifetimes

X-⟨○⟩-(CH$_2$)$_3$-⟨○⟩-X X	Monomer Emission		Excimer Emission		I_D/I_M
	λ_{max},nm	τ^b,ns	λ_{max},nm	τ^b,ns	
H	286	6.35	332	28.9	1.22
Methyl	287	5.03	330	22.4	0.46
Ethyl	287	7.20	330	22.5	0.41
Isopropyl	289	10.4	330	20.2	0.34
1-Methylcyclopropyl	289	8.70	330	29.4	0.07
Vinyl	312	4.73	none		-

a All spectra in nitrogenated cyclohexane solution at room temperature. Concentration of phenyl residues, 1.0×10^{-3} M. b Determined by time-resolved fluorescence spectra recorded under the reduced pressure of 10^{-5} Torr.

It is well known that 1,1-diphenylethylene is not homopolymerized so that the cyclopolymerization of 1,3-bis[p-(1-phenylvinyl)phenyl]propane (DE-C$_3$-DE) is hardly expected (38). The

Table VII TGA Data of Saturated Polymers

Polymer from (⟨○⟩-C(=CH$_2$)-⟨○⟩)$_n$ Z Z	n	Decomp. Temp.,°C		10%-Weight Loss Temp.,°C	
		in air	in N_2	in air	in N_2
-S-	2	420	416	478	480
-(CH$_2$)$_2$-	2	397	429	482	490
-CH$_2$-	2	432	433	492	492
-C(=CH$_2$)Ph-	1	409	441	500	504
-O-	2	429	450	502	508

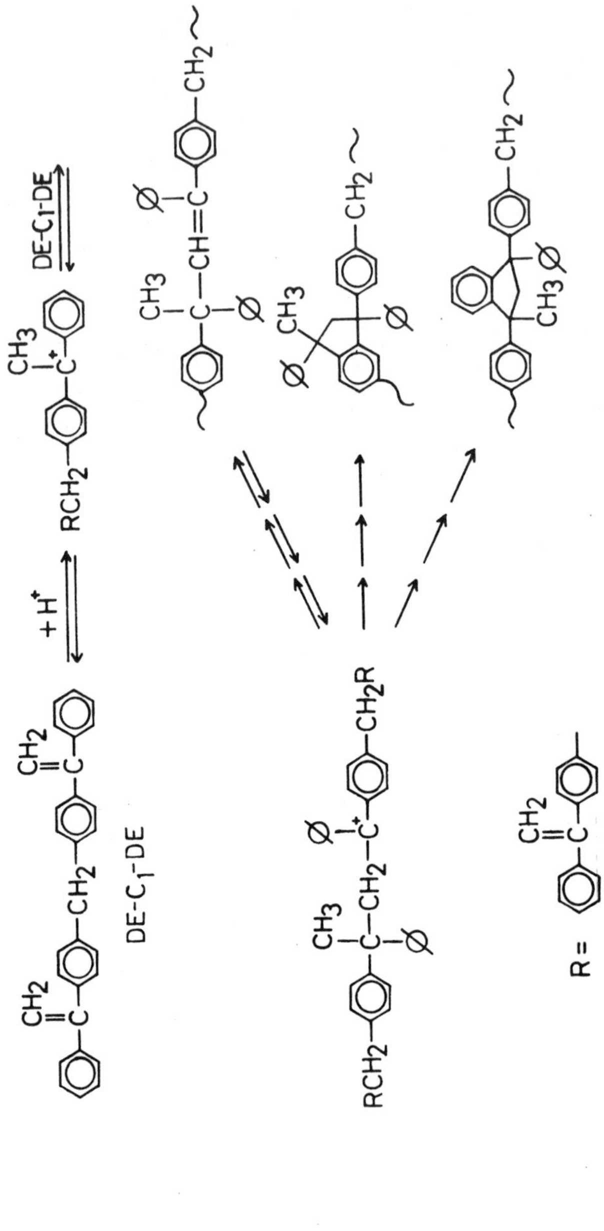

SCHEME I

polymer obtained, either saturated of unsaturated one, by a cationic initiator, was not a cyclopolymer by addition polymerization, but a polymer by stepwise polyaddition as shown in Scheme, using DE-C_1-DE as a representative (39). It is, however, interesting to note that some polymers among saturated ones had relatively good thermostability, as indicated in Table VII (40).

On the effect of the modification of the benzene rings in St-C_3-St on the cyclopolymerization, we have still not had much information. It will be interesting to polymerize monomers substituted at benzene rings symmetrically or unsymmetrically by some groups. At the present moment, we have only one example, the polymerization of VN-C_3-VN, which has more bulky groups but wider π-systems than St-C_3-St (22). Although more detailed works are still needed to compare the data of the cyclopolymerization of VN-C_3-VN with those of St-C_3-St, the monomer seems to be cyclopolymerized equally readily to St-C_3-St: The r_c values, the cyclization constant defined by Aso and his associates (41), for the cyclopolymerizations of St-C_3-St and VN-C_3-VN were 0.15 mol/L (54.0% yield) and 0.14 mol/L (44.6% yield), respectively, when they were polymerized under the same conditions; $[M]_o$ = 0.06 M, at 0 °C in $(CH_2Cl)_2$ by BF_3OEt_2.

Conclusion

After the birth and growth of the cyclopolymerization since 1957 (10), interests to this polymerization now seem to be directed to the studies, where some new properties of the cyclopolymers are seeked. In the literatures (42 - 46), we can find some attempts and successes on this respect.

C3 cyclopolymerization has not yet been finished and has still some interesting unsolved problems as mentioned in this paper. A few of initial objects, however, have been accomplished. That is, it gave the new material which has high π-basicity due to cyclophane units. The future progress is hoped to present further more novel materials produced by the principle of C3 cyclopolymerization where the attractive interaction between cationic site and aromatic π-system functions during the polymerization.

Literature Cited

1. Olah, G. A.; Schleyer, P. v. R., Ed.;"Carbonium Ions, Vol. III"; Wiley Interscience: New York, 1972.
2. Boyd, R. H. Tetrahedron 1966, 22, 119-22.
3. Shieh, C.-F.; McNally, D. C.; Boyd, R. H. Tetrahedron 1969, 25, 3653-65.
4. Boyd, R. H. J. Chem. Phys. 1968, 49, 2574-83.
5. Furukawa, J.; Nishimura, J. J. Polym. Sci. Polym. Lett. Ed. 1976, 14, 85-90.
6. Wiley, R. H.; Mayberry, G. L. J. Polym. Sci. A, 1963, 1, 217-26.

7. Schwachula, G.; Lukas, D. J. Chromato. 1974, 102, 123-30.
8. Traynelis, V. J.; Hergenrother, W. L.; Livingston, J. R.; Valicenti, J. A. J. Org. Chem. 1962, 27, 2377-83.
9. Nishimura, J.; Ishida, Y.; Hashimoto, K.; Shimizu, Y.; Oku, A.; Yamashita, S. Polym. J. 1981, 13, 635-9.
10. Butler, G. B.; Angelo, R. J. J. Am. Chem. Soc. 1957, 79, 3128-31.
11. Butler, G. B.; Crawshaw, A.; Miller, W. L. J. Am. Chem. Soc. 1958, 80, 3615-8.
12. For a review; see Cram, D. J.; Cram, J. M. Account Chem. Res. 1971, 4, 204-13. /704.
13. Cram, D. J.; Steinberg, H. J. Am. Chem. Soc. 1951, 73, 5691-
14. Sheehan, M.; Cram, D. J. J. Am. Chem. Soc. 1969, 91, 3553-8.
15. Cram, D. J.; Allinger, N. L.; Steinberg, H. J. Am. Chem. Soc. 1954, 76, 6132-41.
16. Cram, D. J.; Bauer, R. H. J. Am. Chem. Soc. 1959, 81, 5971-7.
17. Otsubo, T.; Kitasawa, M.; Misumi, S. Bull. Chem. Soc. Jpn. 1979, 52, 1515-20.
18. Melzer, G.; Schweitzer, D.; Hausser, K. H.; Colpa, J. P.; Haenel, M. W. Chem. Phys. 1979, 39, 229-35.
19. Misumi, S. private communication.
20. Nishimura, J.; Yamashita, S. unpublished data.
21. Nishimura, J.; Hashimoto, K.; Oku, A. to be published.
22. Nishimura, J.; Furukawa, M.; Yamashita, S.; Inazu, T.; Yoshino, T. J. Polym. Sci. Polym. Chem. Ed. in press.
23. Kawabata, T.; Shinmyozu, T.; Inazu, T.; Yoshino, T. Chem. Lett. 1979, 315-8.
24. For a review; see Butler, G. B.; Corfield, G. C.; Aso, C. Prog. Polym. Sci. 1975, 4, 71-207.
25. McCormick, C. L.; Butler, G. B. J. Macromol. Sci. Rev. Macromol. Chem. 1972, 8, 201-33. / 70.
26. Nishimura, J.; Ishida, Y.; Mimura, M.; Nakazawa, N.; Yamashita, S. J. Polym. Sci. Polym. Chem. Ed. 1980, 18, 2061-
27. Furukawa, J.; Nishimura, J. J. Polym. Sci. Sym. 1976, 56, 437-46.
28. Nishimura, J.; Mimura, M.; Nakazawa, N.; Yamashita, S. J. Polym. Sci. Polym. Chem. Ed. 1980, 18, 2071-84.
29. Shinmyozu, T.; Kumagae, K.; Inazu, T.; Yoshino, T. Chem. Lett. 1977, 43-6.
30. Tsuji, T.; Hienuki, Y.; Nishida, S. Chem. Lett. 1977, 1015-6.
31. Tsuji, T.; Shibata, T.; Hienuki, Y.; Nishida, S. J. Am. Chem. Soc. 1978, 100, 1806-14.
32. Nishimura, J.; Ishida, Y.; Yamashita, S. unpublished data.
33. Nishimura, J.; Tomisaki, D.; Ohta, M.; Yamashita, S. unpublished data.
34. Longworth, J. W.; Bovey, F. A. Biopolymers, 1966, 4, 1115-29.
35. Nishijima, Y.; Yamamoto, M. Polym. Preprint, 1979, 20(1), 391-4.
36. Nishimura, J.; Yamashita, S. Polym. J. 1979, 11, 619-27.
37. Nishimura, J.; Ouchi, Y.; Yamashita, S. unpublished data.

38. Bywater, S.;" The Chemistry of Cationic Polymerization" ed. by Plesch, P. H.; Pergamon: London, 1963; p307.
39. Nishimura, J.; Tanaka, N.; Hayashi, N.; Yamashita, S. J. Polym. Sci. Polym. Chem. Ed. 1980, 18, 515-26.
40. Nishimura, J.; Tanaka, N.; Yamashita, S. J. Polym. Sci. Polym. Chem. Ed. 1980, 18, 1203-11.
41. Aso, C.; Nawata, T.; Kamao, H. Makromol. Chem. 1963, 68, 1-11.
42. Butler, G. B.; Lien, Q. S. Polym. Preprint, 1981, 22(1), 54-5.
43. Mathias, L. J.; Canterberry, J. B. Polym. Preprint, 1981, 22(1), 38-9.
44. Gibson, H. W.; Bailey, F. C.; Pochan, J. M.; Harbour, J. Polym. Preprint, 1981, 22(1), 35-6.
45. Butler, G. B. J. Polym. Sci. Polym. Sym. 1975, 50, 163-80.
46. Breslow, D. S. Polym. Preprint, 1981, 22(1), 24.

RECEIVED June 1, 1982.

15

Free Radical Polymerization of 2,5-Diphenyl-1,5-hexadiene[1]

LUIGI COSTA and MARINO GUAITA

Istituto di Chimica Macromolecolare dell'Università di Torino, Via G. Bidone 36, 10125 Torino, Italy

The thermal (free radical) polymerization of 2,5-diphenyl-1,5-hexadiene has been investigated at temperatures ranging from 55°C to 150°C. The polymers obtained have mole fractions of residual unsaturation lower than 0.5, indicating that in the competition between intramolecular and intermolecular chain propagation reactions the former ones prevail over the latter ones. The calculated difference between the entropy changes accompanying the two types of propagation reactions, in good agreement with the experimental difference between the activation entropies of the same reactions, supports the interpretation that the intramolecular propagation is favoured by a lower entropy decrease, i.e. a less negative activation entropy.

In previous papers we had shown that the free radical cyclopolymerization of acrylic and methacrylic anhydrides (1,2) and of o-divinylbenzene (3) yields large fractions of cyclic structural units because the ring forming reactions are endowed with a lower activation entropy and a not too higher activation energy than the chain propagation reactions in which linear structural units are formed. The cyclic structures obtained in the polymerization of the two anhydrides are 6-membered rings, and those obtained in the polymerization of o-divinylbenzene are 7-membered rings. Since the balan-

[1] This is Part 20 of a series.

0097-6156/82/0195-0197$06.00/0
© 1982 American Chemical Society

ce between the effects of activation entropy and energy are presumably influenced by the ring size, we thought it of interest to extend the investigation to the free radical cyclopolymerization of 2,5-diphenyl-1,5-hexadiene (DPHD), a process already studied by Marvel and Gall (4), in which 5-membered ring structural units are formed, according to the following mechanism:

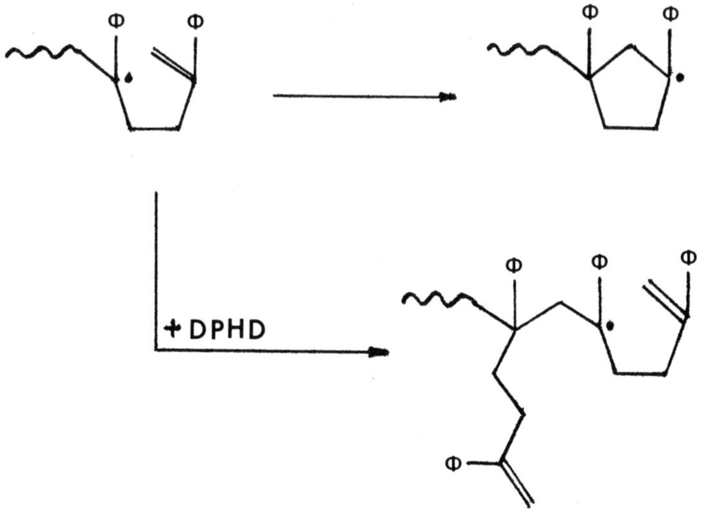

+ DPHD

The results of our investigation are reported and discussed here.

Experimental

DPHD has been prepared according to Marvel (4): the white, shiny crystals had melting point 51°C (lit. (4): 51.0 - 51.8°C); the IR and NMR spectra were consistent with the expected structure.
The polymerizations were carried out in bulk in phials sealed under vacuum, at temperatures ranging from 55°C to 150°C, in the presence of small amounts (approximately 0.1 mol/l) of 2,2'-azobisisobutyronitrile as a free radical initiator. However, taking into account the very long polymerization times, one can assume that most of the polymers obtained were thermally initiated. At a time properly selected to have a monomer-to-polymer conversion of about 10 %, the content of

the phials was dissolved in benzene (insoluble residues have never been found), precipitated in methanol, redissolved and reprecipitated.

The IR spectra of the polymers showed the bands at 1625 cm^{-1} and 895 cm^{-1}, already present in the spectrum of the monomer and attributable to the olefinic double bonds, with an intensity considerably reduced in comparison to the intensity in the spectrum of the monomer. The determination of the mole fraction f_u of residual unsaturation in the polymers were made by IR analysis, measuring the ratio between the absorbance bands at 895 cm^{-1} and 700 cm^{-1} (monosubstituted phenyl ring), having measured the same ratio in the spectra of DPHD and of α-methylstyrene.

Results and discussion

The results relative to the polymerization of DPHD are collected in Table I. The monomer concentrations at the different temperatures have been calculated by measuring in a dilatometer the specific volume of the pure monomer.

It is readily apparent from the data of Table I that the polymerization rate increases with increasing temperature. This is a usual result in the free radical polymerization of many monomers, but it is a little surprising for DPHD, owing to its similarity with α-methylstyrene, a monomer which does not polymerize by free radical mechanism at temperatures higher than 61°C

Table I - Polymerization of DPHD in bulk.

Temperature (°C)	Monomer conc. (mol/l)	Polymerization time (days)	Yield (%)	f_u
55	n.d.	5	0	----
70	4.10	5	4.0	n.d.
70	4.10	36	11.8	0.485
100	3.99	15	12.1	0.400
130	3.89	10	12.9	0.198
150	3.83	8	14.4	0.170

(ceiling temperature (5)). The reason of this is much probably due to the fact that DPHD, undergoing cyclopolymerization, actually yields polymers of different structures at different temperatures. More specifically by increasing the polymerization temperature the cyclic unit content in the polymers increases (as evidenced by the f_u values of Table I), and the cyclic units might be less prone than the linear ones to depolymerize. This might be reasonable, because the depolymerization through the cyclic units involves the breaking of two C–C bonds, as it is shown in the following scheme:

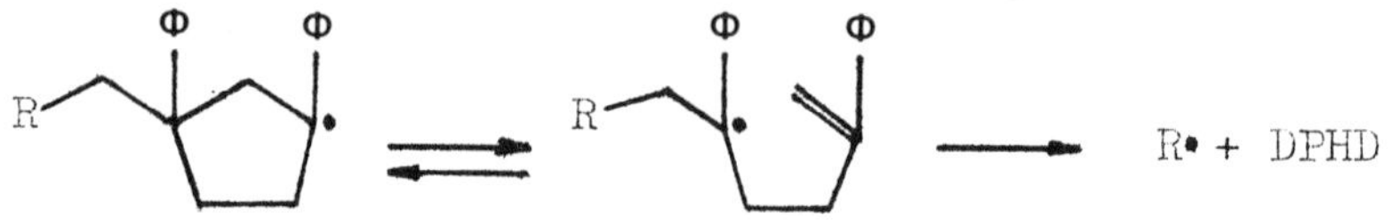

R• + DPHD

However, after the first bond is broken, a double bond remains in the neighborhood of a free radical, and there is a definite probability they meet again to reform the cyclic structure. In other words, the depolymerization involving cyclic structures should be slower than the depolymerization through open structures of otherwise similar constitution. On the other hand, there is no reason to think that a ring forming polymerization would be slower, as far as the chain propagation is concerned, than a polymerization in which similar chemical species are involved, and which exclusively yields open, non cyclic structures. The net effect should then be an increase of the temperature at which polymerization and depolymerization occur at the same rate, that is an increase of the ceiling temperature.

The relation between the mole fraction f_u of unsaturated, non cyclic, polymer structural units and the monomer concentration [M] in the system in which the polymer has been obtained can be expressed in the form (6,7):

$$\frac{1 - f_u}{f_u} = \frac{r_c}{2\,[M]} \qquad (1)$$

where r_c is the cyclization ratio, that is the ratio

between the rate constants for intra- and intermolecular chain propagation, respectively. Since $1 - f_u = f_c$ is the mole fraction of cyclic structural units in the polymer, it is immediately apparent from eq. (1) that the cyclization ratio is twice the monomer concentration at which the polymer with 50 % unsaturated and 50 % cyclic structural units is formed.

By introducing into eq. (1) the f_u and [M] values of Table I, the r_c values collected in Table II are calculated. Taking into account that the monomer concentrations at which the polymerizations of DPHD were carried out are those corresponding to the undiluted monomer, one can conclude that it is practically impossible to obtain polymers from DPHD in which the cyclic units do not prevail over the unsaturated ones.

The cyclization ratio r_c can be expressed in the form (1):

$$r_c = \exp\left[\frac{\Delta S_c^{\ddagger} - \Delta S_p^{\ddagger}}{R}\right] \exp\left[-\frac{E_c - E_p}{RT}\right] \qquad (5)$$

where $\Delta S_c^{\ddagger} - \Delta S_p^{\ddagger}$ is the difference between the activation entropies for the intra- and intermolecular chain propagation respectively, and $E_c - E_p$ is the corresponding difference between the activation energies. These differences can be obtained by plotting $\ln r_c$ as a function of 1/T, as obtained from the data of Table II. Figure 1 shows a good linear dependence of $\ln r_c$ on 1/T: from the slope and the intercept one calculates:

Table II - Cyclization ratios in the polymerization od DPHD.

Temperature (°C)	Monomer concentration (mol/l)	r_c (mol/l)
70	4.10	8.70
100	3.99	11.97
130	3.89	31.12
150	3.83	37.40

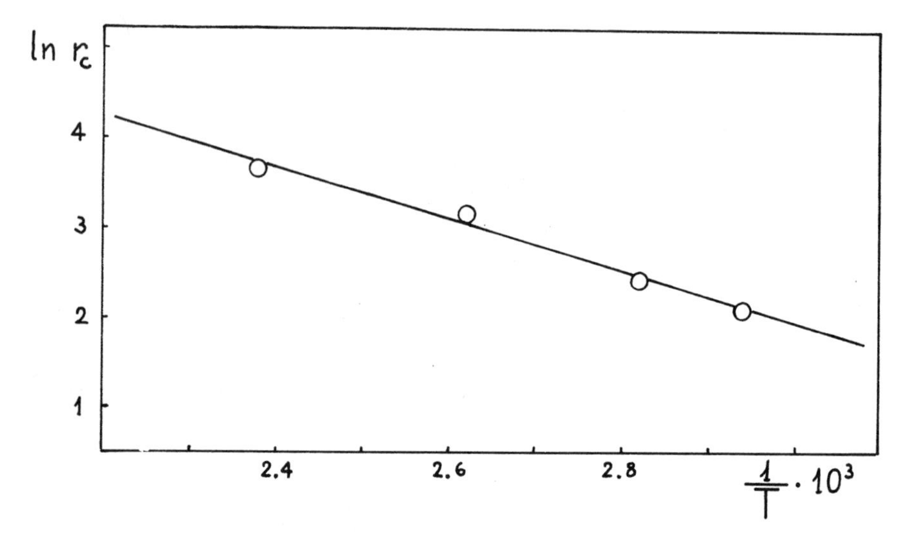

Figure 1. Arrhenius plot of the cyclization ratios of the thermal polymerization of 2,5-diphenyl-1,5-hexadiene.

$$E_c - E_p = 23.7 \text{ kJ/mol}$$
$$\Delta S_c^{\ddagger} - \Delta S_p^{\ddagger} = 86.0 \text{ J/mol K} \qquad (3)$$

The reliability of these results can be checked by calculating (8,1) the entropy changes ΔS_c and ΔS_p associated with the intra- and intermolecular chain propagation reactions, using the assumption, already found to be correct in several cases (8), that the entropy change associated with these reactions in the ideal gas state is close to the experimental activation entropy. The general expression for ΔS_p takes the form:

$$\Delta S_p = S(-CH_2-CH\lessdot) - S(CH_2{=}CH-) + S(conj) + R \ln \sigma \nu \qquad (4)$$

where $S(CH_2-CH\lessdot)$ is the molar entropy of the carbon atoms which, in the intermolecular chain propagation reaction, enter the polymer chain; $S(CH_2{=}CH-)$ is the molar entropy of one double bond of the monomer; $S(conj)$ is the decrement of the molar entropy of the monomer due to conjugation effects which are lost in the reaction; σ is the symmetry number of the monomer molecule and ν is the number of optical isomers which can be formed in the reaction. For the intermolecular chain propagation reaction occurring in the free radical polymerization of DPHD the following values must be used:

$$S(-CH_2-CH\lessdot) = -109.4 \text{ J/mol K}$$
$$S(CH_2{=}CH-) = 64.3 \text{ J/mol K}$$
$$S(conj) = 5.8 \text{ J/mol K}$$

By introducing these values, together with $\sigma = 2$ and $\nu = 2$, into eq. (4) one obtains:

$$\Delta S_p^{press} = -156.4 \text{ J/mol K}$$

where the superscript "press" indicates that the entropy is referred to the standard state of unit pressure (1 atm = 101,325 Pa). In order to convert this value to the value ΔS_p^{conc} corresponding to the standard state of unit concentration (1 mol/l), the entropy change occurring in the isothermal compression of both the reactants and the reaction product from the molar volume at 1 atm to the volume of 1 l has to be taken into account.

It can be immediately shown that, for an ideal gas system at 298.16 K, the following relationship hold:

$$\Delta S^{conc} = \Delta S^{press} - 26.6 \ \Delta n \qquad (5)$$

where Δn is the increment of the number of molecules in the reaction. In the present case $\Delta n = -1$, and one obtains:

$$\Delta S_p^{conc} = -129.9 \ \mathrm{J/mol\ K}$$

The entropy change accompanying the ring closure in the free radical polymerization of DPHD is considered (8,1) as similar to the entropy change accompanying the following reaction carried out in the ideal gas phase:

$$\Phi\text{-}\dot{C}H\text{-}CH_2\text{-}CH_2\text{-}C(=CH_2)\text{-}\Phi \ \ (P\bullet) \longrightarrow \Phi\text{-}CH\text{(-}CH_2\text{-)(-}CH_2\text{-}CH_2\text{-)}\dot{C}\text{-}\Phi \ \ (Q\bullet) \qquad (6)$$

P• and Q• radicals are formally derived from the growing chain radicals of the polymerization of DPHD, before and after an intramolecular chain propagation, by substituting with a H atom the part of the chain which is not directly involved in the reaction, and is assumed to contribute in the same way to the molar entropy of both the reactant and the reaction product. It has been already reported (8,1) that the entropy change ΔS_c occurring in reaction (6) can take the form:

$$\Delta S_c = S_{Q\bullet} - S_{P\bullet} = S_{QH} - S_{PH} + R \ln(\sigma_{QH}\sigma_{P\bullet}/\sigma_{PH}\sigma_{Q\bullet}) \qquad (7)$$

where S_{QH} and S_{PH} are the molar entropies (at 298.16 K in the ideal gas state) of the hydrocarbon molecules QH and PH obtained by saturating with a H atom the free electron of Q• and P• respectively, and σ_{QH}, σ_{PH} $\sigma_{Q\bullet}$ and $\sigma_{P\bullet}$ are the symmetry numbers of the species indicated. In the present case, PH and QH are 2,5-diphenyl-1-pentene and 1,3-diphenylcyclopentane, respectively. The molar entropies of these molecules can be calculated summing up the entropy contributions of the different groups. Specifically, for PH one obtains:

$$S_{PH} = 2\cdot S(\Phi-) + 3\cdot S(-CH_2-) + S(CH_2{=}C<) - S_{conj} =$$
$$= 2\cdot 303.4 + 3\cdot 39.2 + 64.3 - 5.8 = 582.9 \ \mathrm{J/mol\ K.}$$

where the group contributions are those already reported (1) and S_{conj} has been taken as equal to the lowering of the molar entropy of styrene due to the conjugation between phenyl and vinyl groups. In the case of 1,3-diphenylcyclopentane, the molar entropy can be calculated as follows:

$$S_{QH} = S(1,3\text{-DIMETHYLCYCLOPENTANE}) - 2\cdot S(-CH_3) + 2\,S(-\Phi) =$$
$$= 367.1 - 2\cdot 118.8 + 2\cdot 203.4 = 536.3 \text{ J/mol K}$$

By introducing into eq. (7) the values calculated for S_{PH} and S_{QH} together with the symmetry numbers $\sigma_{PH} = \sigma_{P\bullet} = \sigma_{Q\bullet} = 1$ and $\sigma_{QH} = 2$, one obtains:

$$\Delta S_c = -40.8 \text{ J/mol K}$$

(since in the cyclization $\Delta n = 0$, it follows from eq. (5) $\Delta S_c^{conc} = \Delta S_c^{press}$) and, taking into account the value previously calculated for ΔS_p^{conc}:

$$\Delta S_c - \Delta S_p = 89.1 \text{ J/mol K}$$

in very good agreement with the experimental difference $\Delta S_c^{\ddagger} - \Delta S_p^{\ddagger}$. Such an agreement gives support to both the experimentally determined differences between the activation entropies and energies, and one can conclude that in the free radical polymerization of DPHD, as well as in other cyclopolymerization processes (8), the prevalence of the ring forming reactions over the intermolecular chain propagations is accounted for on the basis of the fact that the former reactions, which would be slightly disfavoured as far as the energy changes are involved, are strongly favoured by a lower entropy decrease, which means a less negative activation entropy.

Literature cited

1. Guaita, M. Makromol. Chem. 1972, 157, 111.
2. Chiantore, O; Camino, G.; Chiorino, A.; Guaita, M. Makromol. Chem. 1977, 178, 125.
3. Costa, L.; Chiantore, O.; Guaita, M. Polymer 1978, 19, 202.
4. Marvel, C.S.; Gall, E.J. J. Org. Chem. 1960, 25, 1784.
5. McCormick, H.W. J. Polymer Sci. 1957, 25, 488.

6. Mercier, J.; Smets, G. J. Polymer Sci. 1962, 57, 763.
7. Gibbs, W.E. J. Polymer Sci. 1964, A2, 4815.
8. Guaita, M. "Cyclopolymerization and Polymers with Chain-Ring Structures"; American Chemical Society: Washington, DC, 1982.

RECEIVED March 8, 1982.

16

Synthesis and Cationic Polymerization of 1,4-Dimethylenecyclohexane

L. E. BALL[1], A. SEBENIK[2], and H. J. HARWOOD

University of Akron, Institute of Polymer Science, Akron, OH 44325

The synthesis and polymerization characteristics of 1,4-dimethylenecyclohexane are described. Cationic polymerization of this monomer yields relatively low molecular weight polymers containing appreciable amounts of endocyclic double bonds. In contrast to our earlier claim, 1,4-dimethylenecyclohexane does not seem to cyclopolymerize to a significant extent.

The polymerization of nonconjugated dienes have been investigated extensively during the last 20 years and the scope of monomers that undergo cyclopolymerization has been shown to be very broad (1-4). Trienes, cyclic dienes, and comonomer combinations that can yield cyclic radicals containing pendant unsaturation, etc., form an interesting subclass of those monomers that undergo cyclopolymerization, because they offer the possibility of incorporating bicyclic rings into polymeric backbones (5-9).

This paper concerns the polymerization behavior of 1,4-dimethylenecyclohexane, I, a monomer that was claimed in an earlier report (9) to polymerize cationically to afford a soluble, essentially saturated polymer that was presumed to have the bicyclic repeating structure II.

CH_2=⟨cyclohexane⟩=CH_2 ⟶ [$-CH_2-$ (bicyclic)]$_n$

I II

Subsequent characterization (10,11) of the polymer by infrared, ultraviolet and nmr (^{1}H-) - spectroscopy indicated the presence of

[1] Current address: Sohio Research Center, Cleveland, OH 44128.
[2] Current address: Kemijski Institute "Boris Kidric," Ljubljana, Yugoslavia.

0097-6156/82/0195-0207$06.00/0
© 1982 American Chemical Society

substantial amounts of methyl groups as well as a chromophore absorbing weakly at 260 nm. This information cast doubt on the validity of structure II. Recent cmr studies (12) have clarified the structure of polymer. This paper covers the synthesis of 1,4-dimethylenecyclohexane, its polymerization characteristics and the results of recent ^{1}H- and ^{13}C- NMR studies on the structures of its polymers.

Synthesis and Characterization of 1,4-Dimethylenecyclohexane

1,4-Dimethylenecyclohexane was synthesized in 54 percent yield by the following reaction. The material was shown by gas-liquid chromatography to be better than 99 percent pure.

$$ICH_2-C_6H_{10}-CH_2I \xrightarrow[t\text{-}BuOH]{KOtBu} CH_2=C_6H_8=CH_2$$

It was characterized by chemical analysis, ozonolysis to form formaldehyde, hydrogenation, bromination and by its infrared and ^{1}H-NMR spectra. The diene has also been prepared by pyrolysis of 1,4-bis-(acetoxymethyl)cyclohexane (13,14), from cyclohexanedione by the Wittig reaction (11,15) or by reaction with diazomethane (16), and by the electrochemical or metal reduction of 1,4-dihalobicyclo-[2,2,2-]octanes (17,18) or of 1,1,2,2-tetrakis-(bromomethyl)-cyclobutane (19). It is very stable if stored in ammonia rinsed, dry flasks.

Polymerization Studies

Attempts to polymerize I by free radical or anion initiated polymerization techniques were unsuccessful. Attempts to prepare copolymers of I with SO_2 (10) led only to complex mixtures of low molecular weight products, in contrast to the claims of a patent (20). It proved possible to polymerize I with $TiCl_4/(iBu)_3Al$ catalysts at 25° when Ti/Al ratios of ∿2.7 were employed (21) but substantial portions (25-75%) of the product were crosslinked and the soluble portions contained approximately one double bond for every two monomer units. It seems that the double bonds in I react independently during such polymerizations.

As expected, attempts to polymerize I cationically were more successful. Polymerizations were conducted in methylene chloride or in t-butyl chloride solutions using $AlCl_3$ or gaseous BF_3 as catalysts. The optimum temperature for polymerization in methylene chloride using BF_3 as catalyst was between -30° and -20°C. In this temperature range, it was possible to prepare soluble polymers having molecular weights as high as 10,000 in up to 75 percent yield. At lower temperatures, the yield of polymer and its molecular weight were reduced considerably. The effect of monomer concentration on the yield of polymer obtained at -78° is shown in Figure 1. It is seen that the yield increases with monomer concentration up to about 2 M. It then falls at higher concentrations, presumably because of the influence of monomer on the dielectric constant of the reaction mixture. Table I summarizes the results obtained in studies on the cationic polymerization of 1,4-dimethylenecyclohexane.

Polymer Characterization

A polymer prepared at -30° softened at approximately 175°. Its cryoscopically determined molecular weight was 8400. Unsaturation determinations by bromination and hydrogenation indicated the presence of approximately one double bond per molecule. These determinations were misleading, however, because subsequent nmr studies, especially ^{13}C-NMR studies, indicated the presence of an appreciable amount of unsaturation. The infrared and early ^{1}H-NMR spectra (60 MHz) of the polymer indicated the presence of methyl groups and a small amount of residual unsaturation. The intensity of ultraviolet absorption by the polymer at 258 nm (ε=27,000 per mole of polymer or ε=340/monomer unit) suggested that one diene residue may have been present per polymer chain.

Infrared Studies. Figure 2 shows the relative absorbances at 890 ($>C=CH_2$), 1370 ($-CH_3$), 1460 (CH_2), 1660 (C=C) and 2900 (C-H)cm^{-1} observed for polymers prepared from -20° to -78°. It can be seen that the methyl content of the polymers increases and that the amount of residual exocyclic unsaturation concurrently decreases as polymerization temperature increases.

NMR Studies. Several poly-1,4-DMC samples prepared by BF_3-initiated polymerizations using CH_2Cl_2 as a solvent were analyzed by ^{1}H-NMR (300 MHz) and ^{13}C-NMR (20 MHz) spectroscopy. These are identified as samples 13,16 and 22 in Table I. In addition to the methanol-insoluble products obtained by pouring the reaction mixtures into methanol, oligomeric products obtained by evaporating the mother liquors from the precipitation were also examined. In all cases, the methanol-soluble and methanol-insoluble fractions yielded very similar spectra. Consequently only the spectra of the methanol-insoluble products will be discussed.

Figure 3 shows the ^{1}H-NMR spectrum of sample 22 and Figure 4 shows the ^{13}C-NMR spectra of samples 13, 16 and 22. These spectra indicate the presence of large amounts of methyl groups and of unsaturated structures. The methyl groups are of two types -- methyl groups present on saturated carbons (as indicated by ^{1}H-signals at 0.9 ppm and ^{13}C-signals at 25.0 ppm) and methyl groups on unsaturated carbons (as indicated by ^{1}H-signals at 1.65 and 2.25 ppm and by ^{13}C-signals at 19.8-23.4 ppm). There seem to be four types of unsaturated structures present: (a) residual exocyclic methylene groups, III, are indicated by ^{1}H-signals at 4.5-4.6 ppm and ^{13}C-signals at 105-106 and 146-149.5 ppm; (b) endocyclic unsaturation,

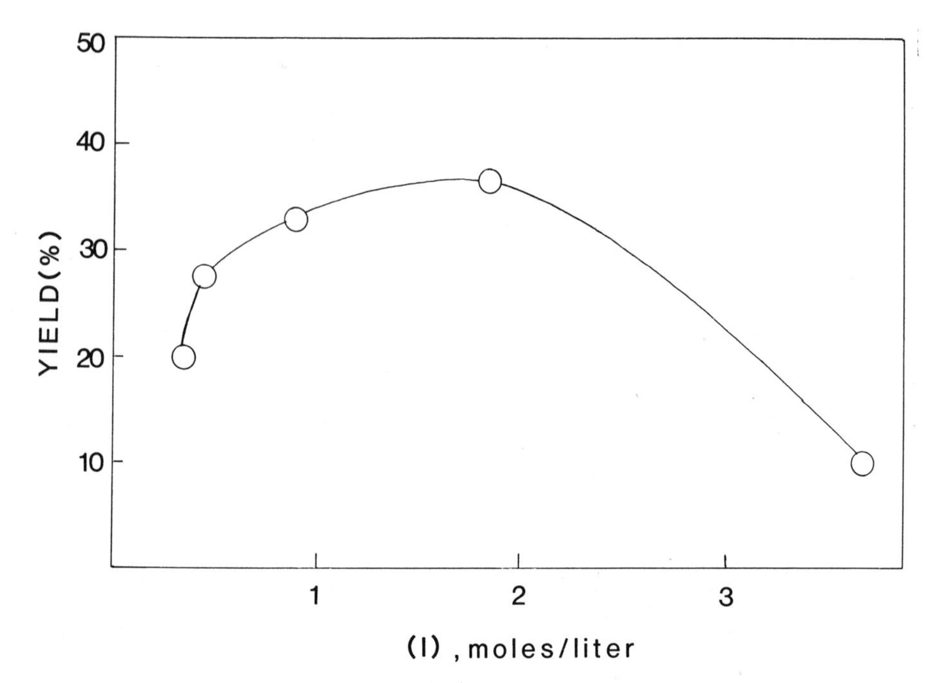

Figure 1. Effect of 1,4-dimethylenecyclohexane concentration on polymer yield for polymerizations conducted at —78°C using CH_2Cl_2 as solvent and BF_3 as initiator.

TABLE I

Studies on the Cationic Polymerization of

1,4-Dimethylenecyclohexane[a]

Exper.	Temp.	Monomer Concen. (wt%)	Reaction Time (Hrs.)	Catalyst	% Yield	[η]	Mol. Wt.
1[a]	-22°C	1.5	5	$AlCl_3$	14	-	-
2	-78°C	1.5	40	$AlCl_3$	12	-	-
3[a]	-22°C	1.5	3	$BF_3 \cdot Et_2O$	0	-	-
4[a]	-78°C	0.2	4	$BF_3 \cdot Et_2O$	0	-	-
5	-78°C	1.3	4	$TiCl_4$	0	-	-
6	-78°C	1.5	2	$SbCl_5$	0	-	-
7	-78°C	0.3	2	BF_3	20	-	-
8	-78°C	0.4	2	BF_3	27.6	-	-
9	-78°C	0.7	2	BF_3	32.5	-	-
10	-78°C	1.3	2	BF_3	36.7	-	-
11	-78°C	2.1	2	BF_3	8.6	-	-
12	-50°C	1.3	2	BF_3	9	-	-
13	-45°C	2.3	6	BF_3	47	-	-
14	-40°C	2.1	2	BF_3	17	0.045	3,700[b]
15	-30°C	1.3	1.5	BF_3	46	0.094	10,400[b]
16	-30°C	1.2	3	BF_3	37	-	-
17	-30°C	1.3	1.5	BF_3	64	0.076	7,900[b]
18	-30°C	1.3	4	BF_3	58	0.058	5,500[c]
19	-30°C	1.2	3	BF_3	55	0.079	8,400[c]
20[a]	-20°C	1.3	7.5	BF_3	2.5	-	-
21	-20°C	1.1	2	BF_3	74	0.092	10,000[c]
22	-15°C	1.2	6	BF_3	45	-	-
23	-0°C	2.1	2	BF_3	0	-	-

[a] CH_2Cl_2 was used as the solvent except for experiments 1 and 3 (t-BuCl), 4 (n-pentane), and 20 (n-hexane).

[b] Calculated from [η] by the following empirical relationship: $[\eta]=9.9\times10^{-3}M^{0.74}$

[c] Determined cryoscopically in benzene.

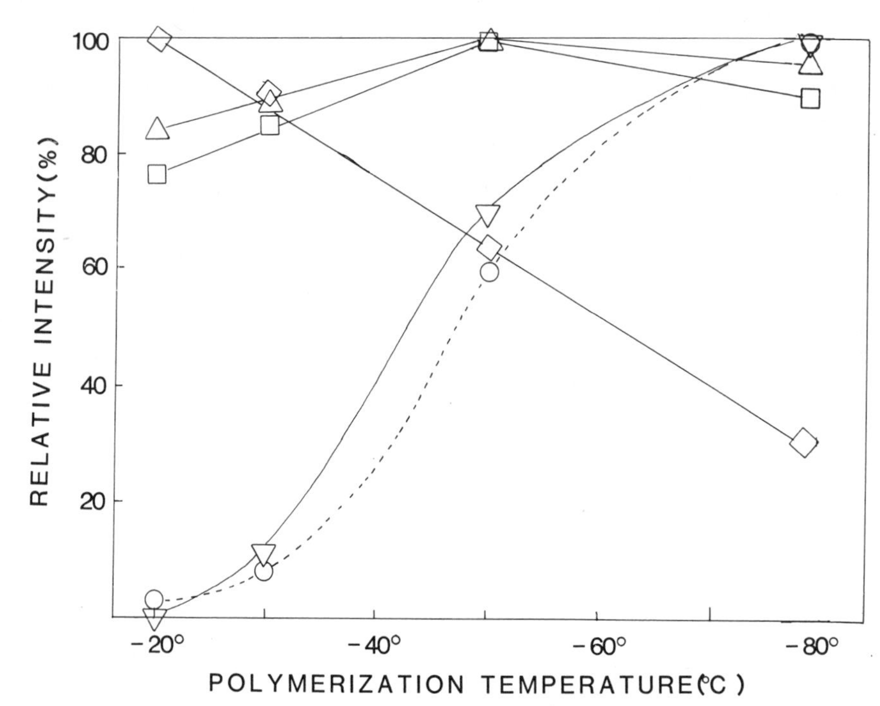

Figure 2. Relative intensities of IR absorptions of poly(1,4-dimethylenecyclohexane) as a function of polymerization temperature. Key: △, C–H stretching at 2900 cm^{-1}; ○, C═C stretching at 1660 cm^{-1}; □, $-CH_2-$ bending at 1460 cm^{-1}; ◇, $-CH_3$ bending at 1370 cm^{-1}; and ▽, $C{=}CH_2$ absorption at 890 cm^{-1}. Relative intensities are based on the maximum intensities observed for individual peaks in the spectra of 5% (w/v) solutions of the polymers in CCl_4.

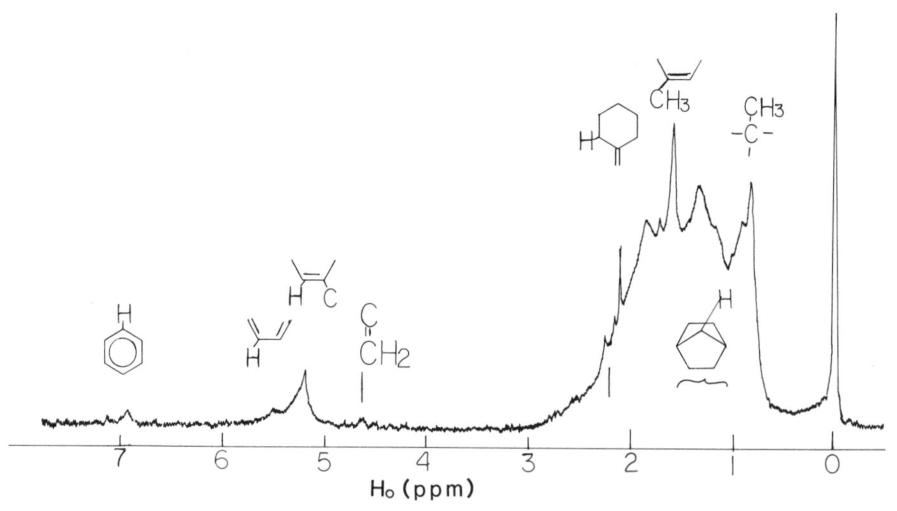

Figure 3. 300 MHz ^1H-NMR spectrum of poly(1,4-dimethylenecyclohexane), Sample 22, in CCl_4 solution.

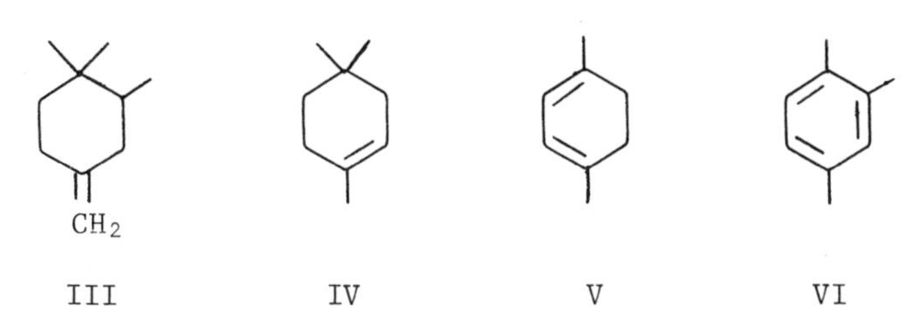

IV, is indicated by ^{1}H-signals at 5.3 ppm and by ^{13}C-signals at 120.0 and 134.0 ppm; (c) conjugated endocyclic diene structures, V, are indicated by ^{1}H-signals at 5.5 ppm and by ^{13}C-signals at 119.0 and 132.5 ppm; and (d) aromatic rings are indicated by ^{1}H-signals at 6.9-7.2 ppm and by ^{13}C-signals at 128-136 ppm. Table II allocates these resonances to specific types of structures believed to be present in the polymers. The assignments are based, in part, on spectra reported for 1-methyl-1,2-cyclohexane (22), trimethylbenzene (23) and for various compounds containing exocyclic double bonds (24, 25, 26).

By making reference to the ^{13}C-NMR spectra reported for a large number of bicyclic hydrocarbons (27), it was possible to predict the chemical shifts expected for any bicyclic structures present in the polymers. These are also listed in Table II. According to these predictions, bridged methylene groups present in (2,2,1)-bicyclic systems should be responsible for absorptions at 45-52 ppm in ^{13}C-NMR spectra and at 1.2-1.6 ppm in ^{1}H spectra. As can be seen in Figures 3 and 4, there is only minor absorption in this region. It should be noted that quaternary carbons can also be responsible for absorption in this region.

Table II contains estimates of the relative amounts of various structures believed to be present in the polymers. It is clear that the polymers have very complicated structures.

Discussion

Although soluble polymers can be prepared by cationic polymerization of 1,4-dimethylenecyclohexane, a number of factors make it unlikely that this monomer undergoes appreciable cyclopolymerization. Infrared and nmr spectroscopy show that the polymers prepared at -78° contain appreciable quantities of exocyclic double bonds, and that polymers prepared at higher temperatures contain appreciable amounts of methyl groups. Carbon and proton magnetic resonance spectroscopy show that the methyl groups are present on both saturated and unsaturated carbon atoms. In addition, the ^{13}C-NMR spectra of the polymers contain strong signals for unsaturated carbon atoms. Polymers prepared at -78° contain mostly exocyclic double bonds and tend to be crosslinked. Polymers prepared at higher temperatures contain mostly endocyclic double bonds and some aromatic rings. These results and the fact that the polymer-

TABLE II

NMR Analysis of Structures in Poly(1,4-Dimethylenecyclohexane)

Structure		Chemical Shifts (ppm) ^1H	Chemical Shifts (ppm) ^{13}C	Average No. Per Repeat Unit -15°	Average No. Per Repeat Unit -45°
$-CH_2-$ (bicyclic)	$-C\lt$	–	40-48	-	-
	ring CH	1.2-1.6	48-52		
$HC=C(CH_3)$	$=C(C)(CH_3)$	-	132.5-134.0	0.5	0.45
	$=C-H$	5.3	120.0		
	$=C(H)-CH_3$	1.65	23.4		
$\gt C=CH_2$	$\gt C=CH_2$	4.5-4.6	105.0-106.0	0.03	0.08
	$\gt C=CH_2$	-	146.0-149.0		
$\gt C(H)CH_3$	$\gt C(H)-CH_3$	0.8-1.0	25.0	0.45	0.40
aromatic ring (H, CH_3)	ring H	7.0	128.5-131.0	0.05	0.08
	CH_3	2.25	19.8-20.8		
	ring C–CH_3	-	133.0-136.0		
cyclohexadiene ring	=C–H	5.5	119.0	0.03	
	=C<	-	132.5		

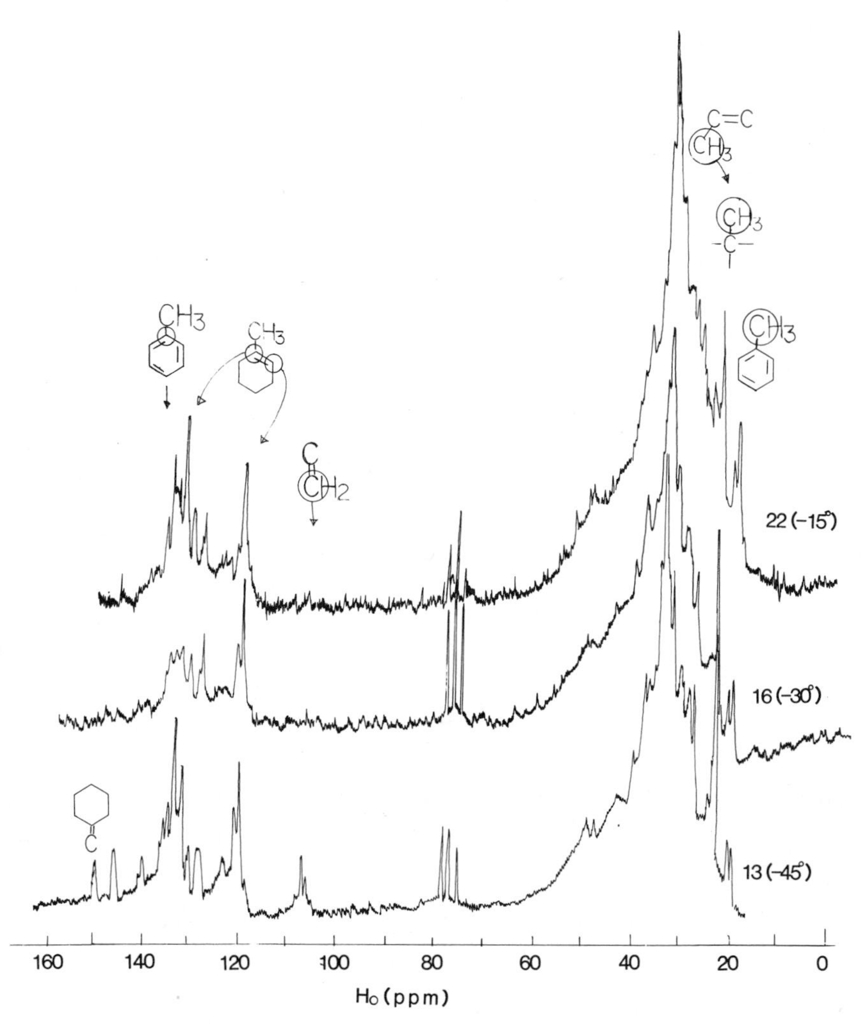

Figure 4. 20 MHz ${}^{13}C$-NMR spectra of poly(1,4-dimethylenecyclohexane), Samples 13, 16, and 22, in CCl_4 solution at room temperature.

izations proceed best between -30° and -15° suggest that proton transfer-rearrangement processes may be a significant feature of cationic 1,4-dimethylenecyclohexane polymerizations. Since double bonds exocyclic to six-membered rings readily rearrange to endocyclic double bonds, it is reasonable to expect the following transformations to occur at the higher polymerization temperatures:

$CH_2=$ I $=CH_2$ $\xrightarrow{H^+}$ $CH_2=$ VII $-CH_3$ $\xrightarrow{H^+}$ CH_3- VIII $-CH_3$

I VII VIII

Monomers I and VII might be responsible for the incorporation of a large number of different structural units in the polymers, viz.

$\sim\sim RCH_2$ $\oplus$ CH_3 → $\sim\sim RCH_2$ $\oplus$ CH_3 III

$\sim\sim R\oplus$ + VII → $\sim\sim RCH_2$ H $\oplus$ CH_3 → $\sim\sim RCH_2$ H $\sim\sim\oplus$ CH_3 IX

$\sim\sim RH$ + [CH_2 $\oplus$ CH_3 ⟷ $CH_2\oplus$ CH_3] → $CH_2\sim\sim\oplus$ CH_3

$\sim\sim R\oplus$ + I → $\sim\sim CH_2$ $\oplus$ → $\sim\sim CH_2$ $\sim\sim\oplus$

$\sim\sim CH_2$ H $\oplus\sim\sim CH_2$ X ← [$\sim\sim CH_2$ H CH_2 $\oplus$ ⟷ $\sim\sim CH_2$ H $\oplus$] → $\sim\sim CH_2$ H $\sim\sim\oplus$ CH_2

The above reactions can explain the incorporation of units with endocyclic unsaturation into the polymers. The highly hindered endocyclic double bonds present in such structures probably have low polymerization ceiling temperatures and would resist polymerization. They could be difficult to brominate or hydrogenate. It is our failure to appreciate the limited reactivity of endocyclic double bonds in structures such as IV, IX and X that caused us to overlook them in our original investigation (9). This limited reactivity also explains why soluble polymers can be formed in 1,4-dimethylenecyclohexane polymerizations even when cyclopolymerization does not seem to occur.

The reactions depicted above also explain the presence of a small amount of conjugated diene units present in the polymers.

The methyl groups present on saturated carbon atoms in the polymers can be explained in a number of ways. Polymerization of endocyclic double bonds is one such possibility. Although it is probably not reasonable to expect such behavior from structures such as III, IV, IX or X, for reasons already discussed, the following reaction seems feasible:

Oxidation-reduction processes involving VIII or related structures and unsaturated structures present in the system (including endocyclic double bonds) are probably an additional source of methyl groups on saturated carbon atoms in the polymers. Such reactions will be driven by the great tendency of VIII and related structures to aromatize, viz.

Alkylation of aromatic structures formed by this process could be responsible for the small amount of aromatic rings incorporated in the polymers.

It thus appears that the principal structural features found in poly(1,4-dimethylenecyclohexane) can be explained by conventional carbonium ion chemistry. There is no indication that cyclopolymerization occurs in these polymerizations and there is much evidence to indicate that the double bonds present in this diene react independently. Some of them are involved in polymerization reactions, but a large proportion isomerize to relatively stable endocyclic double bonds. Under polymerization conditions where isomerization is favorable, soluble, unsaturated polymers having complex structures are obtained. When isomerization reactions are not favorable (low temperatures, use of Ziegler-Natta catalysts), the double bonds polymerize independently and crosslinked products are obtained.

Experimental

1,4-dimethylenecyclohexane - A mixture of potassium metal (59.6g., 1.57 mole) and t-butanol (1-l.) was stirred under reflux in a nitrogen atmosphere for 2 hrs. To the resulting solution of potassium t-butoxide in t-butanol was added trans-1,4-di(iodomethyl)-cyclohexane (28) (182g., 0.5 mole). The mixture was refluxed and stirred under nitrogen for 24 hrs. The mixture was treated with 2400 ml water and the organic layer which separated was washed 4 times with 200 ml portions of water, filtered and dried over anhydrous $MgSO_4$. The liquid was then distilled under reduced pressure to obtain 28.8g. (53% yield) 1,4-dimethylenecyclohexane, b.p. 54-55° (75 mm), n_D^{25} 1.4688, d_4^{25} 0.8133.

Anal. - Calc'd for C_8H_{12} (108.2): C, 88.82; H, 11.18, Bromination Eq. Wt. 54.1. Found: C, 88.86, H, 11.27, Bromination Eq. Wt. 53.6.

The diene quickly absorbed 1.96 moles of hydrogen when hydrogenated at room temperature using platinum as a catalyst. Ozonolysis of the diene in dry pentane at -20°, followed by treatment of the reaction mixture with a dilute solution of dimedon provided a 56 percent yield of the dimedon derivative of formaldehyde, m.p. 188.5-190°

The ^{1}H-NMR spectrum of the diene consisted of two sharp singlets at 2.09 ($-CH_2-$) and 4.66 ($=CH_2$) ppm having relative intensities of 2:1. The diene absorbed only weakly in the ultraviolet region at 262 nm (ε=11.0) and at 234 nm (ε=14.7). In the presence of iodine, it absorbed at 298 nm (ε'=47.7). This is in good agreement with results expected for 1,4-dimethylenecyclohexane based on data reported by Long and Neuzil.(29) Butler and Van Heiningen have also investigated the uv absorption of this monomer (8).

Polymerization of 1,4-Dimethylenecyclohexane with BF_3 - A 100 ml two-neck flask was rinsed with dilute NH_3 solution and dried in air at 150°C. While still hot, the flask was flushed with nitrogen and one of the necks was equipped with a rubber gum seal. After the flask had cooled under N_2 to room temperature, 1,4-dimethylenecyclohexane (6.7 g., 0.062 mole) and methylene chloride (42 ml) were added by syringe. While maintained under nitrogen, the mixture was stirred magnetically and cooled by immersion in a bromobenzene - Dry Ice bath at -30°C for one-half hour. Boron trifluoride was then introduced very slowly into the nitrogen stream flowing over the reaction mixture until a red-orange color developed in the mixture. The mixture was maintained under nitrogen and stirred at -30°C for 3 hours. Triethylamine (1 ml) was then injected into the mixture and the color disappeared. The mixture was filtered through a coarse glass frit into 500 ml of acetone. The polymer which coagulated was filtered and dried to give a finely divided white solid (3.75 g., 55%). The polymer was further purified by dissolving it in hot methylene chloride, filtering the solution through a "medium" glass frit and then adding the filtrate to 500 ml of acetone. The precipitated polymer was collected and dried in a vacuum dessicator. The intrinsic viscosity of the polymer in benzene at 30°C was 0.079. The molecular weight, determined cryoscopically in benzene, was 8,400.

Anal. Calc'd for (C_8H_{12}): C, 88.82; H, 11.18. Found: C, 88.72; 88.67; H, 11.28; 11.12; Bromination Equivalent Wt., 5,200, 6,000; Grams of polymer absorbing one mole of hydrogen, 7,400, 8,200. The polymer was a white powdery solid. It was soluble in most organic solvents including pentane, hexane, benzene, toluene, xylene, methylene chloride, chloroform and carbon tetrachloride. It was insoluble in acetone, methanol and water. The material softened and became fluid between 170-180°C, showing only slight discoloration.

The polymer showed strong methylene absorption at 2900, 2840 and 1450 cm^{-1} in the infrared region. In addition, weak absorptions at 1640 and 888 cm^{-1} indicated the presence of a few terminal olefinic linkages. Absorption at 1380 cm^{-1} indicated the presence of methyl groups.

The ultraviolet spectrum of the polymer, measured in cyclohexane solution, was essentially the same in the presence or absence of iodine. The polymer showed only weak absorption in CH_2Cl_2 solution, having a maximum of 258 nm (ε=340/monomer unit or 27,000/polymer molecule). After treatment with a dilute solution of chlorine in CCl_4, the polymer no longer showed an absorption maxima at 258 nm; instead, tail absorption of an intense band whose maximum would occur below 210 nm was observed.

Polymerizations conducted using other catalysts, temperatures, solvents, and monomer concentrations were conducted according to the procedure discussed above. The results are summarized in Table I.

NMR Measurements - ^{1}H- and ^{13}C- NMR spectra of the polymers in $CDCl_3$ solution at ambient temperature were recorded at 300 and 20 MHz, respectively, using Varian HR-300 (CW mode) and CFT-20 (FT mode) spectrometers. The ^{13}C-NMR spectra were the result of over 100 K accumulations using a pulse width of 10μ sec. Tetramethylsilane was used as an internal standard.

Acknowledgement

We are pleased to acknowledge that the participation of one of us in this study was made possible by a grant from the Yugoslavian Academy of Sciences.

Literature Cited

1. Butler, G.B. J. Polym. Sci., Polymer Symposium 1978, 64, 71.
2. Corfield, C.G. Chem. Soc. Rev. 1972, 1(4), 523.
3. McCormick, C.L.; Butler, G.B. J. Macromol. Sci., Rev. Macromol. Chem. 1972, C8, 201.
4. Corfield, C.G.; Aso , C.; Butler, G.B. "Progress in Polymer Science"; A.D. Jenkins, Ed.; Pergamon Press; Oxford, 1975, Vol. 4, pp 71-207.
5. Butler, G.B.; Miles, M.L.; Brey, W.C. J. Polym Sci. 1965, 3A, 723.
6. Butler, G.B.; Miles, M.L. J. Polymer Sci. 1965, 3A, 1609.
7. Frazer, A.H.; O'Neill, W.P. J. Am. Chem. Soc. 1963, 85, 2613.
8. Butler, G.B.; Van Heiningen, J.J. J. Macromol. Sci. 1974, A8, 1139, 1175.
9. Ball, L.E.; Harwood, H.J. Am. Chem. Soc., Div. Polym. Chem., Preprints 1961, 2(1), 59.
10. Ball, L.E. Ph.D. Dissertation, University of Akron 1961; Dissertation Abstr. 1961, 22, 1404; Chem. Abstr. 1962, 56, 5846 h.
11. Shannon, F.D. Ph.D. Dissertation, University of Akron 1964; Dissertation Abstr. 1965, 25, 3849; Chem. Abstr. 1965, 62, 13249c.
12. Sebenik, A; Harwood, H.J. Amer. Chem. Soc., Div. Polym. Chem., Preprints 1981, 22, 15.
13. Lautenschlaeger, F.; Wright, G.F. Can. J. Chem. 1963, 41, 1972.
14. Fr. Patent 1,323,328 (April 5, 1963); Chem. Abstr. 1963, 59, 9784.
15. St-Jacques, M.; Bernard, M. Can. J. Chem. 1969, 47, 2911.
16. Acker, D.S.; Blomstrom, D.C. U.S. Patent 3,115,506 (Dec. 24, 1963); Chem. Abstr. 1964, 60, 14647d.
17. Wiberg, K.B.; Epling, G.A.; Jason, M. J. Am. Chem. Soc. 1974, 96, 912.
18. Wiberg, K.B.; Burgmaier, G.J. J. Am. Chem. Soc. 1972, 94, 7396.
19. Buchta, E.; Kröniger, A. Chimia 1969, 23, 225.
20. Spainhour, J.D. U.S. Patent 3,331,819 (July 18, 1967); Chem. Abstr. 1967, 67, 74012k.
21. Denkowski, G.S. M.S. Dissertation, University of Akron 1965.
22. Marshall, J.L.; Miller, D.E., Org. Magn. Reson. 1974, 6, 395.

23. Stothers, J.B. "Carbon-13 NMR Spectroscopy, Academic Press, New York, N.Y., 1972,
24. Johnson, L.F.; Jankowski, W.C., "Carbon-13 NMR Spectra", Wiley-Interscience, New York, 1972.
25. Pouchert, C.Y.; Campbell, J.R., "The Aldrich Library of NMR Spectra", Aldrich Chemical Co., Inc., Milwaukee, Wisconsin, 1974.
26. Olah, G.A.; and White, A.M. J. Am. Chem. Soc. 1969, 91, 3954.
27. Lippmaa, E.; Pehk, T.; Paasivirta, J.; Belikova, N.; Plate, A. Org. Magn. Reson. 1970, 2, 581.
28. Haggis, G.H.; Owen, L.N. J. Chem. Soc. 1953, 404.
29. Long, R.D.; Neuzil, R.W. Anal. Chem. 1955, 27, 1110.

RECEIVED February 1, 1982.

17

Acetylene-Terminated Fluorocarbon Ether Bibenzoxazole Oligomers

ROBERT C. EVERS and GEORGE J. MOORE
Air Force Wright Aeronautical Laboratories, Materials Laboratory, Wright-Patterson Air Force Base, OH 45433

JERALD L. BURKETT
University of Dayton Research Institute, Dayton, OH 45469

Acetylene terminated fluorocarbon ether bibenzoxazole (FEB) oligomers were prepared by the polycyclocondensation of fluorocarbon ether diimidate esters with fluorocarbon ether bis(o-aminophenol) monomers followed by subsequent reaction with a new endcapping agent, 2-amino-4-ethynylphenol. The average number of repeat units in the oligomers was controlled by the stoichiometry of the oligomerization reaction. Oligomer structure was verified by elemental and infrared spectral analyses. Onset of curing of the acetylene terminated FEB oligomers, as determined by differential scanning calorimetry, occurred in the 180-200°C range with reaction maxima being reached above 280°C. The resultant cured oligomers exhibited glass transition temperatures as low as -55°C. No crystalline melt temperatures were observed. Onset of breakdown of the reddish, transparent rubbers during thermogravimetric analysis in an air atmosphere took place in the 375-400°C range; complete weight loss occurred at approximately 600°C.

Thermooxidatively stable fluorocarbon ether bibenzoxazole (FEB) polymers with glass transition temperatures (Tg) as low as -58°C can be synthesized through the acetic acid-promoted polycyclocondensation of fluorocarbon ether diimidate or -dithioimidate esters with fluorocarbon ether bis(o-aminophenol) monomers. (1) By judicious modification of polymer structure, greater hydrolytic stability and more ready curability can be achieved. (2) The latter property is effected through incorporation into the polymer backbone of hydrocarbon cure sites reactive to radical-induced cure reactions. The present article describes the synthesis of acetylene terminated FEB oligomers that can be cured by thermal reactions of acetylene end groups. These oligomers can be prepared as shown in Scheme I by polycyclocondensation of

0097-6156/82/0195-0223$06.00/0
© 1982 American Chemical Society

fluorocarbon ether diimidate esters with fluorocarbon ether bis(o-aminophenol) monomers followed by endcapping with an endcapping agent, 2-amino-4-ethynylphenol. The average number of repeat units, n, is controlled by the stoichiometry of the reaction.

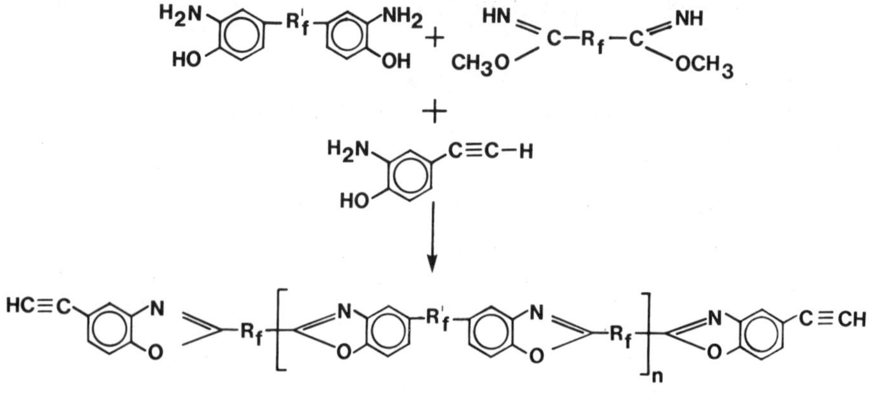

n=0, 1, 2

$R_f = CF_2(OCF_2CF_2)_xO(CF_2)_5O(CF_2CF_2O)_yCF_2$ $x+y=5, 6$

$R'_f = (CF_2)_2O(CF_2)_5O(CF_2)_2$

Scheme I

Synthesis of these oligomers and the requisite endcapping agent is given below. Oligomer characterization and preliminary cure studies are also described.

Results and Discussion

Synthesis of Endcapping Agent. 2-Amino-4-ethynylphenol was prepared as shown in Scheme II. 3-Nitro-4-hydroxyacetophenone was treated with acetic anhydride to protect the phenolic group. The key reaction in the synthetic sequence entailed treatment of the acetate ester with a Vilsmeier reagent generated from phosphorous oxychloride and N,N-dimethylformamide. 4-Acetoxy-3-nitro-β-chlorocinnamaldehyde was obtained in 43% yield. Subsequent treatment of the aldehyde with sodium hydroxide solution gave 4-ethynyl-2-nitrophenol. Alternatively, this reaction scheme could be carried out with a p-tosyl group protecting the phenolic function. However, both formation of the tosylate ester and subsequent regeneration of the free phenol were more difficult. After formation of the acetylene group, the nitro group was selectively reduced with sodium dithionite in an alkaline medium to yield the desired 2-amino-4-ethynyl-phenol. Since this compound was somewhat sensitive to both heat and light, it was stored in the cold shielded from light.

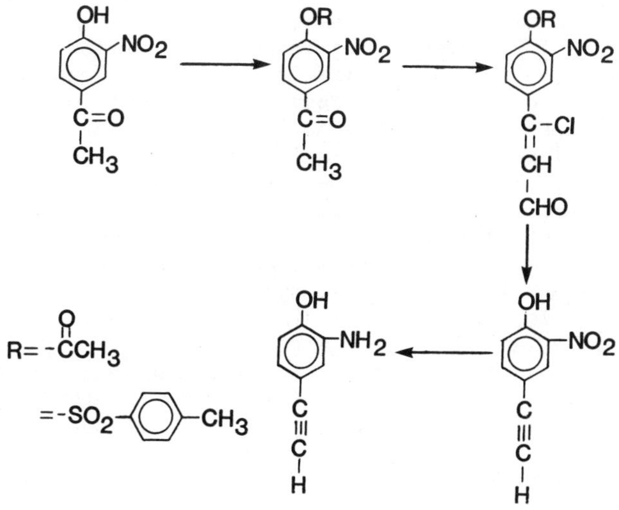

Scheme II

An alternate synthesis as shown in Scheme III of the intermediate 4-ethynyl-2-nitrophenol was briefly investigated. The

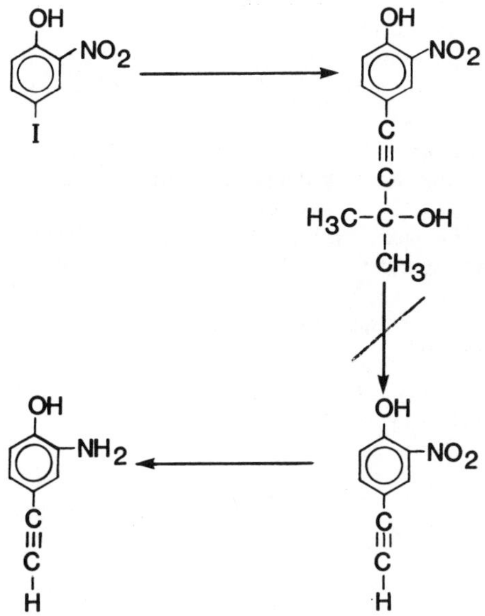

Scheme III

palladium-catalyzed reaction(3) of 4-iodo-2-nitrophenol with 2-methyl-3-butyn-2-ol led to a 41% yield of the butynol adduct. Potassium hydroxide cleavage of acetone(3) to give the desired 4-ethynyl-2-nitrophenol did not proceed well in either toluene or water and gave predominantly unchanged starting material.

Synthesis of Oligomers. Initial synthesis efforts were directed toward establishing the reactivity of the endcapping agent, 2-amino-4-ethynylphenol. Model reactions between this o-aminophenol and fluorocarbon ether diimidate esters were carried out under reaction conditions normally used for the synthesis of FEB polymers.(1) The acetic acid-promoted condensations in hexafluoroisopropanol (HFIP) led to blood-red reaction mixtures and products whose infrared spectra were inconsistent with the model compound structure. Further investigation indicated that the deep coloration was attributable to decomposition of the endcapping agent in HFIP. Repetition of the syntheses (Table I, Trials 1 and 2) in a methylene chloride reaction medium led to high yields of the desired products. Syntheses of the acetylene terminated FEB oligomers were subsequently carried out in a methylene chloride-Freon 113 (1,1,2-trichloro-1,2,2-trifluoroethane) reaction medium. The fluorocarbon solvent was necessary to permit ready solubility of the oligomers throughout the entire reaction period. The average number of repeat units, n, in the oligomers was governed by the reaction stoichiometry (see Table I). Acetic acid-promoted condensation of fluorocarbon ether diimidate ester and fluorocarbon ether bis(o-aminophenol) at approximately 40°C was monitored by infrared spectra and yielded oligomers terminated with imidate ester groups. When this reaction was judged to be complete, addition and subsequent reaction of 2-amino-4-ethynylphenol led to high (85-90%) yields of the acetylene terminated oligomers. Total reaction times were 36 to 48 hours.

Attempts to improve the hydrolytic stability of the FEB oligomers by extending the synthetic sequence described above to fluorocarbon ether dithiomidate esters (2) derived from hexafluoropropylene oxide were unsuccessful. The dithioimidate esters proved to be unreactive in the methylene chloride-Freon 113 reaction medium. A search for a solvent in which the dithioimidate ester was sufficiently reactive and the endcapping agent was stable was unsuccessful.

Characterization and Curing of Oligomers. While the acetylene terminated model compounds were white waxy solids, the oligo-

Table I
Synthesis of Model Compounds and Oligomers

Trial No.	Diimidate Ester-mmol	Bis(o-amino-phenol)-mmol	4-amino-2-ethynyl-phenol-mmol	Acetic Acid-mmol	Elemental Analysis - Calc'd (Found) C	H	N
1 n=0, x+y=5	0.50	0	1.10	2.10	33.73 (33.10)	0.65 (0.34)	2.25 (2.37)
2 n=0, x+y=6	0.50	0	1.11	2.07	32.62 (32.74)	0.59 (0.22)	2.06 (2.09)
3 n=1, x+y=5	1.00	0.50	1.10	4.10	30.80 (30.82)	0.48 (0.47)	1.91 (2.15)
4 n=1, x+y=6	3.00	1.50	3.30	12.60	30.05 (30.20)	0.47 (0.10)	1.77 (1.68)
5 n=2, x+y=5	0.75	0.50	0.55	5.79	30.00 (29.42)	0.44 (0.21)	1.82 (2.08)
6 n=2, x+y=6	1.50	1.00	1.10	6.20	29.35 (29.37)	0.41 (0.13)	1.70 (1.77)

mers were pale yellow, very viscous syrups. Model compound and oligomer structure were determined by elemental and infrared spectral analyses. Elemental analysis values are given in Table I. The infrared spectra exhibited in all cases a sharp band at 3310 cm^{-1}, indicative of a terminal acetylene group, (4) and absorptions attributable to the FEB polymer structure.(5) Absent were absorptions in the 3400-3500 cm^{-1} region, indicative of unreacted o-aminophenol and at 1680 cm^{-1}, indicative of unreacted imidate ester. A representative infrared spectrum is shown in Figure 1.

The thermal behavior of the acetylene terminated FEB model compounds and oligomers was studied by differential scanning calorimetry (DSC). The results of the thermoanalytical measurements are summarized in Table II. The model compounds (Trial Nos. 1 and 2) exhibited only crystalline melting points (Tm) while the oligomers exhibited Tg's. In addition to showing initial strong baseline shifts attributed to the oligomer Tg's, the DSC scans (ΔT=20°C/min) of the oligomers exhibited strong exotherms beginning in the 180-200°C range and reaching maxima at above 280°C. These exotherms were indicative of reaction of the terminal acetylene groups and took place at higher temperatures for oligomers in which the average number of repeat units was 2. A DSC curve for an uncured FEB oligomer is shown in Figure 2. Tg's

Table II
Curing of Model Compounds and Oligomers

Trial No.	Initial Tg- °C	Initial Tm- °C	Cure Onset- °C	Max. Rate Cure- °C	Tg After Cure- °C
1 n=0, x+y=5	-	49	175	268	4
2 n=0, x+y=6	-	41	175	268	-
3 n=1, x+y=5	-56	22	180	285	-45
4 n=1, x+y=6	-60	-	195	280	-46
5 n=2, x+y=5	-54	-	190	318	-45
6 n=2, x+y=6	-61	-	200	318	-55

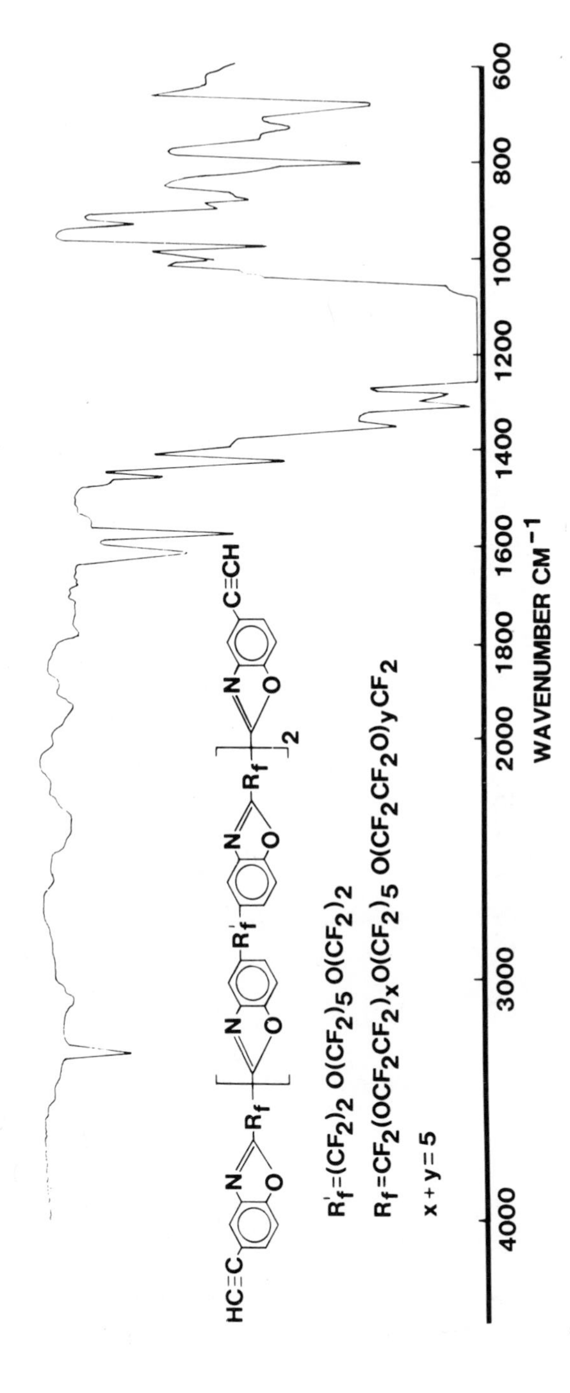

Figure 1. IR spectrum of fluorocarbon ether bibenzoxazole oligomer.

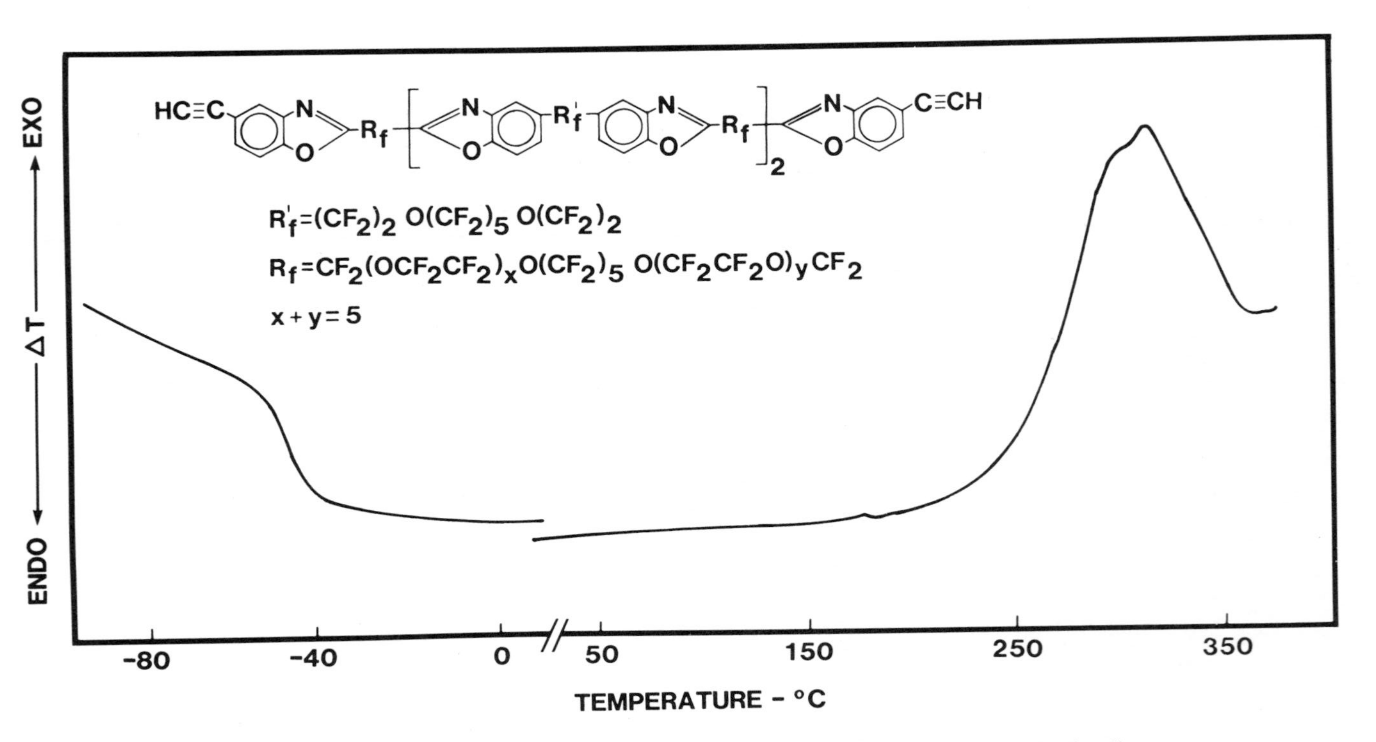

Figure 2. Differential scanning calorimetry curve of fluorocarbon ether bibenzoxazole oligomer.

of the cured oligomers were determined after heating of the oligomers to approximately 370°C during the DSC scans and subsequent cooling. No Tm's were observed.

A sample of oligomer (Trial No. 4) was heated in a steel mold at 160°C/16 hours and then at 185°C/24 hours to give a transparent, reddish rubber. Additional heating at 240°C/3 hours did not further advance the cure. The soft, rather weak rubber exhibited a Tg of -45°C which was almost identical to the value of -46°C obtained during the DSC measurements.

Thermooxidative stability of a cured sample (Trial No. 3) was evaluated by thermogravimetric analysis in air. The thermogram (Figure 3) indicated onset of breakdown in the 375-400°C range and was quite similar to that observed for an FEB polymer having the same repeat unit structure.(1)

Experimental

2-Amino-4-ethynylphenol. A mixture of 4-hydroxy-3-nitroacetophenone(6) (50.0g, 0.28 mole) and acetic anhydride (100.0g, 0.98 mole) was stirred at reflux for 45 minutes to form a clear, light yellow solution. The excess acetic anhydride was stripped off under reduced pressure to give an oil which gradually solidified. Recrystallization from 500 ml of isopropanol (charcoal) yielded 43.0g (63%) of light yellow crystals of 4-acetoxy-3-nitroacetophenone, mp 62-63°C.

Anal. Calcd for $C_{10}H_9NO_5$: C, 53.80; H, 4.08; N, 6.27; mol. wt., 223. Found: C, 53.54; H, 3.93; N, 6.45; mol. wt., 223 (mass spectrum).

Phosphorus oxychloride (45.0 ml, 0.49 mole) was added dropwise with stirring over a 30 minute period to spectral grade N, N-dimethylformamide (150 ml, 1.94 mole) at 20°C. The solution had taken on a blood-red color by the time addition was complete, which indicated the presence of the Vilsmeier reagent. To the solution at room temperature was added powdered 4-acetoxy-3-nitroacetophenone (43.0g, 0.19 mole). The reaction temperature was maintained at 55 to 60°C for three hours at which time the reaction mixture was poured into 1500 ml of ice water. The resultant light tan solution was neutralized with sodium bicarbonate and the precipitate which formed was isolated by filtration. It was allowed to air dry on the frit to yield 34g of crude product which was subsequently dissolved in 400 ml of methylene chloride. The cloudy solution was treated with charcoal, filtered, and an additional 500 ml of methylene chloride and 500 ml of heptane were added. This solution was treated with charcoal, filtered and reduced in volume to approximately 1700 ml. Upon cooling, an oil separated on the sides of the beaker. The supernatant was decanted, reduced in volume to 1500 ml, and poured into 200 ml of hexane. Yellow crystals separated from the cooled solution to yield 22g (43%) of 4-acetoxy-3-nitro-β-chlorocinnamaldehyde, mp 128-130°C.

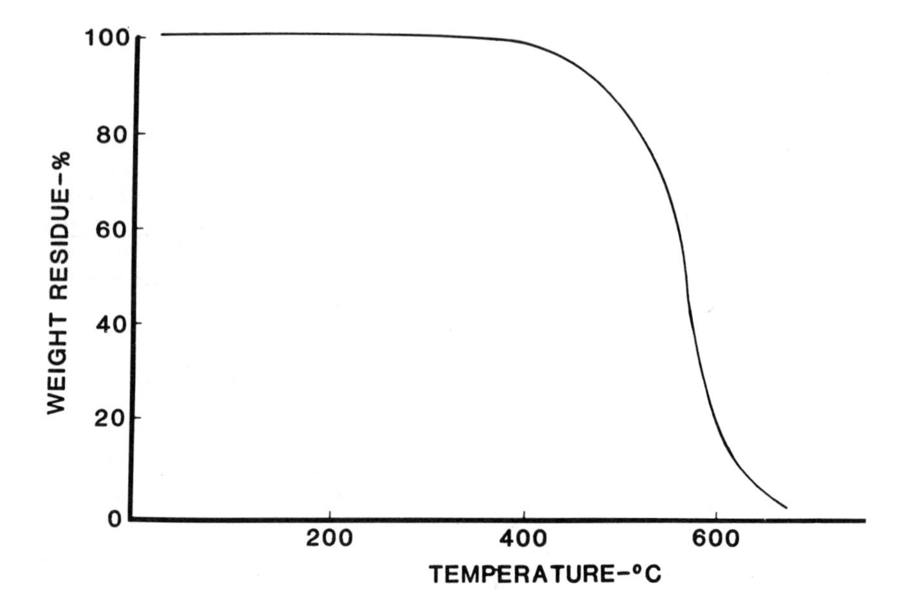

Figure 3. Thermogravimetric analysis thermogram of cured fluorocarbon ether bibenzoxazole oligomer.

Anal. Calcd for $C_{11}H_8ClNO_5$: C, 48.98; H, 3.00; N, 5.19; mol. wt., 269. Found: C, 48.96; H, 2.80; N, 5.06; mol. wt., 269 (mass spectrum).

4-acetoxy-3-nitro-β-chlorocinnamaldehyde (22.0g, 0.08 mole) was added to a stirred solution of sodium hydroxide (25g, 0.63 mole) in 1100 ml of water. The reaction mixture was maintained at 60 to 70°C for 20 minutes and was then cooled to 0°C. A small amount of insolubles was filtered off and the clear solution was neutralized with glacial acetic acid to give a creamy-white precipitate which was isolated by filtration. After being allowed to air dry on the frit, the crude product was dissolved in 400 ml of methylene chloride. This solution was treated with charcoal, filtered, and reduced in volume to 200 ml to yield, upon cooling, 9.6g (59%) of 4-ethynyl-2-nitrophenol as yellow crystals, mp 115-116°C.

Anal. Calcd for $C_8H_5NO_3$: C, 58.92; H, 3.07; N, 8.58; mol. wt. 163. Found: C, 58.85; H, 2.94; N, 8.82; mol. wt., 163 (mass spectrum).

4-ethynyl-2-nitrophenol (3.7g, 0.02 mole) was dissolved in a solution of potassium carbonate (3.7g, 0.03 mole) in 250 ml of water. Then sodium dithionite (30.0g, 0.17 mole) was added and the resultant solution stirred at room temperature for one hour. The aqueous solution was extracted repeatedly with methylene chloride (total volume of 1500 ml). The methylene chloride solution was treated with anhydrous magnesium sulfate and then with charcoal. After the filtered solution was reduced in volume to 500 ml, 1500 ml of hexane was added. The resultant solution was reduced in volume to 1200 ml at which time the product began to crystallize out of solution. Cooling to ice water temperature yielded 0.90g (34%) of 2-amino-4-ethynylphenol as white crystals, mp > 105°C (vigorous decomposition).

Anal. Calcd for C_8H_7NO: C, 72.14; H, 5.33; N, 10.51; mol. wt., 133. Found: C, 72.47; H, 5.66; N, 9.95; mol. wt., 133 (mass spectrum).

4-Ethynyl-2-nitrophenol (Attempted Alternate Synthesis). To a stirred solution of 4-iodo-2-nitrophenol (14.0g, 0.053 mole)(7) in 150 ml of tetrachloroethylene was added bis(triphenylphosphine) palladium dichloride (0.74g, 0.001 mole), cuprous iodide (0.40g, 0.002 mole), triphenyl phosphine (0.03g, 0.001 mole), and 50 ml of dry triethylamine. After the reactants were thoroughly mixed, 2-methyl-3-butyn-2-ol (5.4g, 0.064 mole) was added. The stirred reaction mixture was heated slowly to 70-75°C and maintained at that temperature for 16 hours. The tetrachloroethylene and triethylamine were removed under reduced pressure and the red gum-like residue was extracted with hot heptane. The heptane solution was treated with charcoal and the heptane was stripped off to give a yellow solid (9.5g). The solid was dissolved in benzene and was placed in a silica gel column. Elution with 1:1 benzene/hexane yielded 1.8g of product. The remainder of the product was removed by elution with 1:1 methylene chloride/hexane. Recrystallization from hexane gave 5.0g (41%) of the yellow butynol adduct mp 80-82°C.

Anal. Calcd for $C_{11}H_{10}NO_4$: C, 59.72; H, 5.01; N, 6.33; mol. wt., 221. Found: C, 59.61; H, 5.04; N, 6.27; mol. wt., 221 (mass spectrum).

The butynol adduct (1.0g, 0.005 mole) was dissolved in 200 ml of dry toluene. Potassium hydroxide (0.5g, 0.009 mole) was added and the stirred reaction mixture was refluxed under nitrogen for eight hours. The toluene was gradually distilled over to remove any acetone which resulted from cleavage of the butynol adduct. The volume of the reaction mixture was kept constant by the addition of fresh toluene. The reaction mixture was filtered and the toluene removed under reduced pressure to give 0.90g of yellow solid, mp 65-70°C. Infrared and mass spectroscopy as well as thin layer chromatography revealed this to be almost exclusively starting material.

The butynol adduct (1.0g, 0.005 mole) was dissolved in a basic solution of potassium hydroxide (0.5g, 0.009 mole) in 200 ml of distilled water. With the volume being kept constant, the orange solution was slowly distilled under nitrogen for ten hours to remove any acetone which was formed. The cooled reddish solution was made acidic by the addition of dilute sulfuric acid to give 0.80g of gummy yellow material. Recrystallization from hexane gave 0.50g of almost pure starting material, mp 79-81°C.

<u>Acetylene Terminated FEB Oligomer (Trial No. 4)</u>. The fluorocarbon ether diimidate ester (3.586g, 0.0030 mole) and glacial acetic acid (0.380g, 0.0063 mole) were dissolved in a mixture of 60 ml of Freon 113 and 45 ml of methylene chloride at 45°C. To this clear solution was added the fluorocarbon ether bis(o-aminophenol) (1.048g, 0.0015 mole) in four potions over the course of an hour. After the reaction mixture was stirred under nitrogen at 40°C for 22 hours, 2-amino-4-ethynylphenol (0.445g, 0.0033 mole) and additional glacial acetic acid (0.380g, 0.0063 mole) in 15 ml of methylene chloride were added to the clear, pale yellow solution. The reaction was continued for an additional 20 hours at which time the clear reaction mixture was transferred to a separatory funnel. It was washed successively with water, dilute sodium bicarbonate solution, and again with water. The organic layer was then dried over anhydrous magnesium sulfate, treated with charcoal, and reduced in volume to approximately 15 ml. The clear concentrated solution was transferred to a vial and the remaining organic solvents were carefully removed at 58°C and 0.01 mm Hg to yield 4.36g (86%) of viscous, pale yellow oligomer.

Anal. Calcd for $C_{79}H_{14}N_4F_{94}O_{22}$: C, 30.05; H, 0.44; N, 1.77. Found: C, 30.20; H, 0.10; N, 1.68.

Acknowledgement

The authors thank Mrs. Paula Beck for assistance in the synthesis work. The DSC and thermooxidative stability data were contributed by Mr. E. J. Soloski and Dr. G. F. L. Ehlers.

Literature Cited

1. Evers, R. C., J. Polym. Sci., Polym. Chem. Ed., 1978, 16, 2833.
2. Evers, R. C.; Abraham, T., Am. Chem. Soc., Prepr. Polym. Div., 1979, 20(1), 478.
3. Sobourin, E. T., Am. Chem. Soc., Prepr. Petr. Div., 1979, 24(1), 233.
4. Bellamy, L. J., "The Infrared Spectra of Complex Molecules;" Second Edition, Wiley:New York, 1958; p. 58.
5. Evers, R. C., J. Polym. Sci., Polym. Chem. Ed., 1978, 16, 2815.
6. Pope, Frank G., Proc. Chem. Soc., 1912, 28, 2933; Chem. Abstr. 1912, 7, 2933.
7. Keimatsu, S., J. Pharm. Soc. Japan, 1924, 507, 319; Chem. Abstr., 1924, 18, 2503.

RECEIVED April 28, 1982.

18

Initiation of an Acetylene-Terminated Phenyl Quinoxaline

M. A. LUCARELLI, W. B. JONES, JR., L. G. PICKLESIMER, and T. E. HELMINIAK

Air Force Wright Aeronautical Laboratories, Materials Laboratory, Wright-Patterson Air Force Base, OH 45433

A series of heterogeneous mixtures was made by adding various amounts of bis(triphenyl phosphine) nickel II chloride, as an initiator, to an acetylene-terminated phenylquinoxaline. The series was then evaluated using the Differential Scanning Calorimeter (DSC) and the shifts in the exothermic reaction with temperature were associated with the initiator concentrations. From these data, the initiator concentration of 1.0%, by weight, was selected for further investigation. Control samples of the polymer sans initiator were also carried through the investigation. Heats of reaction were measured for these materials in both air and nitrogen atmospheres. Specimens were then interrupted at various stages of cure at 177°C (350°F). Subsequently, these specimens were placed in the DSC and the residual exotherms were measured. The data were interpreted in terms of the extent of cure and used in conjunction with Kinetic theory to estimate the maximum amount of cure that could be expected at 177°C (350°F). The methods and results are discussed herein.

A family of acetylene-terminated phenyl quinoxalines have been synthesized by the Polymer Branch of the Materials Laboratory.(1) These phenyl quinoxalines are remarkable for their thermooxidative stability and resistance to moisture. These materials have potential for structural applications as adhesives or composite matrix resins.(2) The feature of moisture resistance makes the materials especially attractive for bonding aluminum. However, problems arise from the fact that aircraft aluminum alloys (and their surface oxides) are altered by exposures to temperatures above 177°C (350°F) and this is much lower than the polymerization temperatures of the acetylene-terminated oligomers.

This chapter not subject to U.S. copyright.
Published, 1982, American Chemical Society

In order to overcome this problem, it was decided to select one of the phenylquinoxalines and investigate the feasibility of initiating the cure reactions at 177°C (350°F).

Physical and mechanical characteristics have been measured and catalogued for the phenyl quinoxaline family in a related internal research project. From these data, 3,3'-Bis(4-[3-ethynyl phenoxy] phenyl)-2,2'-diphenyl-6,6' biquinoxaline (or BA-DAB-BA for short) was selected. Bis-(triphenyl phosphine) nickel II chloride was chosen as the initiator.

Sample Preparation Method

Although it is desirable to have a homogeneous blend of initiator and polymer, a practical method for obtaining this is not known. Several methods were tried, and the one that reduced the peak reaction temperature the greatest amount (as-measured, using the DSC) was selected. Initially, a small sample of the initiator and polymer, both powders at room temperature, were ground together with mortar and pestle. The blend was tested in the DSC. Trying another method, the initiator was dissolved in methanol and allowed to precipitate on to the polymer (nonsoluble) as the methanol evaporated. The DSC scans showed a significantly greater reduction in the exotherm peak temperature.

A third method was tried that gave undesirable results. The initiator and monomer were dissolved in tetrahydrofuran, a common solvent. In attempting to strip off the THF in a roto-vac at room temperature, it was found that polymerization of the BA-DAB-BA was initiated and advanced, even while there was still retained solvent. Since these factors contribute to processing difficulties and undesirable porosity in cured specimens, this method was discarded, and the methanol precipitation method was selected.

Influence of Initiator Concentration

A dilute solution was prepared by dissolving one gram of initiator in 100ml of methanol. The solution was then precipitated on to weighed amounts of BA-DAB-BA so that the final concentrations were 0.01, 0.05, 0.10, 0.25, 0.5, 1.00, 2.40, 5.0, and 7.50% by weight. Sufficient methanol was added in order to wet the BA-DAB-BA. The mixture was mechanically stirred while the methanol was allowed to evaporate. Once the mixture was dry in air it was held under vacuum overnight to remove any residual methanol. The solid mixtures were then scanned in the DSC from 20 to 400°C at a rate of 20°C/minute. The exotherm onset and peak temperatures were noted and are shown in Table I. Also, indicated in Table I is the effect of particle size in the heterogeneous blend. From these data, the initiator concentration of 1% was selected for further investigation.

Table I. Exotherm Peak Temperature as a Function of Particle Size.

MESH	SIZE	GROUND BADABBA	Tp°C	PRECIPITATED BADABBA	Tp°C
35	500µ	1	242.5	9	175
60	250µ	2	203	10	182
80	177µ	3	210	11	162
120	125µ	4	200	12	180
200	74µ	5	175	13	180
250	58µ	6	-	14	200
325	44µ	7	145	15	230
>325		8	198	16	175

µ = 10^{-4}cm

Curing Studies

Approximately 60 grams of BA-DAB-BA was synthesized using the Hedberg procedure.[1] One-half of this material was evaluated in its pure form, while the other half was first blended to 1% concentration with the initiator using the methanol precipitation method, and then evaluated. Both samples were subjected to DSC scans at 20^oC/min. The results are shown in Figure 1.

Rheology

The viscosities of pure and initiated BA-DAB-BA samples were measured during 175^oC exposure. Specimens were prepared by pressing the uncured powder at 2×10^8 n/m^2 (40,000 psi) into 12mm dia x 2mm thick pellets. The pellets weighed 0.4 gm each. The pellets were placed between preheated (175^oC) parallel plates in a Rheometrics RMS-7200 Mechanical Spectrometer and subjected to low frequency (Sinusoidal Shearing) viscosity measurements. The results, shown in Figure 2, clearly show the initiating influence of the nickel compound. Since plate slippage was experienced above 10^8 poise, an alternate method was used to evaluate the latter stages of cure.

Residual Exotherm

The reaction exotherms were used to indicate the degree of cure for the polymer samples. The areas under the DSC curves (see Fig 1), between 20 and 350^oC, were measured and taken to be the total heats of reaction. Additional specimens were prepared and placed in an air oven at 177^oC (350^oF). Periodically, specimens were withdrawn from the oven, placed in the DSC and scanned from 20 to 350^oC at 20^oC/min; and the residual exotherms were measured. These tests were repeated in a nitrogen environment. The results of the DSC scans are indicated in Figure 3.

From the DSC scans, the percent of reaction that had occurred was determined from the following equation:

$$R = \frac{(\Delta H_T - \Delta H)}{\Delta H_T}$$

where ΔH_T - Total Heat of Reaction; 92.63 cal/gm BA-DAB-BA and 66.18 cal/gm for BA-DAB-BA + initiator.

ΔH - Area under DSC curve after x hrs at 177^oC.

Plotting log R versus 1/time (see Fig 4) and extrapolating the curve to intersect the ordinate, indicates the air cured samples would approach 89% cure completion for pure BA-DAB-BA and 92% for the initiated BA-DAB-BA. The data indicate both materials would approach 95% cure completion in a nitrogen atmosphere.

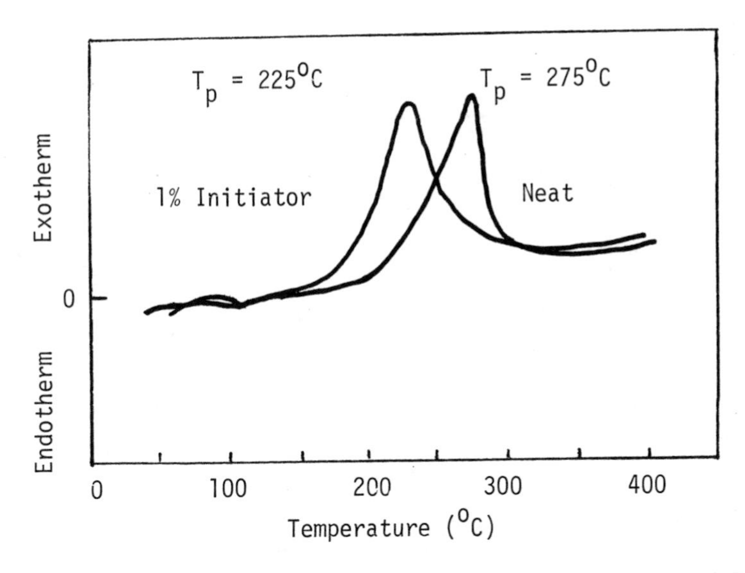

Figure 1. Differential scanning calorimetry for BA-DAB-BA (neat) and BA-DAB-BA (with 1% initiator).

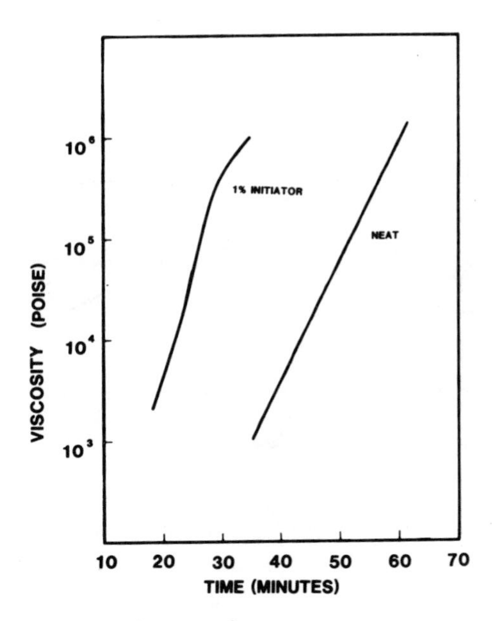

Figure 2. Viscosity for BA-DAB-BA (neat) and BA-DAB-BA (with 1% initiator) at 175°C isotherm.

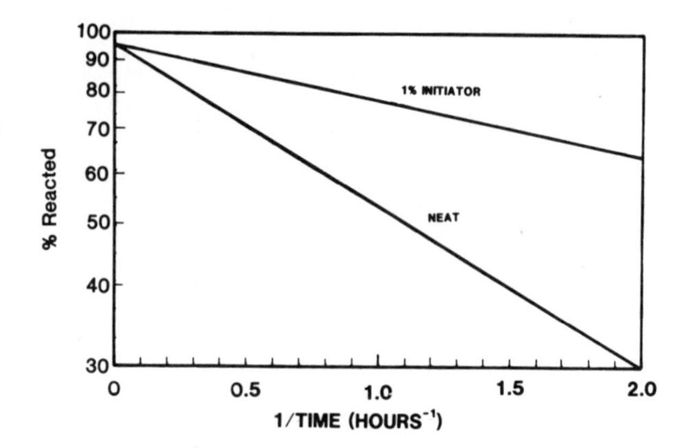

Figure 3. Percent reaction vs. time for BA-DAB-BA (neat) and BA-DAB-BA (with 1% initiator) at 177°C, isothermal air cure.

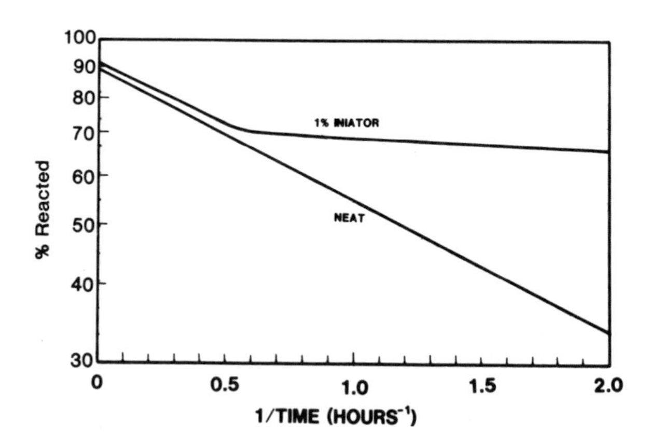

Figure 4. Percent reaction vs. time for BA-DAB-BA (neat) and BA-DAB-BA (with 1% initiator) at 177°C, isothermal nitrogen cure.

Discussion of Results

Progress has been made in initiating the polymerization reaction of an acetylene terminated phenyl quinoxaline at lower temperatures. While this result in itself is of practical significance to the Air Force, the investigation was embellished to provide generic foundations for subsequent studies. Observations were also made that may be of interest to other investigators. These are as follows: (a) At low concentrations (less than 1%), increasing amounts of initiator progressively lower both the onset and peak exotherm temperatures (as measured using the DSC); (b) Above 1%, the DSC curve begins to split into two peaks, probably a chemically initiated peak and the normal thermally initiated peak; (c) High initiator concentrations are not necessary or effective since uncured BA-DAB-BA has a softening or melting point at about 110°C and polymerization would not advance in the crystalline or glassy state. Reactions can, of course, take place in solution (e.g., THF) at lower temperatures;[4-11] (d) A practical method has been found to mix the initiator and polymer. The mix is heterogeneous, as the initiator simply coats the monomer particles, and is sensitive to particle size; (e) Although there is no direct evidence either way, it is considered possible that the initiated polymerization may result in a different molecular structure, since the polymerization temperature and heat or reactions are both lower for the initiated BA-DAB-BA. Neither cure reactions go to completion at 177°C (350°F); (f) The effort reported herein is a part of an on-going project; another portion of which is the mechanical response and fracture spectra of the subject materials. It is hypothesized that the fracture/mechanics behavior will be particularly decisive in determining possible differences in molecular structure.

Literature Cited

1. Hedberg, F.L. and Arnold, F.E., J. Appl. Polym. Sci., (1979) Vol. 24, 73-739.
2. Maximovich, M.G., Locherly, S.C., Kovac, R.F., and Arnold, F.E., Adhesives Age, (1977), 11, 40.
3. Eddy, S.R., Lucarelli, M.A., Helminiak, T.E., Jones, W.B., and Picklesimer, L.G., Adhesives Age, (1980), 23, 2.
4. Venanzi, J. Am. Chem. Soc., (1958), 219.
5. Daniels, J. Inorg. Chem., (1964), 23, 2936.
6. Luttinger and Colthup, J. Org. Chem., (1962), 27, 3752.
7. Colthup and Meriwether, J. Org. Chem., (1962), 27, 5169.
8. Meriwether, Leto, Colthup, Kennerly, J. Org. Chem., (1962), 27, 1591.
9. Luttinger, J. Org. Chem., (1962), 27, 1591.
10. Tsonis and Farona, J. Poly. Sci., (1979), 17, 1779.
11. Chalk and Gilbert, J. Poly. Sci., (1972), 10, 2033.

RECEIVED May 11, 1982.

19

Fracture Behavior of an Acetylene-Terminated Polyimide

W. B. JONES, JR., T. E. HELMINIAK, C. C. KANG, M. A. LUCARELLI, and L. G. PICKLESIMER

Air Force Wright Aeronautical Laboratories, Materials Laboratory, Wright-Patterson Air Force Base, OH 45433

The fracture behavior of an acetylene-terminated polyimide has been experimentally investigated. Engineering fracture methods were adapted and applied to small quantities of this material. Miniature compact-tension (CT) specimens were fabricated and subjected to a programmed series of post-cure cycles. Fracture tests were then conducted at room temperature and at several elevated temperatures in a nitrogen purged environment. The fracture toughness was found to vary with temperature and to depend upon post-cure. These Fracture tests clearly discriminate between the specimens resulting from the different cure cycles with only a small investment of material.

In the evaluation of new materials for potential structural applications, it is important that the structural characteristics be measured as early as possible in the evolution cycle of the material. In response to a growing awareness of the risk inherent in developing totally new and "unknown" materials, the Air Force is developing an evaluation process for structural polymers that reduces the risk with concomitant increase in investment. Interim progress reports have been given on the evaluation process, along with example characteristics measured for some polymers. [1-2] During the adjustments in emphasis and balance of the evaluation methodologies, it has become clear that one of the most desirable characteristics of a structural material is resistance to flaws and damage. Tough materials are preferred for virtually all structural applications.

At present, there is a paucity of fracture toughness data for polymers, and data for new polymers is rare. This is due in part to an unawareness of the value of such data, and in part to a lack of test methods that use only small quantities of materials.

This chapter not subject to U.S. copyright.
Published, 1982, American Chemical Society

In an attempt to remove the latter constraint, effort has been initiated to develop miniature fracture specimens that can be used to characterize limited quantities of new polymers. Hopefully, these efforts will provide the "tool" required to characterize new polymers and will be complimentary to the contributions of other investigators so that a meaningful fracture data base will be generated for all polymers.

The fracture characteristics of a material are important in at least two considerations. One use is as a figure of merit for estimating the material's structural capability in design applications. Another use is to give the physical chemist insight into the mechanisms by which the molecular network formed during polymerization can resist mechanical damage and fracture. Hypothetically, this insight can then provide guidelines for the synthesis of improved fracture resistant structural polymers.

The Material

A relatively large quantity (approximately 200 gm) of an aromatic heterocyclic material was available and was used to develop a fracture test method that could be used to characterize materials limited in available quantities. This material is an acetylene-terminated polyimide, known as Thermid 600*, which has the molecular structure shown in Figure 1. Although the material is a bit difficult to work with in the 100% solid form, this form was used to fabricate specimens. Cure kinetics and rheological data were measured and used to guide specimen fabrication.

Specimen Fabrications

The TH-600, in the form of a finely precipitated powder, was held in a vacuum oven at 100°C for several days to remove solvent and absorbed moisture. Preweighed quantities were then introduced into a preheated (235°C) mold and 100 psi pressure was immediately applied using a hydraulic press. After dwelling for 1/2 hour at these conditions, a consolidated and partially cured bar 0.1" x 0.5" x 3" was removed from the mold. Bars made in this manner were later postcured to the conditions shown in Table 1.

Using a diamond wafering saw, each bar was cut into four equilength rectangles. Holes were then carefully drilled and a central slot with a chevron tip was cut in each rectangle as shown in Figure 2. The chevron tip was then precracked by carefully tapping a razor blade in the slot. It is noted that the specimen is patterned after the ASTM compact tension (CT) specimen used to test metals.[3]

Fracture Testing

Stiff linkage and clevice type grips were fabricated and connected in an Instron testing machine. For each test, the

*Gulf

Figure 1. Molecular structure of Thermid 600.

TABLE 1. Postcure Conditions for TH-600 Fracture Specimens:

CURE	TEMP	TIME	ATMOSPHERE
1	290C	40 hrs	Air
2	340C	40 hrs	Air
3	372C	3½ hrs	Air
4	372C	1 hr	N_2
5	372C	40 hrs	N_2

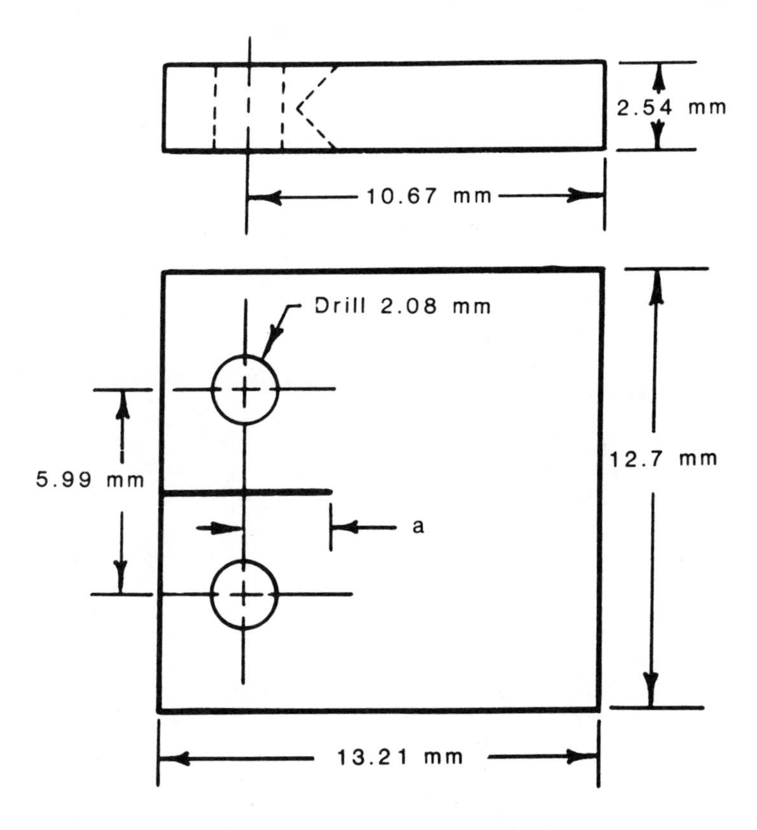

Figure 2. Compact tension specimen used in fracture tests.

specimen was pinned through the holes to the clevices using drill rods as pins. A small Linear Variable Differential Transformer (LVDT) was attached to measure the relative displacement between the clevices. For tests at elevated temperatures, a split furnace was placed around the specimen and stabilized at the test temperature. The Instron was run at 0.002"/min. cross-head rate, and the data were recorded as Force vs. LVDT displacement on an X-Y plotter. Typical data taken at three temperatures are shown in Figure 3. Although effort was made to have a stiff loading frame and linkage, there is enough compliance to permit several increments of crack growth in the specimens at 20°C and 200°C. This is evidenced by the sawtooth patterns. Post-test examination showed striations (See Fig. 4) on the fracture surfaces that correspond to each interruption in crack growth. These features were used to determine the crack lengths that were then used to calculate the fracture toughness K_{IC}. Complications arose at temperatures between 250 and 300°C where the material displayed slow crack growth and consequently did not exhibit initiation/ arrest or surface markings. Since the loading was limited to constant displacement rate on the Instron, fracture was initiated in the high temperature specimens by tapping intermittantly during the steady crack growth process. This usually produced small surface markings with associated down-steps in load.

The fracture data were reduced according to the equation:[3]

$$K_{IC} = \frac{P_{CR}}{b\sqrt{w}} \times f(a/w)$$

$$f(a/w) = \frac{(2+a/w)(0.886+4.64a/w-13.32\ a^2/w^2 + 14.72a^3/w^3-5.6a^4/w^4)}{(1-a/w)^{3/2}}$$

where: a - crack length
w - width
b - thickness
P_{CR} - failure load

The calculated fracture toughness values are tabulated in Table II, and graphically displayed in Figure 5.

Among all the data measured during this study, there are varying levels of confidence placed on the data, depending upon the post-cure and test temperatures. Duplicate specimens were tested at each of the twenty conditions and multiple data points were obtained from each specimen. For Cure #3 tested at 200°C, twenty-four (24) data points were obtained from the two specimens and the scatter was within ± 10% of the average. Moderately high confidence is placed in these data. In other cases only four or five data points make up the average. There is more scatter in these data (up to ±20%) thus these averages are marked (*) in Table II, and we are less confident of these results. Superimposed on the data scatter, which supposedly could be resolved with

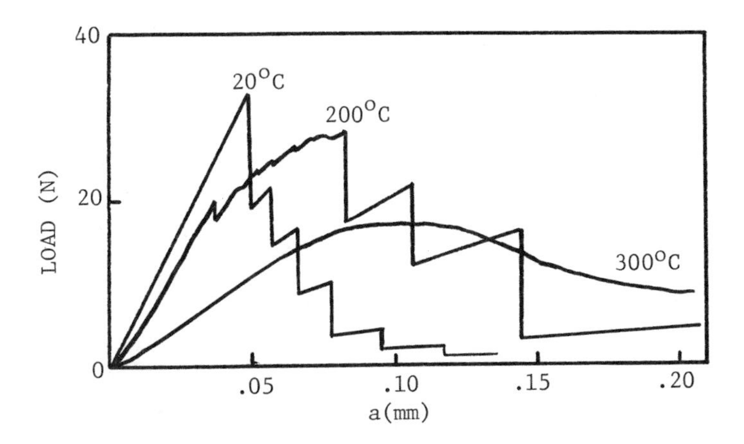

Figure 3. Data taken at three temperatures measuring the relative displacement between clevices in acetylene-terminated polyimides.

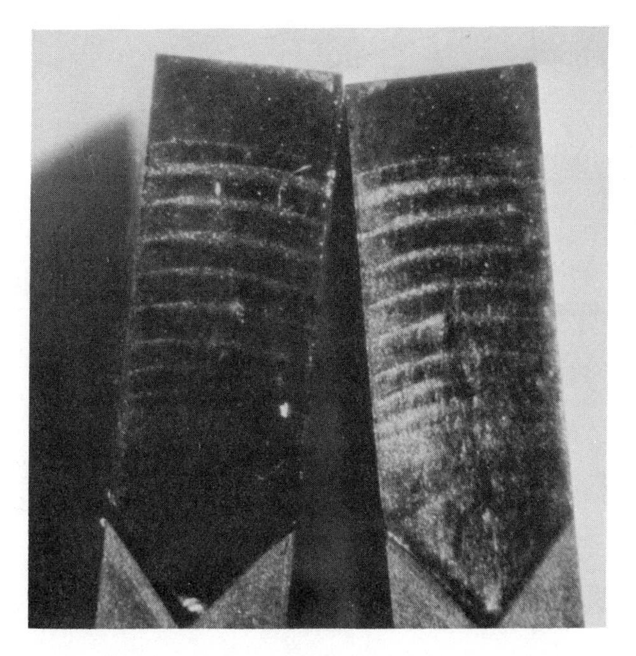

Figure 4. Surface marks left by crack initiation/arrest.

TABLE II. Fracture Toughness, K_{IC} MPa $\sqrt{m}$ Measured for TH-600 After Various Post-Cures

TEMP	20°C	200°C	250°C	300°C
CURE				
1	880	680	660*	230*+
2	710*	490	440	270
3	660*	460	530*	270
4	890	800*	420*	160
5	600	380	360*	330

* less than six data points in average
\+ tested at 290°C

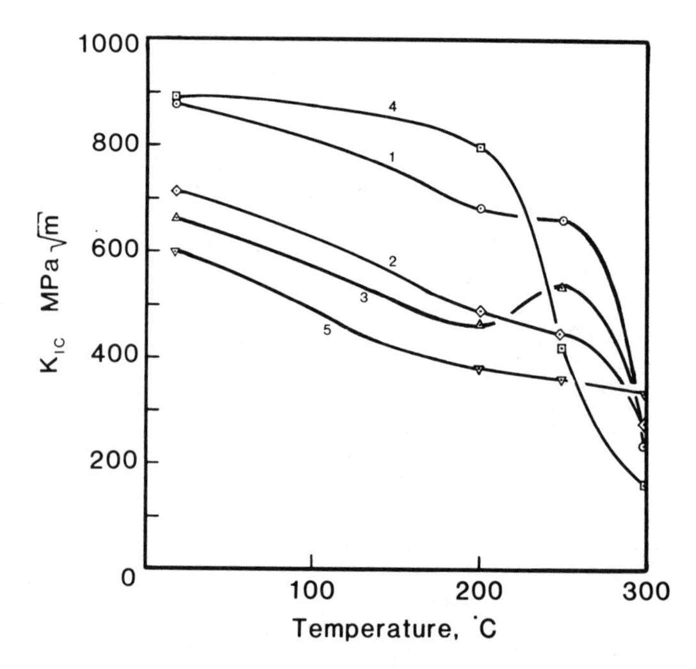

Figure 5. Fracture toughness vs. temperature for Thermid 600 after various post-cures.

supplemental testing, another effect was noted that becomes more pronounced at the elevated temperatures. For a specimen repeatedly tested at constant temperature the fracture toughness was high for short crack lengths, and became progressively lower as the crack length increased. This effect was most pronounced for Cure #4 tested at 300°C, with resulting data being shown below:

a, mm	K_{IC}, MPa $\sqrt{m}$
4.0	230
4.6	220
5.3	170
6.4	70

This effect was not observed for Cure #5 specimens, which incidentally were the most extensively post-cured. Other investigators(4) have reported similar results and have developed methods that would seem to deal effectively with this behavior. The effect stems from the crack length indirectly and depends rather on the time-dependence of the material behavior. When the cracks are short, the specimen is stiff, and hence loads to failure quickly under the constant displacement rate. For the longer crack lengths, the more compliant specimen displays a longer time to failure, with associated lower fracture toughness. Further inquiry in this area is planned.

Even while the above issues are only partially resolved, some conclusions can be drawn from the data displayed in Figure 5. Clearly, the fracture toughness varies with temperature and is influenced by post-cure. Equally as clear, the advancing post-cure conditions evaluated in this study progressively reduced the fracture toughness of the material.

Since post-cure is popularly used to more fully react the initial reactants, develop higher glass transition temperature, and hopefully develop the material's ability to withstand higher use temperatures, the data in Figure 5 shown caution should be exercised, at least for this material. The small improvement in the 300°C properties for Cure #5 material came with significant loss in toughness at temperatures below 250°C.

The miniature Compact Tension fracture specimen is new, and as such, needs thorough evaluation before the resulting data can be considered to be intrinsic material behavior.

In the interim, it is believed the test data will be extremely useful to delineate between materials having varying degrees of toughness, for evaluating a material's sensitivity to processing variations, and to serve as a tool for exploring the Fracture Process itself as an intermediate step to developing tougher materials.

In summary, a fracture test has been successfully applied to a high temperature polymer. Further developments of this test are required and are already underway.

Literature Cited

1. Eddy, S., Lucarelli, M., Helminiak, T., Jones, W., Picklesimer, L.; "Evaluating New Polymers as Structural Adhesives," Adhesives Age, Feb 1980, Vol. 23, No. 2.
2. Jones, W. B., et. al., "Evaluation of an Acetylene Terminated Sulfone Oligomer," presented at American Chemical Society Meeting, March 1980, Houston, TX.
3. ASTM-Standard E3999-78a, Paragraph 9.1.4, 1978.
4. Knauss, W. G., "Fracture of Solids Possing Deformation Rate Sensitive Material Properties," chapter in The Mechanics of Fracture, ed. by F. Erodogan, AMD - Vol 19, Amer. Soc. of Mech. Engineers, NY 1976.

RECEIVED December 31, 1981.

20

Synthesis and Characteristics of Poly(bisdichloromaleimides)[1]

INDRA K. VARMA,[2] GEORGE M. FOHLEN, and JOHN A. PARKER

Ames Research Center, NASA, Moffett Field, CA 94035

Six bisdichloromaleimides having different aromatic structures were prepared by reacting 2,3-dichloromaleic anhydride with (a) 1,5-diaminonaththalene, (b) bis(m-aminophenyl)methylphosphine oxide, (c) 2,5-bis(p-aminophenyl) 1,3,4-oxadiazole, (d) 3,3-bis(p-aminophenyl)phthalide, (e) 9,9-bis(p-aminophenyl) fluorene, (f) 10,10-bis(p-aminophenyl)anthrone. The effect of structural differences on thermal characteristics of these compounds was investigated by dynamic thermogravimetry.

Polymerization of these bisdichloromaleimides was carried by nucleophilic displacement of chlorine with 9,9-bis(p-aminophenyl)fluorene. The resulting polymers were characterized by IR spectroscopy and reduced viscosity measurements. Anaerobic char yields of these polymers at 800°C ranged from 55-60%. In presence of air, a complete loss of weight was observed between 600-650°C. Thermal cross-linking of these polymers was also investigated.

The high-temperature capabilities and flame resistance of polymers may be influenced by the incorporation of aromatic or heterocyclic rings in the backbone (1). We have earlier reported high char yields in bismaleimides and biscitraconimides having fluorene, anthrone, and phthalein groups as bridging groups (2). Incorporation of certain elements such as phosphorus, chlorine, nitrogen, boron, and bromine in synthetic polymers leads to an improvement in flame retardation. Phosphorus-containing bisimide resins with good thermal stability have been reported (3). An

[1] This is Part II in a series.

[2] Current address: Indian Institute of Technology, Centre of Materials Science & Technology, Delhi, Hauz Khas, New Delhi 110016, India.

0097-6156/82/0195-0253$06.00/0
© 1982 American Chemical Society

improvement in limiting oxygen index has been reported in chlorinated aromatic polyamides (4). Chlorine substitution, however, did not alter the char yields. It would be of interest to investigate the influence of chlorine on char yields and on the flame resistance of bisimides.

In this paper we report the synthesis and characterization of bisdichloromaleimides having aromatic and heterocyclic groups in the backbone. Six bisdichloromaleimides with aromatic rings as bridging units were prepared by reacting 2,3-dichloromaleic anhydride with (a) 1,5-diaminonaphthalene, (b) bis(m-aminophenyl) methylphosphine oxide, (c) 2,5-bis(p-aminophenyl) 1,3,4-oxadiazole, (d) 3,3-bis(p-aminophenyl)phthalide, (e) 9,9-bis(p-aminophenyl) fluorene, and (f) 10,10-bis(p-aminophenyl)anthrone. Throughout the text, bisdichloromaleimide monomers and polymers are referred to by the letter designations (a-e) used above and shown in the accompanying sketch.

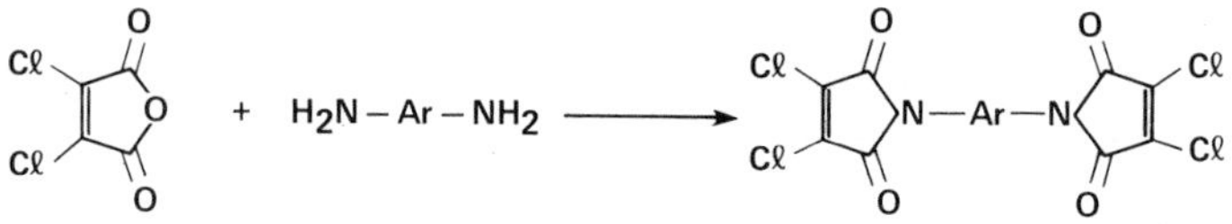

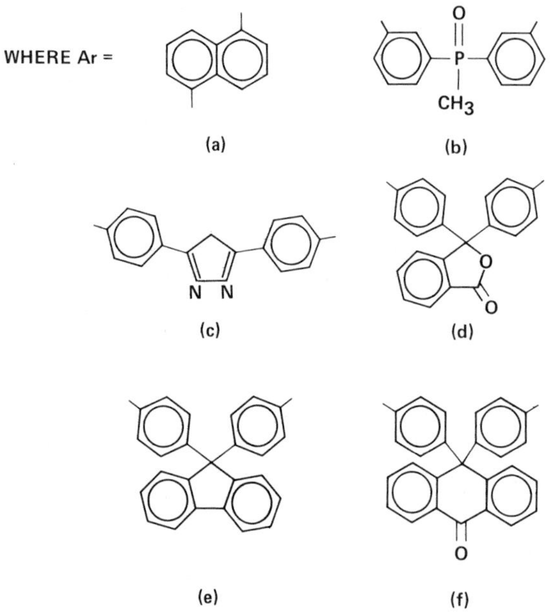

$H_2N-Ar-NH_2$

WHERE Ar =

(a) (b)

(c) (d)

(e) (f)

The bisdichloromaleimides thus obtained may be polymerized thermally by the opening of the double bond. Such cross-linking reactions have been extensively investigated in bismaleimides. Displacement of chlorine by appropriate nucleophiles is yet another way of preparing polymers from bisdichloromaleimides.

Nucleophilic substitution reactions have been effectively exploited in the synthesis of several classes of polymers. However, only scanty reports are available on the application of such reactions to the displacement of chlorine in bisdichloromaleimides, although several examples of displacement of chloride by nucleophiles in N-substituted dichloromaleimides exist in literature. Amines (5), phenols (6), and alcohols and thiols (7) have been used as nucleophiles in such reactions.

Relles and Schluenz (8) have reported the preparation of poly(maleimide-ethers) by using bisphenols and bisdichloromaleimides. These polymers were soluble in dimethylformamide (DMF) and dimethylsulfoxide (DMSO) and gave tough flexible films. Phosphorus-containing poly(maleimide-amines) have been reported to have poor thermal stability and little flame retardation (9).

In the present work, condensation polymerization of bisdichloromaleimides was carried out with 9,9-bis(p-aminophenyl) fluorene.

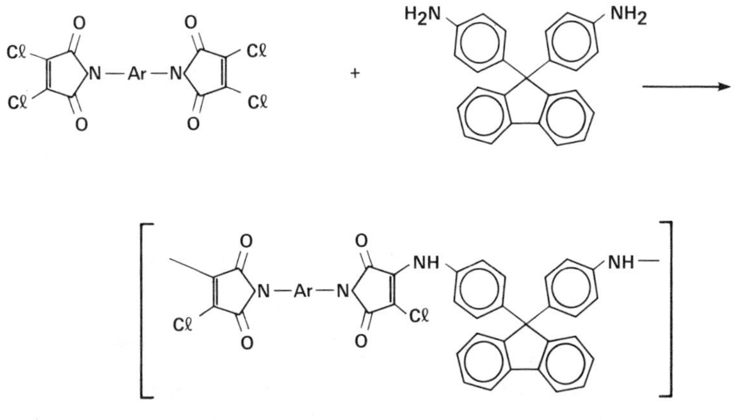

The reaction of diamines and bisdichloromaleimides proceeds with the displacement of one chlorine. The resulting poly(maleimide-amine) contain double bonds as well as chlorine in the backbone. The presence of double bonds in the backbone of these polymers also makes them susceptible to cross-linking reactions which may be initiated by heat.

Experimental

Starting Materials. The diamines such as 1,5-diaminonaththalene or 9,9-bis(p-aminophenyl)fluorene (BAF) were purified by crystallization or precipitation. Melting points (mp) of the dried samples were found to be 187°C–188°C and 236°C–237°C, respectively. Bis(m-aminophenyl)methylphosphine oxide (10) (mp 146°C–149°C); 3,3-bis(p-aminophenyl)phthalide (11) (mp 204°C–205°C); 10,10-bis(p-aminophenyl)anthrone (12) (mp 305°C–306°C) and 2,5-bis(p-aminophenyl) 1,3,4-oxadiazole (13) (mp 260°C–261°C) were prepared according to the methods reported in the literature. The 2,3-dichloromaleic anhydride was available commercially (Aldrich Co.).

Preparation of Bisdichloromaleimides. The method of Martin et al. (14) was used for the preparation of bisdichloromaleimides. Diamines (0.005 mole) and 2,3-dichloromaleic anhydride (0.01 mole) were separately dissolved in glacial acetic acid and the solutions were then mixed and heated slowly to reflux temperature. After refluxing for 2 h, the solutions were cooled, and bisdichloromaleimides were precipitated in water. The products were filtered and washed with water until free of acetic acid. After drying, the products were recrystallized from toluene/hexane.

Preparation of Polymers. The polymerization of the above prepared bisdichloromaleimides was carried out according to the method of Relles and Schluenz (8). The 9,9-bis(p-aminophenyl) fluorene (0.001 mole) was dissolved in 10 ml of freshly distilled DMF containing 0.002 mole of triethylamine as an acid acceptor. To this well-stirred solution, 0.001 mole of a bisdichloromaleimide (a–f) was added and the reaction mixture stirred for 2 h. The polymer was precipitated by pouring the mixture into an excess of water. The precipitate after boiling in acetone was dried in vacuum at 60°C–70°C.

Characterization. Infrared spectra of bisdichloromaleimide monomers and polymers in KBr pellets were recorded, using a Perkin-Elmer 180 spectrophotometer. Elemental analyses were provided by Huffman Laboratories. Mass spectra were recorded at 70 eV on a Hewlett-Packard MS 5980 instrument by the direct inlet procedure. A DuPont 990 thermal analyzer was used to evaluate thermal behavior of bisdichloromaleimide monomers and polymers. Reduced viscosity of the polymers was determined in DMF at 30°C with a Cannon viscometer. Thermal polymerization was studied by heating a known weight of the material from room temperature to the desired temperature in a glass tube. The extent of curing was evaluated by extraction with DMF at room temperature.

Results and Discussion

Characterization of Bisdichloromaleimides. Physical description, molecular weight, and analytical data of the monomers are given in Table I. Melting points of these monomers were above 250°C. Thus, in compounds (b), (c), and (f) in Table I, endothermic transition associated with melting was observed in DSC at

Table I. Characterization of Bisdichloromaleimides

Sample	Formula	Elemental analysis (%)				$M^{+\bullet}$*	Color
		C	H	N	Cl		
(a)	$C_{18}H_6O_4N_2Cl_4$	47.60 (47.57)	1.43 (1.32)	6.14 (6.16)	31.04 (30.83)	454	Light brown
(b)	$C_{21}H_{11}O_5N_2Cl_4P$	47.02 (46.49)	2.23 (2.03)	5.43) (5.17)	25.10 (25.83)	542	Yellow
(c)	$C_{22}H_8O_5N_4Cl_4$	49.06 (48.17)	1.67 (1.46)	10.21 (10.22)	25.95 (25.55)	548	Yellow
(d)	$C_{28}H_{12}O_6N_2Cl_4$	56.12 (54.9)	2.20 (1.96)	4.44 (4.57)	24.11 (22.87)	612	Yellow
(e)	$C_{33}H_{16}O_4N_2Cl_4$	63.03 (61.49)	2.75 (2.48)	4.39 (4.34)	20.63 (21.74)	644	Yellow
(f)	$C_{34}H_{16}O_5N_2Cl_4$	62.23 (60.53)	2.73 (2.37)	4.23 (4.15)	20.74 (20.77)	672	Light green

*Represents the molecular ion in the 70 eV mass spectra of bisdichloromaleimides; figures in parentheses are calculated values.

265°C, 369°C, and 300°C, respectively. Elemental analyses agree well with the assigned structures of the monomers.

Electron-impact-induced fragmentation patterns of 2,5-bis(p-dichloromaleimidophenyl) 1,3,4-oxadiazole (compound c); 3,3-bis(p-dichloromaleimidophenyl)phthalide (compound d); and 10,10-bis(p-dichloromaleimidophenyl)anthrone (compound f) are given in Figure 1. The common features present in the mass spectra of these compounds are summarized in Table II. A low-intensity molecular ion peak was observed in most of these monomers, with the exception of compounds (a) and (b). Loss of carbon dioxide from the molecular ion (M-44) was indicated only in compound (d), which contains a phthalide group. Carbon dioxide may be eliminated from this site. Similar fragmentation patterns have earlier been obtained with corresponding bismaleimides and biscitraconimides (15).

The other fragment ions observed in the mass spectra indicate the loss of the dichloromaleimidyl group $(M-164)^+$ (compounds a, b, and c) and dichloromaleimidophenyl group $(M-240)^+$ (compounds b, d, e, and f) from the molecular ion. Cleavage of the dichloromaleimidyl group is responsible for the appearance of fragment ions at m/z 122 $(C_3Cl_2O)^+$ and 87 $(C_3ClO)^+$. However, the corresponding $(M-122)^+$ ion was absent in b, c, d, and f. Relles and Schluenz (16), while working the N-aryl dichloromaleimides, have observed that the tendency of charge to remain in the isocyanate half on

N=C=O^{+} + $C_3Cl_2O^{\cdot}$
M-122
OR
N=C=O$^{\cdot}$ + $C_3Cl_2O^+$
M/Z 122
M-240
M-164

cleavage depends on the electron donating or electron withdrawing effect of the substituent. Compounds (a) and (c), in which such an ion was observed, contain a naphthalene or fluorene ring, whereas in other cases electron-attracting groups (carbonyl, oxadiazole, or phosphine oxide) are present in the molecule. A fragment ion at m/z $(M-148)^+$ was observed in some compounds. This could arise by the loss of the $(C_4Cl_2)ON$ neutral fragment from the maleimide group. Appearance of the corresponding $(M-80)^+$ and $(M-94)^+$ fragments has earlier been observed in bismaleimides and biscitraconimides (15). Simultaneous loss of CO_2 (from the phthalide group) and of the dichloromaleimidyl group may account for the fragment ion at $(M-208)^+$ in compound (d). In compound (b), fragment ions of the high intensity (40% of base peak) were

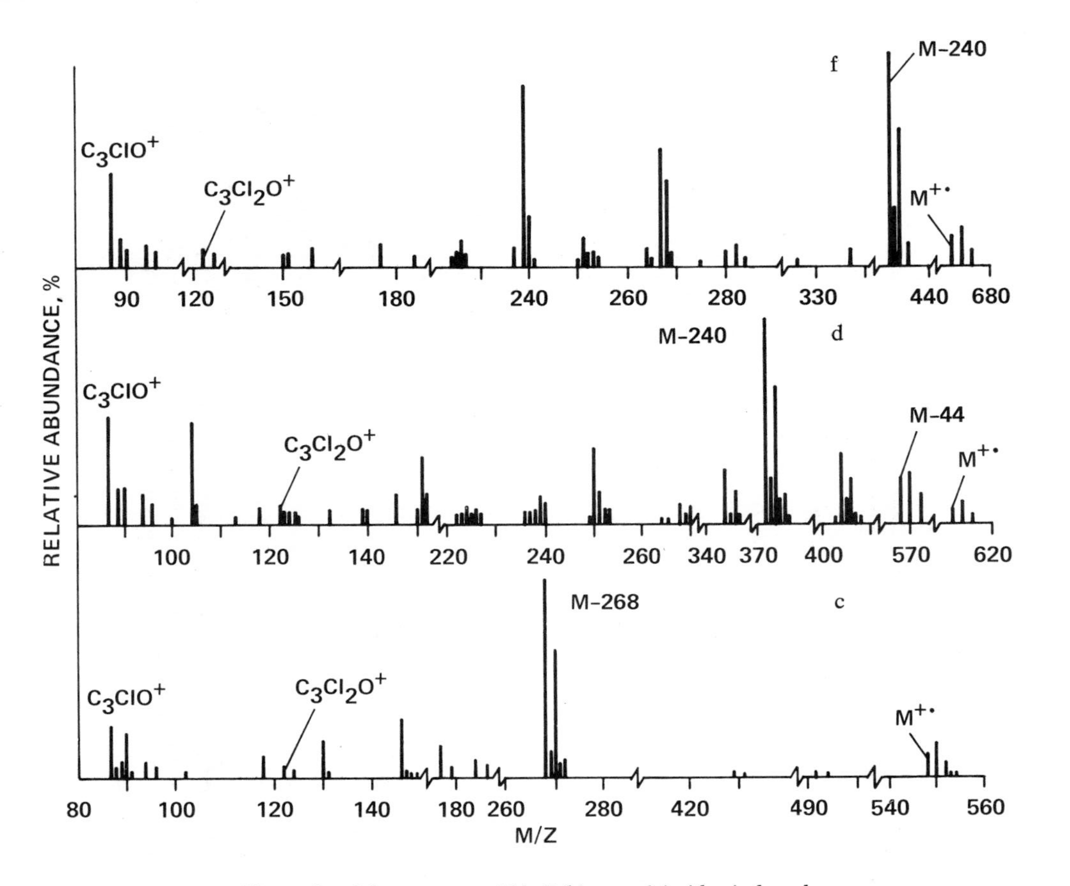

Figure 1. Mass spectra of bisdichloromaleimides f, d, and c.

Table II. Relative Abundances (% Base Peak) of Some Fragment Ions in the 70 eV Mass Spectra of Bisimides*

	Sample					
	a	b	c	d	e	f
(Mol wt)	(454)	(542)	(548)	(612)	(644)	(672)
$(M)^+$	65.8 (4)	66 (4)	14.7 (4)	9.2 (3)	59.6 (3)	16.4 (3)
$(M-35)^+$	2.2	47.0	-	-	-	-
$(M-44)^+$	-	-	-	22.1 (3)	-	-
$(M-63)^+$	8.5	40.9	-	-	-	-
$(M-122)^+$	36.8 (3)		-	-	11.7 (2)	-
$(M-148)^+$	14.1		-	-	-	-
$(M-164)^+$	5.8	23.2 (2)	-	-	20.2	-
$(M-208)^+$	-		-	33.6 (3)	-	-
$(M-240)^+$	-	16.6 (2)	-	100 (3)	76.6 (2)	100 (3)
$(M-244)^+$	100		-	-	-	-
$(M-362)^+$	-		-	-	100	9.5
$(87)^+$	67.6 (2)	100 (2)	25.1 (2)	53.0 (2)	59.6 (2)	47.6 (2)
$(122)^+$	11.5 (2)	33.7 (3)	6.6 (2)	10.6 (3)	13.8	11.6 (2)
$(164)^+$	3.4 (2)	8.8 (2)	-	5.5	9.6	7.4
$(239)^+$	-	7.7	-	16.6	98.9	93.7
$(240)^+$	-	11.6	14.7	12.9	41.5	26.5
$(268)^+$	-		100	12.9	-	47.1

*The numbers in parenthesis next to intensity values indicate the number of peaks associated with the chlorine isotopic clusters for each ion. Intensities given are for the all ^{35}Cl peaks of each cluster.

observed at M-1, M-15, M-35, and M-63. The M-1 ion may arise by loss of hydrogen and formation of bridged phosphafluorenyl ion (17). The scheme in Sketch 4 may account for the appearance of these ions.

The IR spectra of the compounds showed characteristic absorptions owing to the imide group (at 1790 cm^{-1} and 1725 ±5 cm^{-1}), the double bond (at 1620 ±10 cm^{-1}), and C-Cl (at 880 cm^{-1}) (Fig. 2a). In monomer (d) absorption owing to the phthalide group is observed at 1730 cm^{-1} (Fig. 2b). In monomer (f) the carbonyl group absorption appears at 1660 cm^{-1}. In monomers (b), absorptions owing to P = O (at 1190 cm^{-1}), P-C_6H_5 (at 1425 cm^{-1}), C_6H_5 (at 1480 cm^{-1}), and C-N (at 1380 cm^{-1}) were also observed.

Curing of Bisdichloromaleimides. Differential scanning calorimetry was used as a tool in characterizing the curing behavior of bisdichloromaleimides. A broad exotherm in the temperature range of 270°C-450°C was observed in these samples. This may be attributed to thermal polymerization by opening of the double bonds. It may also arise by thermal decomposition of bisdichloromaleimides. Thermogravimetric analysis of these monomers did indicate a loss in weight above 300°C.

It was, therefore, decided that we would study thermal polymerization of bisdichloromaleimides at 300°C for 30 min. The resulting product was soluble in DMF to a great extent (Table III) with the exception of compound (b). This indicates the absence of thermal polymerization under these conditions. Anaerobic char yields of these thermally treated bisdichloromaleimides depended on their backbone structure; a very low value was obtained in compounds (a) and (c); compound (b), which contained phosphorus, was most stable. Condensed phase reactions are influenced by the presence of phosphorus in these polymers. An almost linear relationship is observed between anerobic char yields at 800°C and bridge formula weight of bisdichloromaleimide (Fig. 3).

Characterization of Bisdichloromaleimide-Amine Polymers. The structures of the polymers (a) to (f) were confirmed by their IR spectra; absorptions owing to the secondary amine were observed at ~3300 cm^{-1} and 1509-1515 cm^{-1}, and absorption owing to the carbonyl group of the imide ring was observed at 1720 cm^{-1}. An additional band was observed around 1660 ±2 cm^{-1} (Fig. 4). This has been assigned to the carbonyl group conjugated with the nitrogen of secondary amine (6, 9). The intensity of the C-Cl band at 880 cm^{-1} was reduced in the polymers, relative to that of the monomers. Elemental analyses of these polymers agreed with the assigned structures (Table IV). The reduced viscosities of the polymers were in the range of 0.1-0.72 dl/g (Table V).

Thermal Behavior of Polymers. Thermogravimetric analyses of the polymers were carried out both in an air and nitrogen atmosphere (flow rate of 100 cm^3/min) at a heating rate of 10°C/min.

SKETCH 4

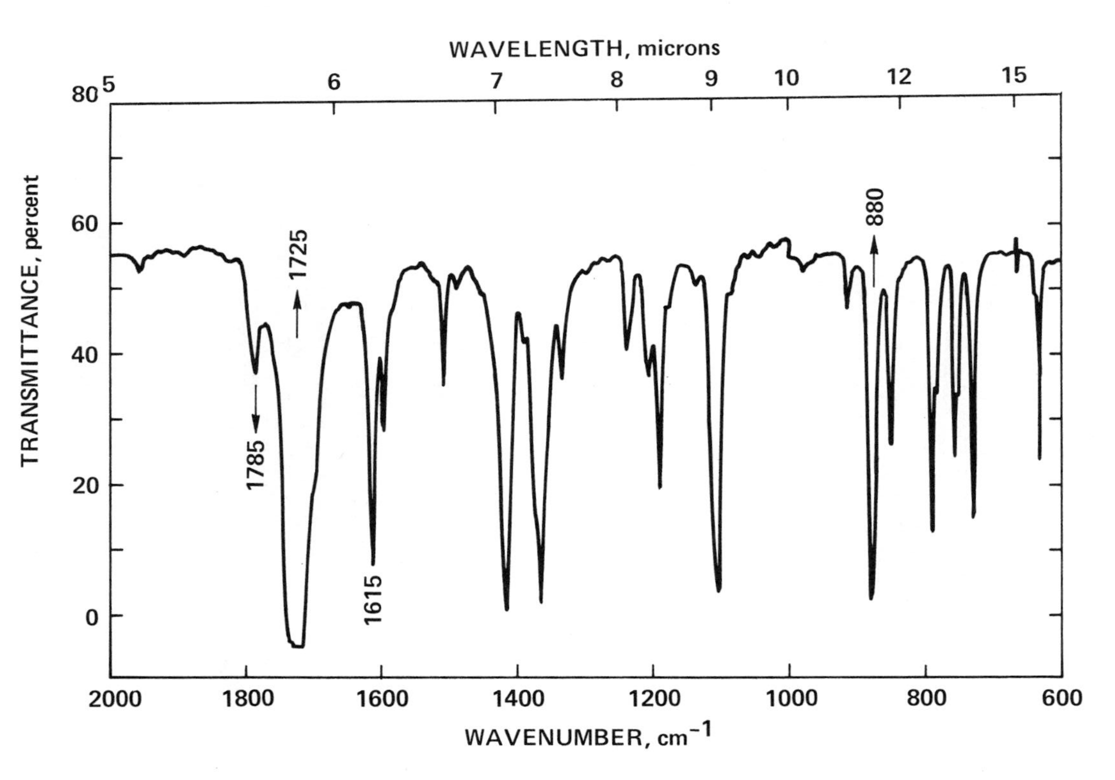

Figure 2a. IR spectra of 1,5-bisdichloromaleimido naphthalene.

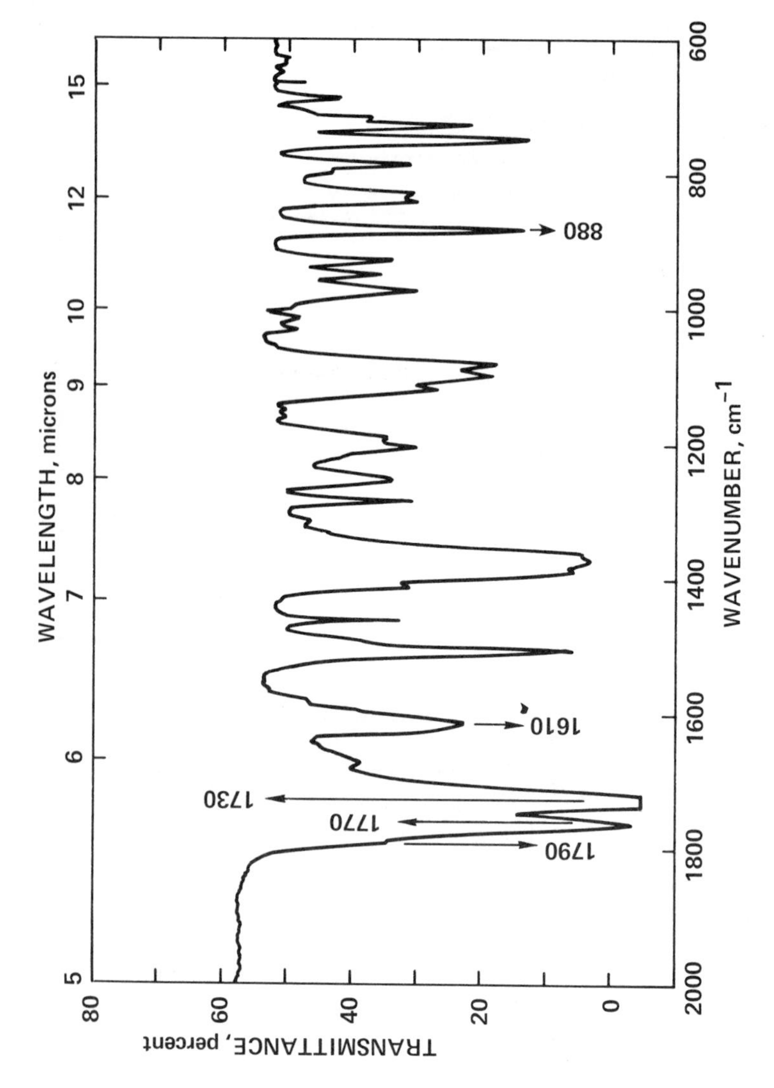

Figure 2b. *IR spectra of 3,3-bis(*p*-dichloromaleimidophenyl)phthalide.*

Table III. Anerobic Char Yields of Bisdichloromaleimides Heated at 300°C for 30 min

Sample	Bridge formula weight	Y_c (%)	Solubility in DMF
(a)	126	10	100
(b)	214	66	6
(c)	220	34	62.3
(d)	284	52.5	86
(e)	316	55.5	100
(f)	344	57	100

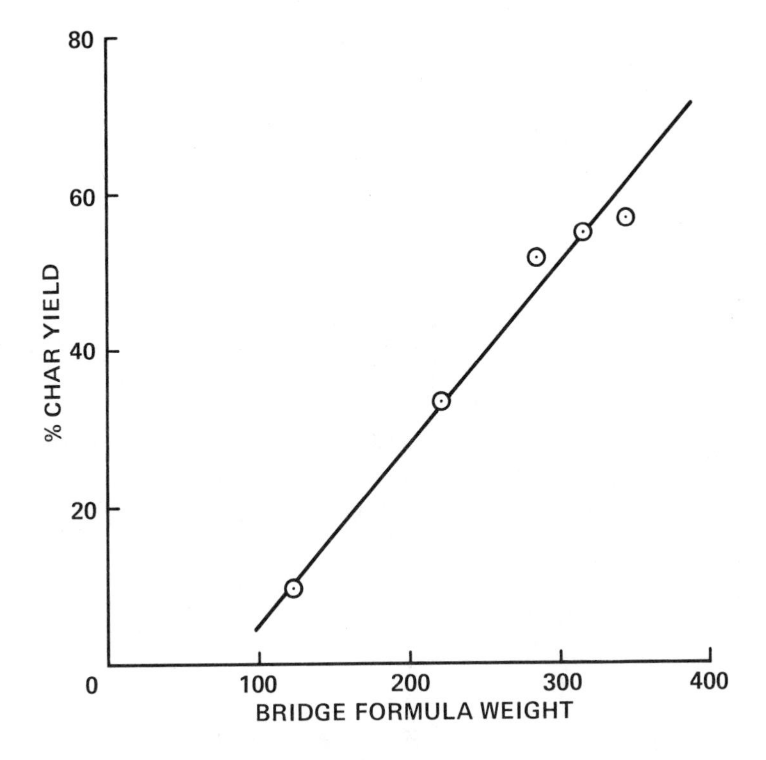

Figure 3. Effect of bridge formula weight on anaerobic char yields (at 800°C) of bisdichloromaleimides.

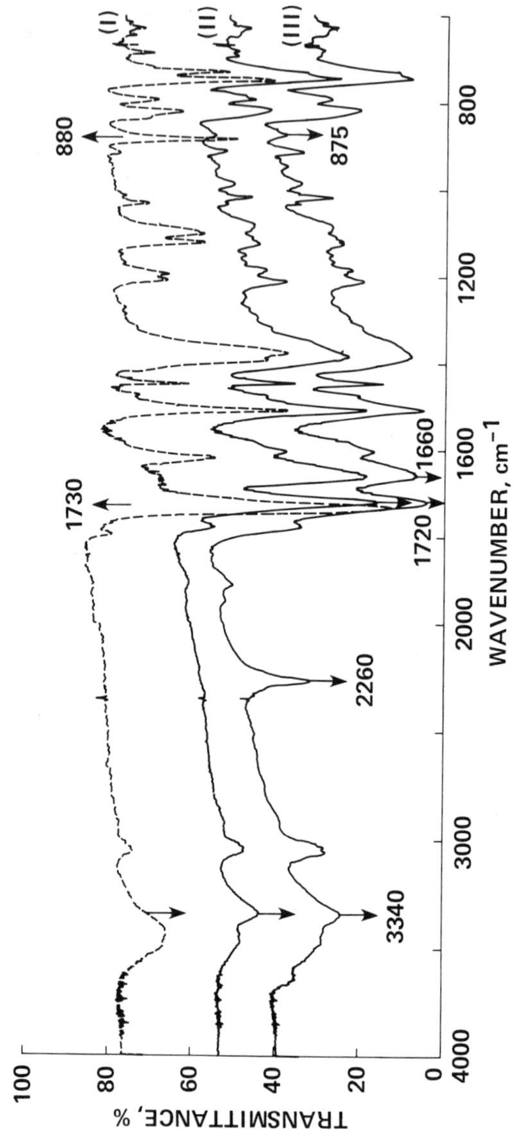

Figure 4. IR spectra of monomer e (I), polymer e (II), and polymer e, heated at 300°C for 30 min (III).

Table IV. Characterization of Poly(maleimide-amines)

Sample	Formula of repeat unit	Elemental analysis* (%)			
		C	H	N	Cl
(a)	$C_{43}H_{24}N_4O_4Cl_2$	69.39 (70.6)	3.45 (3.2)	7.35 (7.67)	10.84 (9.5)
(b)	$C_{46}H_{29}N_4O_5Cl_2$	66.24 (67.48)	3.78 (3.54)	6.81 (6.84)	8.21 (8.56)
(c)	$C_{47}H_{26}N_6O_5Cl_2$	66.95 (68.4)	3.17 (3.15)	10.34 (10.19)	9.05 (8.4)
(d)	$C_{53}H_{30}N_4O_6Cl_2$	71.34 (71.6)	3.62 (3.37)	6.49 (6.3)	6.71 (7.88)
(e)	$C_{58}H_{34}N_4O_4Cl_2$	75.65 (75.65)	3.85 (3.69)	6.01 (6.08)	7.39 (7.6)
(f)	$C_{59}H_{34}N_4O_5Cl_2$	73.76 (75.47)	3.64 (3.6)	5.80 (5.97)	6.04 (7.4)

*Figures in parentheses are calculated values.

Table V. Anaerobic Char Yields and Decomposition Temperatures of Bisdichloromaleimide-Amine Polymers

Polymer	Formula weight of repeat unit	$\eta sp/c$* (dl/g)	T_i (°C)†		T_{max} (°C)‡		Y_c** (%)
			Air	Nitrogen	Air	Nitrogen	
(a)	730	0.56	335	350	500	580	61
(b)	818	0.13	310	340	480	550	55
(c)	824	0.22	330	335	520	540	56
(d)	888	0.72	340	400	525	525	60
(e)	920	0.27	360	400	510	580	60
(f)	948	0.41	350	390	525	550	60

*In DMF at 30°C at a concentration of 0.2 ±0.02 g/dl.
**Anaerobic char yield at 800°C.
†T_i = initial decomposition temperature.
‡T_{max} = temperature of maximum rate of decomposition.

The polymers were stable up to 300°C in air and nitrogen but started losing weight at higher temperatures (Fig. 5). The anaerobic char yields of the polymers at 800°C were in the range of 55%-61% (Table V). Presence of fused aromatic rings in these polymers apparently did not influence the char formation tendency. The temperature of the maximum rate of decomposition was reduced in the presence of air, and an almost complete loss in weight occurred at about 600°C (Table V).

Cross-Linking Reactions of Polymers. Bisdichloromaleimide-amine polymers contain (a) a double bond in the maleimidyl group, (b) chlorine, and (c) secondary amine group (-NH-). It may be possible to cross-link them either by the opening of the double bond (thermal polymerization) or by the nucleophilic displacement of chlorine by the secondary amine. The representative reaction scheme for such reactions is shown in Figure 6. The extent of such reactions may be evaluated by solubility measurements in dimethylformamide.

Thermal cross-linking reactions were evaluated by heating polymers (a), (b), (d), and (f) at 300°C for 30 min. Solubility of the resulting product in DMF was considerably reduced (<0.8%). In the IR spectra of thermally treated polymers an additional band was observed at 2260 cm^{-1} in polymers (e) and (f) (Fig. 4). Such a band may indicate the formation of nitrile group or isocyanate group. In polymers (a), (b), and (d) no such new bands were observed in the IR spectra.

A four-ply graphite cloth laminate based on polymer (e) was fabricated. Limiting oxygen index of this laminate was determined according to ASTM D2863-74 and was found to be 71%. Similar values for the graphite cloth laminates fabricated from 9,9-bis(p-maleimidophenyl)fluorene have been obtained earlier (2).

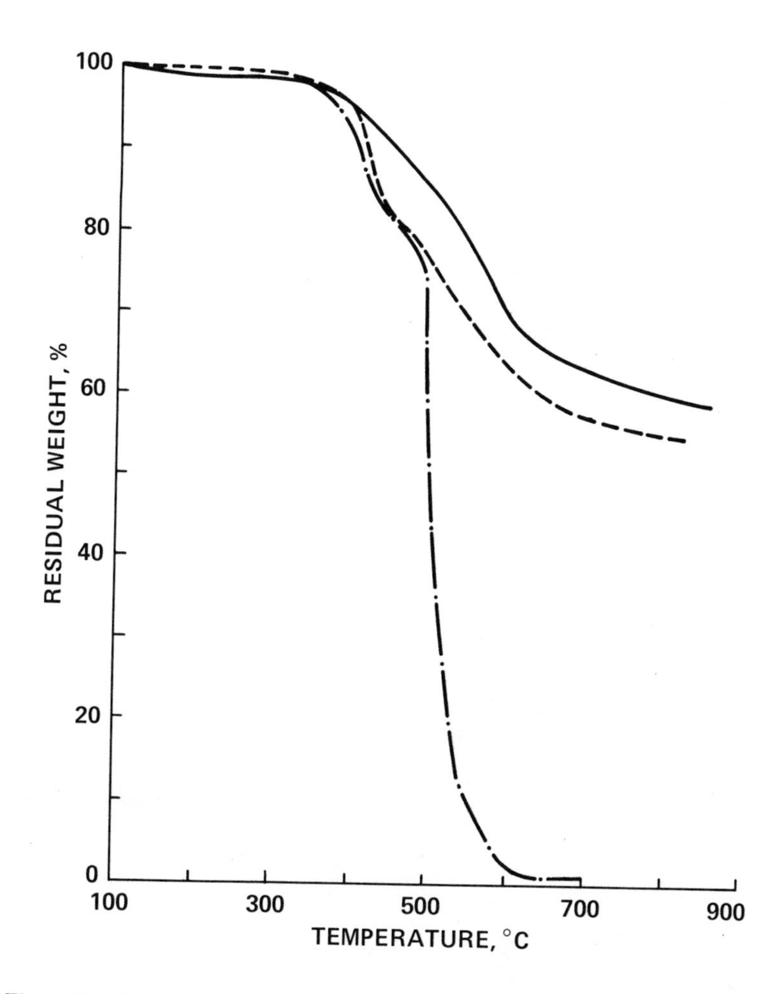

Figure 5. Primary thermograms of polymer e in nitrogen (——), polymer e in air (– · –), and monomer e heated at 300°C (– – –).

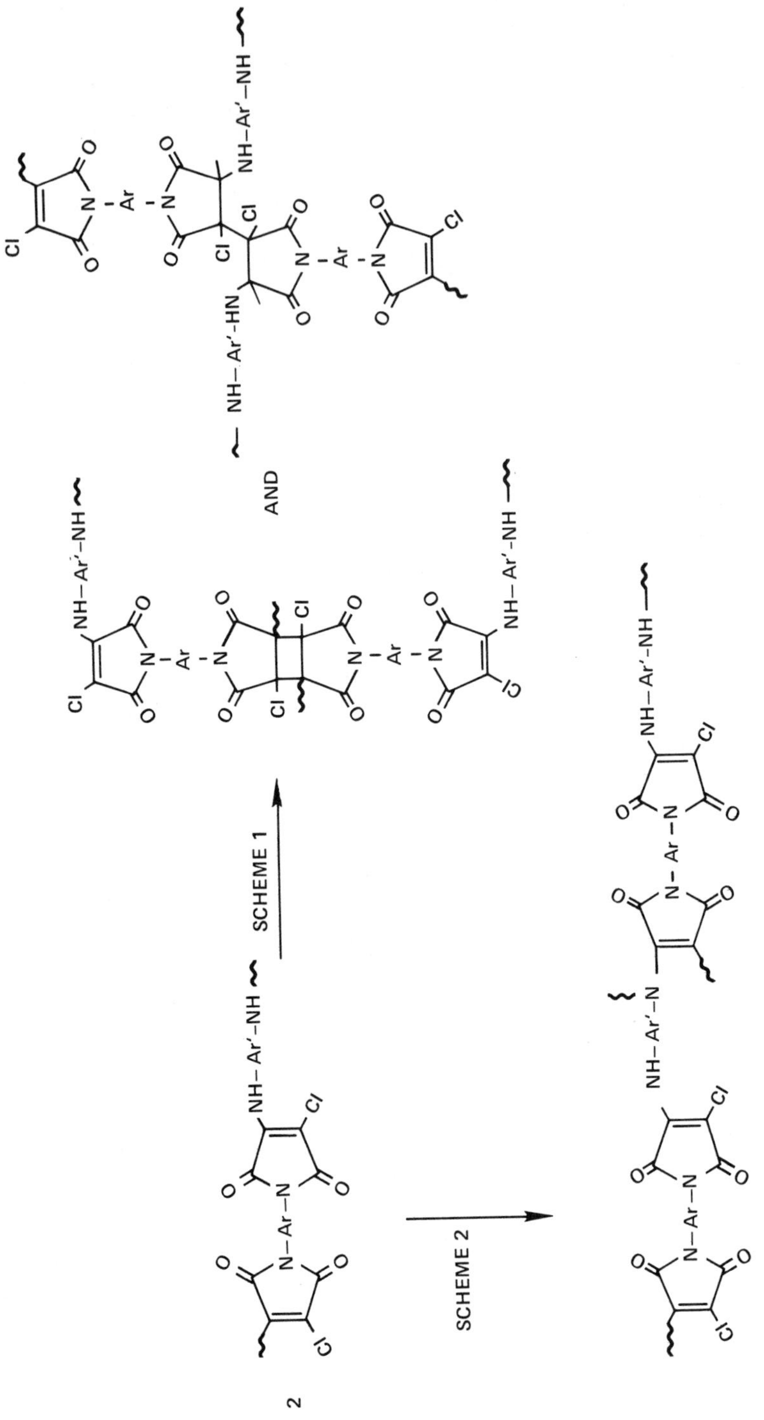

Figure 6. Representative scheme of cross-linking reactions.

Acknowledgment

The authors thank Dr. Ming-ta Hsu for mass spectral data.

Literature Cited

1. Van Krevelen, D. W. Polymer 1975, 16, 1615.
2. Varma, I. K.; Fohlen, G. M.; and Parker, J. A. Polymer Preprints 1980, 21, 1972.
3. Varma, I. K.; Fohlen, G. M.; and Parker, J. A. U.S. Patent 4,276,344, June 1981.
4. Khanna, Y. P. and Pearce, E. M. Organic Coatings and Plastics Chemistry 1980, 43, 94.
5. Oda, R.; Hayashi, Y.; and Takai, T. Tetrahedron 1968, 24, 4051.
6. Relles, H. M. and Schluenz, R. W. J. Org. Chem. 1972, 37, 3637.
7. Fickentscher, R. Tetrahedron Lett. 1969, 4273.
8. Relles, H. M. and Schluenz, R. W. J. Polym. Sci., Polym. Chem. Ed. 1973, 11 561.
9. Kondo, M.; Sato, M.; and Yokoyama, M. Europ. Polym. J. 1980, 16, 537.
10. Kourtides, D. A.; Parker, J. A.; Giants, T. W.; Bilow, N.; and Hsu, M-t. Proceedings of the Adhesives for Industry Conference, El Segundo, Calif., June 24-25, 1980.
11. Etienne, A. and Arios, J. C. Bull. Soc. Chim. France 1951, 727.
12. Hubacher, M. H. J. Am. Chem. Soc. 1951, 73, 5885.
13. Preston, J.; Bach, M. C.; and Clements, J. B. "Condensation Monomers," J. K. Stille and T. W. Campbell, Eds., Wiley, New York, 1972, p. 576.
14. Martin, E. L.; Dickinson, C. L.; and Rolands, J. A. J. Org. Chem. 1961, 26, 2032.
15. Varma, I. K.; Fohlen, G. M.; Hsu, M-t.; and Parker, J. A. Org. Mass Spectrom. 1981, 16, 145.
16. Relles, H. M. and Schluenz, R. W. J. Org. Chem. 1972, 37, 1742.
17. Williams, D. H.; Ward, R. S.; and Cooks, R. G. J. Am. Chem. Soc. 1968, 90, 966.

RECEIVED November 30, 1981.

21

Thermally Stable Polyimides from Pyrrole Dicarboxylic Acid Anhydride Monomers

R. W. STACKMAN

Celanese Research Company, Summit, NJ 07091

Arylene-bis-(pyrrole dicarboxylic acid anhydrides) were prepared by the condensation of two moles of diethyl diacetyl succinate with one mole of aromatic diamine, followed by hydrolysis and dehydration. Condensation of these novel dianhydrides with various aromatic diamines resulted in the formation of poly (amic acids) which were further condensed to polyimides. If the diethyl diacetyl succinate and aromatic diamine were reacted in equimolar quantities an N-(amino aryl) pyrrole diester was formed which can be further condensed to give polyimide directly. The initially prepared poly (amic acids) lose water in the range of 150-230°C to form the polyimide which shows thermal stability up to 400°C. Heating of the N-(amino aryl) pyrrole diester monomer under vacuum at 200-250°C was found to be sufficient for conversion to the polyimide. These pyrrole derivative monomers therefore are easily obtained from relatively inexpensive starting materials and appear to offer promise of being an attractive route to thermally stable polymers at reasonable cost.

Aromatic polyimides possess outstanding thermal stability as well as being unusually high melting, intractable and insoluble (1). Polyimides are prepared either by polyamide salt techniques, by condensation of dianhydrides with diisocyanates (2) or by reaction of an aromatic diamine with a dianhydride to give a poly(amic acid) followed by dehydration to give the polyimide. The polyimides from a variety of diamines have been reported and the dianhydride unit has been varied widely (1).

This report deals with polyimides prepared from novel dianhydrides containing a bispyrrole structure.

Dicarboxy pyrroles were first reported in the literature many years ago (3). These compounds are prepared by the

0097-6156/82/0195-0273$06.00/0
© 1982 American Chemical Society

reaction of ethyl diacetyl succinate with an amine in glacial acetic acid as shown by the following equation:

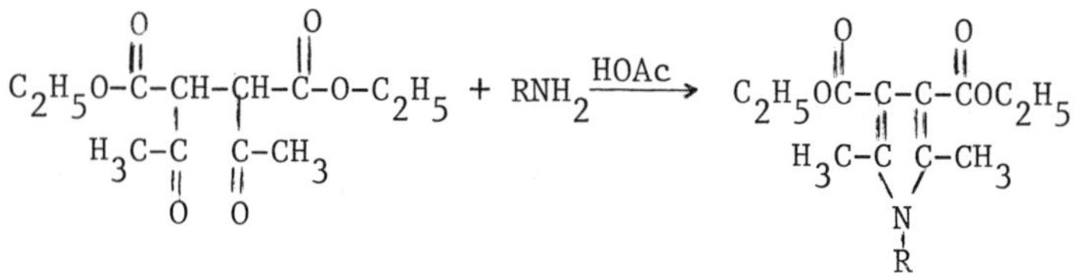

O O
|| ||
$C_2H_5O\text{-}C\text{-}CH\text{-}CH\text{-}C\text{-}O\text{-}C_2H_5$ + RNH_2 $\xrightarrow{HOAc}$ $C_2H_5OC\text{-}C\text{-}C\text{-}COC_2H_5$
$H_3C\text{-}C$ $C\text{-}CH_3$ $H_3C\text{-}C$ $C\text{-}CH_3$
O O N
R

By extending this reaction to the use of aromatic diamines one can obtain tetracarboxylic acids which may be subsequently converted to the dianhydride by the following series of reactions:

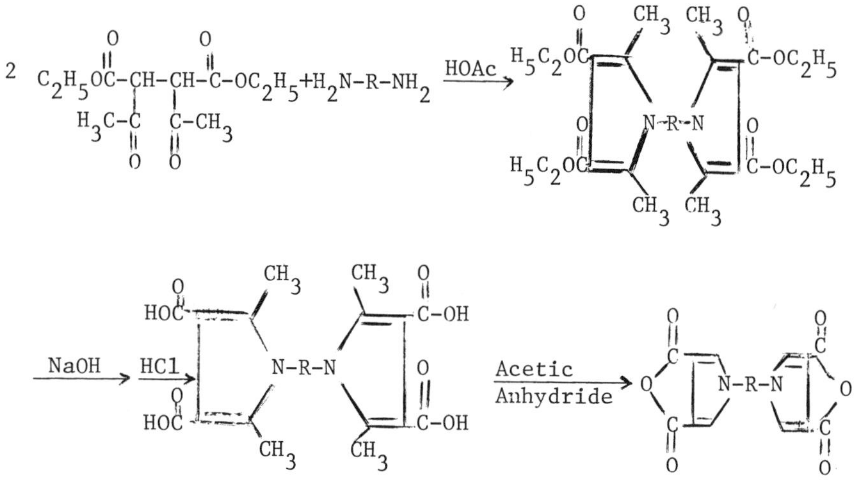

2 $C_2H_5OC\text{-}CH\text{-}CH\text{-}C\text{-}OC_2H_5 + H_2N\text{-}R\text{-}NH_2$ $\xrightarrow{HOAc}$
$H_3C\text{-}C$ $C\text{-}CH_3$
O O

O CH_3 CH_3 O
H_5C_2OC $C\text{-}OC_2H_5$
O N-R-N O
H_5C_2OC $C\text{-}OC_2H_5$
CH_3 CH_3

$\xrightarrow{NaOH}$ $\xrightarrow{HCl}$
O CH_3 CH_3 O
HOC C-OH
N-R-N
O O
HOC C-OH
CH_3 CH_3

$\xrightarrow{\text{Acetic Anhydride}}$
O O
C C
O N-R-N O
C C
O O

The dianhydrides may then be reacted with aromatic diamines to give a poly(amic acid) which is converted to the polyimide.

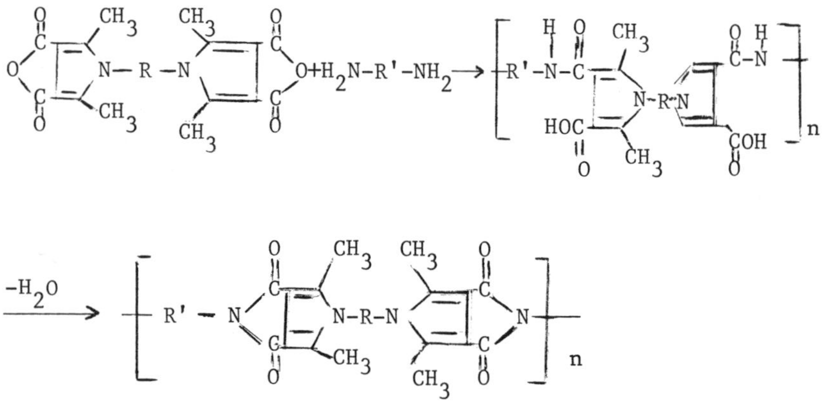

O CH_3 CH_3 O
C C
O N— R —N O $+ H_2N\text{-}R'\text{-}NH_2 \longrightarrow$
C C
O CH_3 CH_3 O

H O CH_3 O H
$R'\text{-}N\text{-}C$ C-N
N-R-N
HOC COH
O CH_3 O
n

$\xrightarrow{-H_2O}$
O CH_3 CH_3 O
C C
R' — N N-R-N N
C C
O CH_3 CH_3 O
n

One modification of this procedure has been to react the ethyl diacetyl succinate with diamine on an equimolar basis to give an N(aminophenyl)pyrrole diester which can then undergo further condensation to give the polyimide directly.

C_2H_5-C(=O)-C-CH(C(=O)CH_3)-CH(C(=O)CH_3)-C(=O)-OC_2H_5 + H_2N-R-NH_2

↓

H_5C_2-O-C(=O)- / H_5C_2-O-C(=O)- (pyrrole ring, CH_3, CH_3), N-R-NH_2

↓

(imide ring: C=O, N, C=O fused to pyrrole ring with CH_3, CH_3, N -R)$_n$

Discussion

<u>Synthesis of bis(pyrrole) monomers</u>. The condensation of ethyl diacetyl succinate with amines to give dialkyl esters of N substituted pyrroles proceeds in near quantitative yields to the bis(dicarboethoxy dimethyl pyrrole). The conditions for this reaction are the same as those used by earlier workers to prepare mono dicarboxy pyrrole derivatives (3-8). Upon cooling of the acetic acid reaction solution the product, bis pyrrole derivative, crystallizes out and is recovered in sufficient purity to be used without further purification. Attempts to hydrolyze the bis(dicarboethoxy pyrrole) derivatives in aqueous NaOH proved unsuccessful due to the high degree of insolubility. If, however, the hydrolysis was begun in alcoholic media, the reaction progressed quite rapidly to give the bis(dicarboxy pyrrole). Anhydride formation has been accomplished by refluxing the tetra acid in acetic anhydride. Here again, the product in high purity can be obtained merely by cooling the reaction mixture.

As can be seen from Table I, the yields of the various bis pyrrole monomers are all very high, even though no efforts were made to optimize the yields for these syntheses.

The melting points of some of the derivatives are summarized in Table II. Other analyses performed on these samples (NMR, Infrared, Elemental Analyses) are all consistent with the proposed structures.

Polyimides. The preparation of the polyimides proceeds through poly(amic acid) intermediates. The poly(amic acid) can be isolated providing the temperature of the condensation is properly controlled. At 50°C the structure formed is almost entirely poly(amic acid) while at the solvent reflux temperature almost 100% polyimide is formed.

The presence of the amic acid is also clearly shown by infrared and TGA curves. Figure 1 shows the weight loss of the amic acid structure at 150-230°C. Fifteen percent of the weight of the sample is lost during this period. This loss appears to be of two types, possibly from solvent loss as well as dehydration. Although quantitative estimates have not been made, it appears as though in certain samples, a portion of the weight loss corresponds to the theoretical loss for water in the cyclization (5% for Sample 34b). DSC scans of the poly(amic acid) show a broad endotherm at 160-175°C which probably corresponds with a ring closure to the imide. After that point decomposition can be seen past 400°C. Preparation of the polymer at a higher temperature (160°C) gives almost exclusively polyimide. The results of polymerizations using several dianhydrides and diamines are summarized in Table III. Departure from the theoretical carbon and hydrogen analyses (Table IV) for samples C and D probably indicates incomplete cyclization of the imide structure. No free amic or carboxylic acid functionality is found in the IR spectra of these samples. The TGA curves (Figures 2 & 3) show little or no initial weight loss up to 400°C.

As is common for aromatic polyimides these polymers are all brown to gold in color and are soluble in 97% H_2SO_4. Inherent viscosities as shown in Table IV are low, being less than 0.5 dl/g., indicating low molecular weight. While films could be prepared from the poly(amic acids), which were converted to polyimides on heat treatment (200-250°C for several hours), they were very brittle and broke on handling. This is not surprising in view of the low inherent viscosity (<0.5) for these polymers. Wallach (9) has reported that polyimides with inherent viscosities of less than 1.0 have negligible physical property levels.

Synthesis of N-(amino aryl) pyrrole dicarboxylate monomer. This type of polyimide precursor was prepared by the method of Knorr, who first described these preparations almost 100 years ago. The preparation proceeds readily by refluxing of ethyl diacetyl succinate with

TABLE I

PREPARATION OF PYRROLE MONOMERS

DEDAS (moles)	DIAMINE (moles) $H_2N-R-NH_2$	TETRA ESTER	TETRA ACID	DI-ANHYDRIDE
0.023	R = -(O)-(O)- (0.01)	80%	96%	87%
0.175	-(O)-(O)- (0.08)	93%	94%	95%
0.10	-(O)- (0.05)	82%	100%	98%
0.10	-(O)-O-(O)- (0.05)	92%	100%	70.2%
0.10	(O)- (0.05)	95%	98%	95%
0.10	-(O)-SO_2-(O)- (0.05)	87%	95%	98%

TABLE II

PROPERTIES OF PYRROLE DERIVATIVES

X-C(=O), X-C(=O), CH_3, CH_3 — N - R - N — CH_3, CH_3, C(=O)-X, C(=O)-X

R	X (MPT)°C		
	ESTER	ACID	ANHYDRIDE
-(O)-	140-143	225-230(DECO)	
-(O)-	--	220-225 DECO	194-204 MELT/ DECOMP
-(O)-(O)-	210(230 DECO)	--	--
-(O)-O-(O)-	84-85°	~270 (DECO)	~310°(DECO)
-(O)- SO_2-(O)-	--	--	--

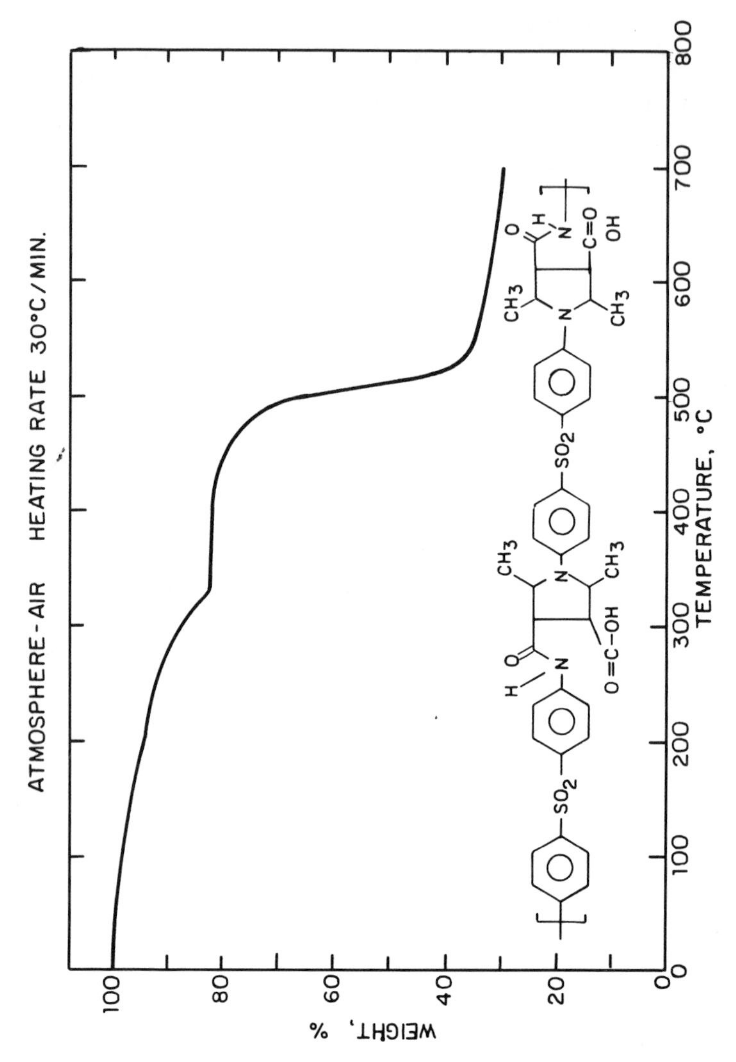

Figure 1. Thermogravimetric analysis of typical polyamic acid.

TABLE III

PREPARATION OF POLYIMIDES

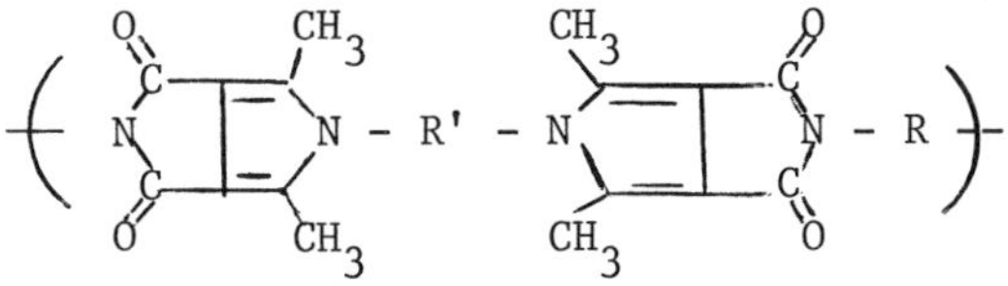

EX NO.	R = R[1]	SOLVENT	TEMP., °C	CONV., %
A	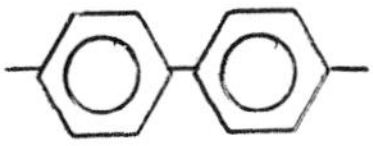 biphenylene	N,N-DIMETHYL ACETAMIDE	160	89
B	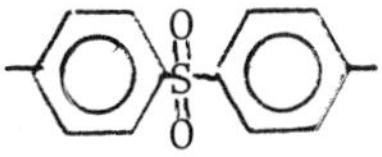 diphenylsulfone	DIMETHYL FORMAMIDE	150	96
C	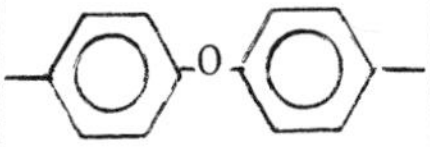 diphenyl ether	N,N-DIMETHYL ACETAMIDE	130	98
D	 phenylene	N,N-DIMETHYL ACETAMIDE	130	87

TABLE IV

PROPERTIES OF POLYIMIDES

Polyimide Analysis			Composition of Polyimides					
			Calculated			Found		
Ex No.	I.V.[a] dl/g	TGA Analysis (5% Wt. Loss)	C	H	N	C	H	N
A	0.29	470°C	76.4	4.5	8.93	76.5	5.72	8.87
B	0.38	455	63.4	3.7	7.41	62.8	4.1	7.23
C	0.48	440	72.7	4.3	8.48	68	4.35	8.05
D	0.38	420	70.5	4.2	11.58	67.7	5.42	10.4

[a]0.1% Conc. in 99% H_2SO_4

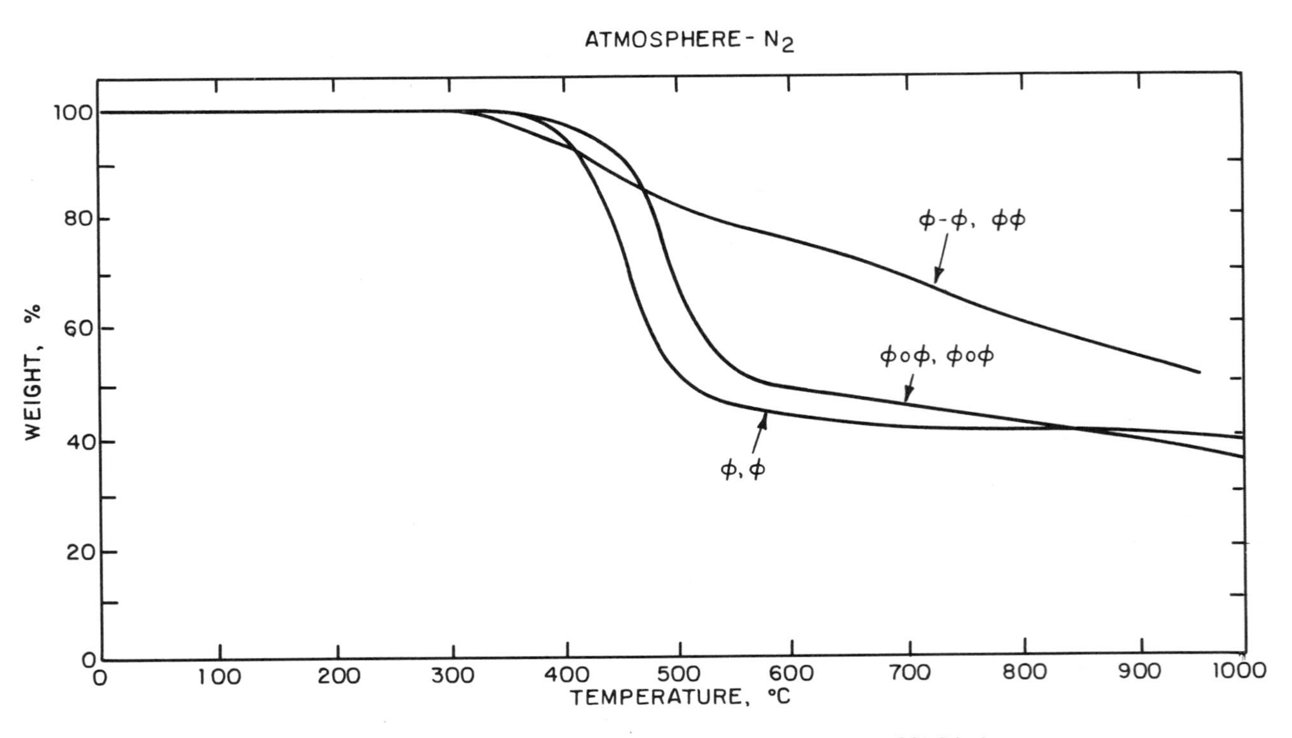

Figure 2. Thermogravimetric analysis of polyimides, heating rate 30°C/min.

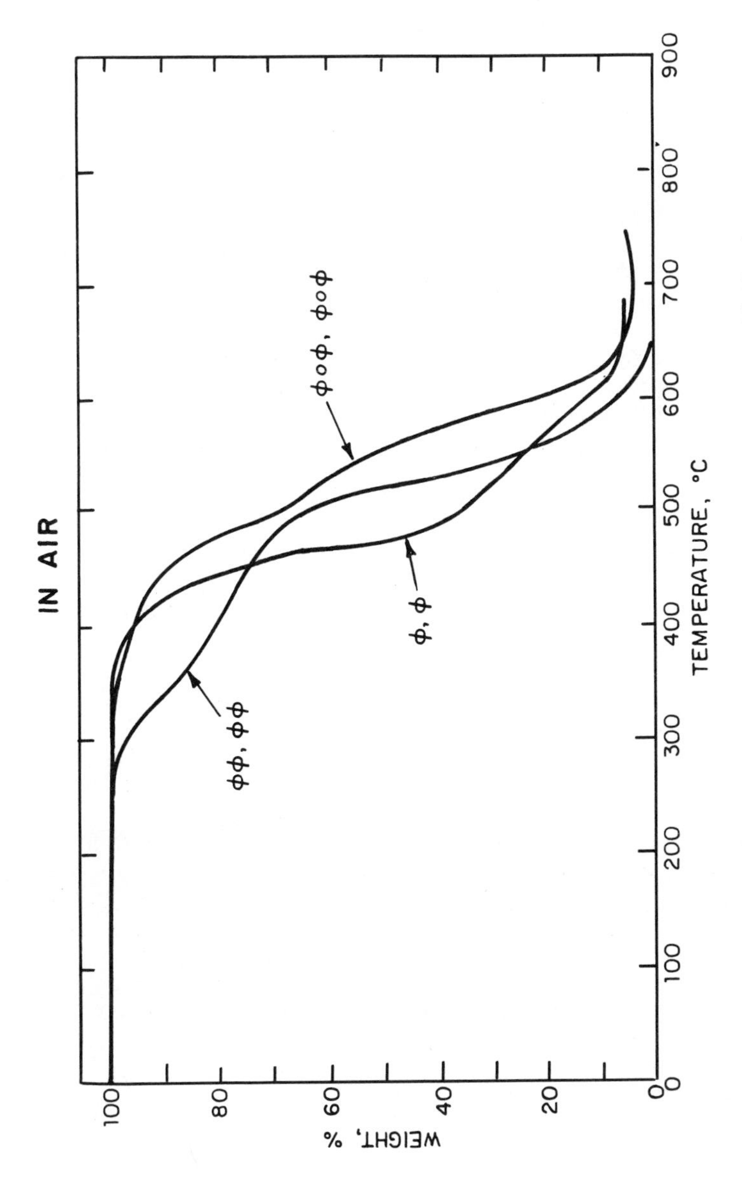

Figure 3. Thermogravimetric analysis of polyimides, heating rate 30°C/min.

aromatic diamine in acetic acid solution. On pouring the acid solution into water and neutralization with sodium bicarbonate the N-(amino aryl)pyrrole diester is recovered as a gold-brown solid. Infrared and NMR spectra are consistent with the proposed structure. This monomer can be converted readily to the polyimide directly by either heating under vacuum at 200-250°C or by refluxing in DMAc solution for a few hours. The infrared spectra of polymers prepared in either manner show only a slight indicating of acid or ester functionality and TGA curves show only minor (<5%) weight losses up to 435°C.

While no variation of the diamine used was carried out in this program, the properties of the polyimide can undoubtedly be affected by use of various diamines. Flexibility could be introduced by longer chain length diamines, and anisotropy may be possible through the use of thermotropic or lyotropic liquid crystalline diamines.

In addition to the polyimide structures discussed, dianhydride monomers have also been found to be useful in the preparation of polypyrroles. Reaction of dianhydrides with tetramines in polyphosphoric acid leads to these thermally stable polymers as indicated below.

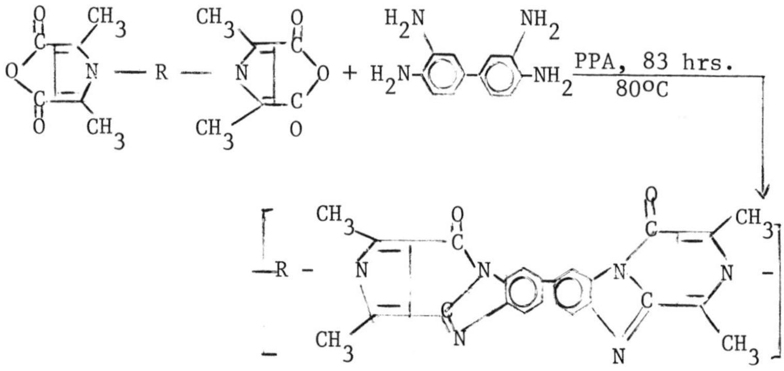

PPA, 83 hrs. 80°C

Condensation of the 4,4' biphenyl bispyrrole dianhydride with 3,4,3',4' tetramino biphenyl gave a dark brown-red polymer which shows exceptional thermal stability in nitrogen as indicated by the TGA curve (Figure 4)--the polymer, maintaining 80% of its original weight up to 700°C. While only a very limited number of experiments were performed in quest of these ladder polymers, it appears that the bis pyrrole dianhydrides do in fact lead to polymers of high thermal stability.

Conclusion. While the mechanical properties of the polyimides prepared in this study are insufficiently high to be of commercial interest, these materials do show

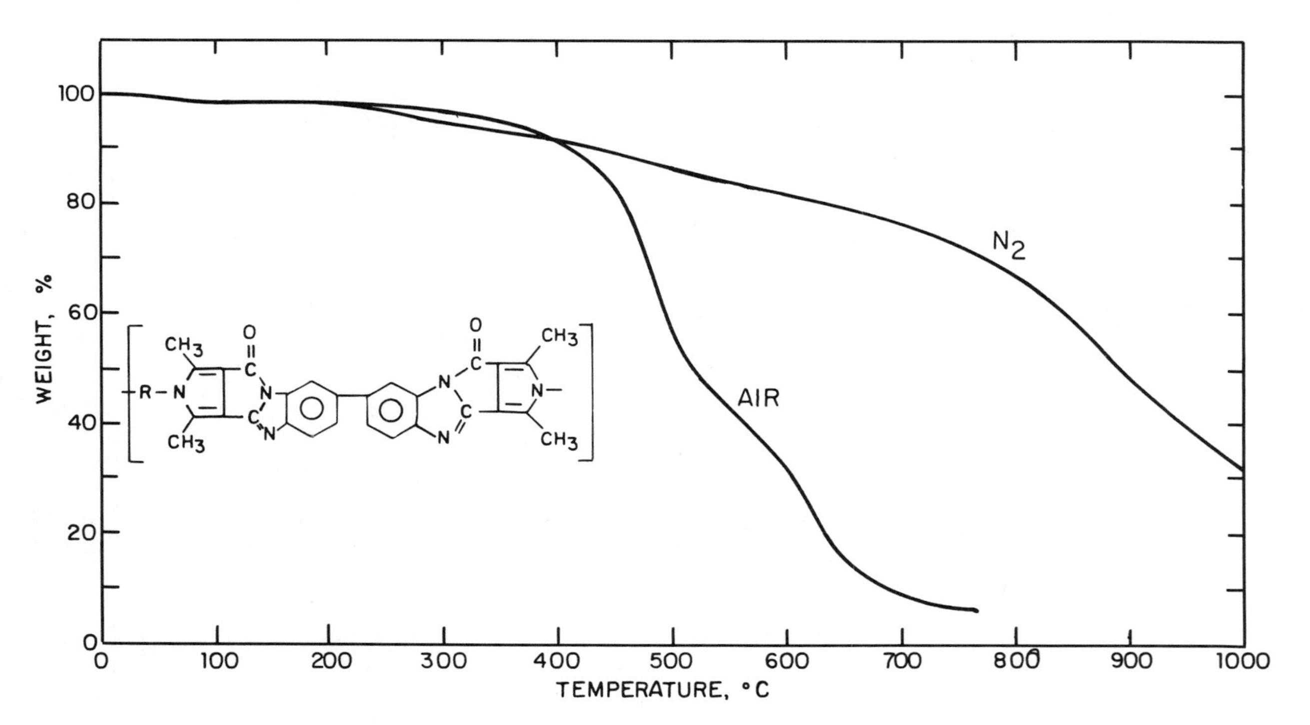

Figure 4. Thermogravimetric analysis of polypyrrone, heating rate 30°C/min.

excellent thermal stability. It seems likely that with additional purification of the monomers and selection of the proper polymerization conditions high molecular weight materials should be easily obtainable. The potentially readily available and economically attractive starting materials for these polyimides, together with the simple reaction conditions and excellent yields obtained in the preparation of the pyrrole precursors, make these attractive monomers for thermally stable polymers.

Experimental

Monomer Preparation. Diethyl diacetyl succinate was prepared by the method of Knorr (8). Condensation of this compound with aromatic diamines was accomplished by the method of Helferich and Pietach (3) to give the bis(3,4 dicarboethoxy-2.5 dimethyl pyrrole) derivatives. These dicarboethoxy pyrroles were hydrolyzed to the acids in aqueous ethanol with NaOH and converted to the dianhydrides by treatment with acetic anhydride at the reflux temperature.

The N-(aminophenyl) pyrrole diester was prepared by a similar procedure using equal molar quantities of ethyl diacetyl succinate and an aromatic diamine.

Preparation of poly(amic acids). Prepared by mixing the dianhydrides with equimolar amounts of diamines for 5 hours at 50°C. A golden brown solid, recovered by precipitation into water showed the presence of both amide and acid functionality in the infrared spectrum.

Preparation of polyimides. A similar procedure was employed with the exception that the dianhydride and diamines were stirred in refluxing DMAc (160°C) for 5 hours. The recovered polymers show no free acid functionality in the infrared spectra, and TGA curves show only small weight losses (<5%) below 470°C.

Literature Cited

1. Sroog, C.E., J. Polym. Sci., Macromolecular Reviews, 1976, 11, 161-208.
2. Meyers, R.A., J. Polym.Sci., 1969, A-1, 7 2757
3. Knorr, L., Annalen der Chemie, 1886, 238, 296
4. Helferich, B. and Pietach, G., J. Prakt Chem. 1962, 17, 213
5. Sargent, D.E., U.S. Pat. 2,453,671, Nov. 9 1948 to American Cyanamid.
6. Sickels, J.P. and Sargent, D.E., U.S. Pat. 2,453,675, Nov. 9, 1948 to American Cyanamid.
7. Sickels, J.P., U.S. Pat. 2,453,676, Nov. 9, 1948 to American Cyanamid.
8. Scholz, T.F., U.S. Pat. 2,453,674, Nov. 9, 1948 to American Cyanamid.
9. Wallach, M.L., Polymer Preprints 8 (1) 656 (April, 1967)

RECEIVED December 7, 1981.

22

Triazine Ring Cross-linked Polyimides and Refractory Materials Derived from Them

LI-CHEN HSU and WARREN H. PHILIPP

NASA, Lewis Research Center, M.S. 106-1, Cleveland, OH 44135

Catalytic cyclopolymerization or polycyclotrimerization of imide oligomers containing nitrile groups with or without graphite fibers is described. The graphite fiber reinforced triaryl-s-triazine ring (TSTR) cross-linked polyimides with ring-chain structures have good mechanical properties at elevated temperatures. On pyrolysis, the TSTR cross-linked polyimides were converted to refractory type materials which are believed to be graphitic type ladder polymers containing some nitrogen in their cyclic structures.

The polyimide class of polymers are known to possess a high degree of thermal stability. They decompose in an inert atmosphere around 500°C and in air about 400°C as indicated by thermogravimetric analysis (1). Because of the great thermal stability of 2,4,6-triphenyltriazine (2) (decomposes above 486°C), copolymerization of this compound with imides should lead to enhanced heat resistant materials in the form of triaryl-s-triazine ring (TSTR) cross-linked (XL) polyimides.

This paper describes the chemistry, reaction mechanism, and method of preparation of these high temperature materials. Their possible structures and refractory properties are also discussed.

Processable Aromatic Nitrile-Modified Imide Precursors

The first step in making TSTR cross-linked polyimides is to synthesize the processable aromatic nitrile-modified imide precursors of the type (I), (II) and (III)

This chapter not subject to U.S. copyright.
Published, 1982, American Chemical Society

(I)

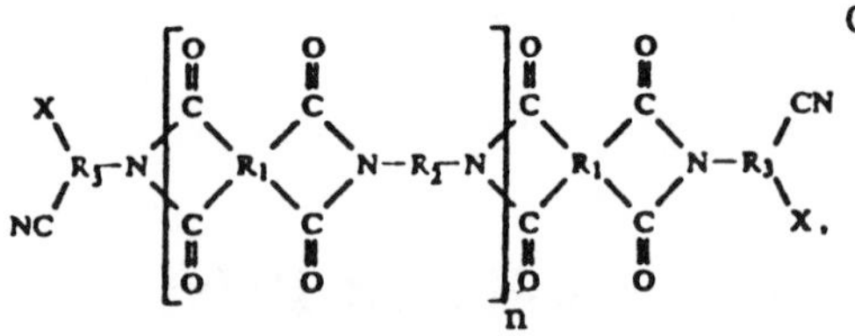

(II)

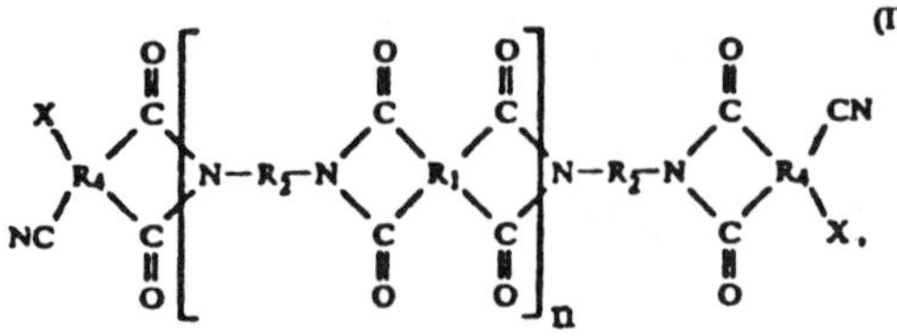

(III)

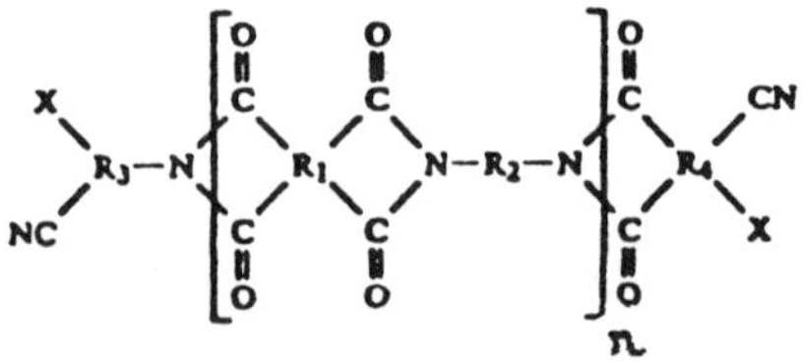

from a tetracarboxylic acid dianhydride or its derivatives, and a diamine having the structural formula,

$$H_2N - R_2 - NH_2$$

with a nitrile cross-linking agent having one of the following structural formulas:

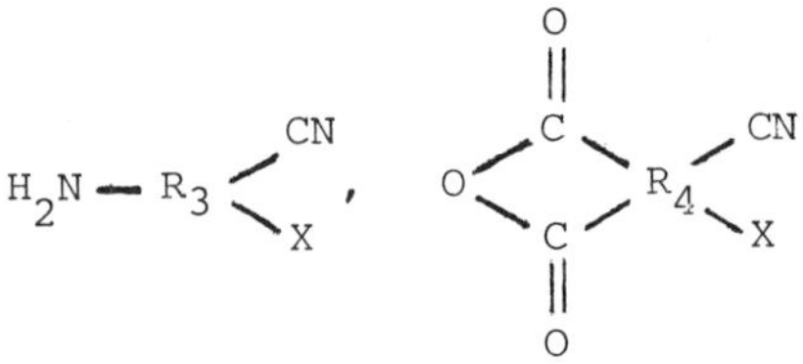

n is an integer usually from 0 to 3. R_1, R_2, R_3, and R_4 are aryl radicals. X is a monoradical including H (3).

Cyclopolymerization or Polycyclotrimerization

Cross-linking involving cyclopolymerization or polycyclotrimerization of the aromatic nitrile in the precursor is accomplished by thermal curing as illustrated by the following reaction:

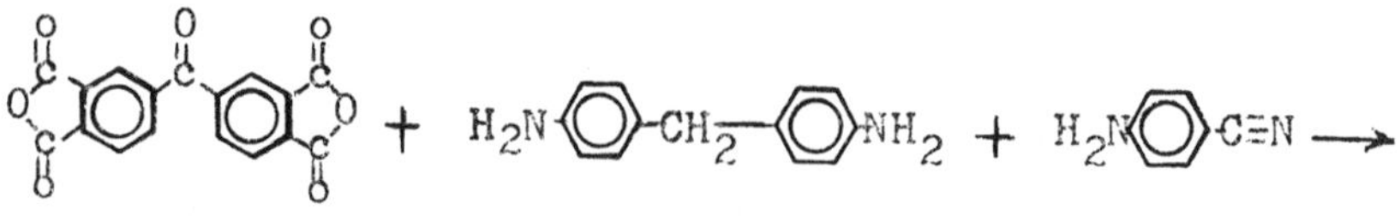

BTDA + MDA + PABN →

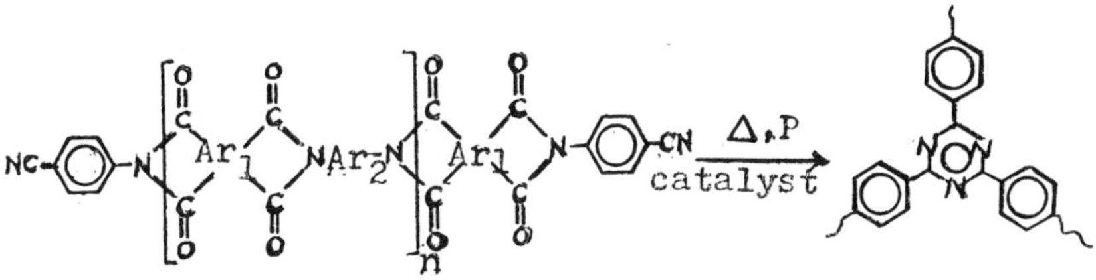

The key to the success of this process is a combination of specific catalysts and optimum reaction conditions. Effective catalysts for trimerization of aromatic nitriles are listed in Table I (4). Optimum reaction conditions for processing the aromatic nitrile-modified imide precursors depend on the chemical structure and characteristic property of the individual precursor of concern. In general, yield of the polymeric products increases with the increase of reaction temperature, pressure, time, and concentration of catalyst within the range of practical experimental limits (5).

Proposed Reaction Mechanism

Fig. 1 illustrates the "Push-Pull" mechanism suggested by H. Smith Broadbent (6) for catalytic trimerization of aromatic nitriles in general. This mechanism should apply to the aromatic nitrile-modified imide precursors as well: (a) The tautomeric catalyst, for example 2-hydroxypyridine, approaches a nitrile molecule and donates a proton to it through a six-member ring environment. (b) This activated molecule (cation) attracts a second nitrile molecule and transfers the positive charge to it. (c) Because of the influence of the resonance-stabilized catalyst anion, a third nitrile molecule will join to complete a triazine ring with the release of the proton rather than forming a linear configuration.

TSTR Polyimides with Ring-Chain Structures

Although the key factors for the preparation of TSTR containing polyimides by cyclopolymerization or polycyclotrimerization of aromatic nitrile-modified imide precursors are the specific catalysts and optimum processing conditions used, tailored design of polymer structure still plays an important role so far as desirable properties and reproducibility are concerned. We believe that cyclopolymerization of aromatic nitrile-modified imide precursors having structures (I), (II), or (III) would result in TSTR cross-linked polyimides with the Ring-Chain Ring type structure shown in Fig.2. If an excess of aromatic nitrile cross-linking agent is used, the TSTR polyimides would have the Ring-Chain Chain type structure shown in Fig.3. Since both types of TSTR containing polyimides are composed of a Ring-Chain

Table I. EFFECTIVE CATALYSTS FOR TRIMERIZING AROMATIC NITRILES

CATALYST	FORMULA
ORGANIC SULFONIC AND SULFINIC ACIDS	
p-TOLUENE SULFONIC ACID	$CH_3C_6H_4SO_2OH \cdot H_2O$
BENZENE SULFONIC ACID	$C_6H_5SO_2OH \cdot H_2O$
1-NAPHTHALENE SULFONIC ACID	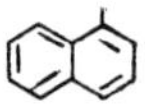 $SO_2OH \cdot 2H_2O$
2-NAPHTHALENE SULFONIC ACID	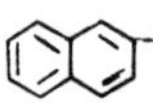 SO_2OH
BENZENESULFINIC ACID	SO_2H
ORGANIC PHOSPHONIC AND PHOSPHINIC ACIDS	
TRICHLOROMETHYL PHOSPHONIC ACID	$CCl_3PO(OH)_2$
PHENYL PHOSPHONIC ACID	$C_6H_5PO(OH)_2$
PHENYL PHOSPHINIC ACID	$C_6H_5PO(OH)H$
METALLIC ACETYLACETONATES	
FERRIC ACETYLACETONATE	$(C_5H_7O_2)_3Fe$
ZINC ACETYLACETONATE	$(C_5H_7O_2)_2Zn \cdot 2H_2O$
HETEROCYCLIC AMIDES	
2-HYDROXYPYRIDINE	 N —OH ⇌ NH =O
2-PYRIMIDINOL	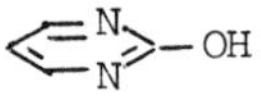 N N —OH
2-QUINOLINOL	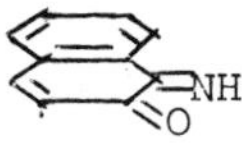 NH O

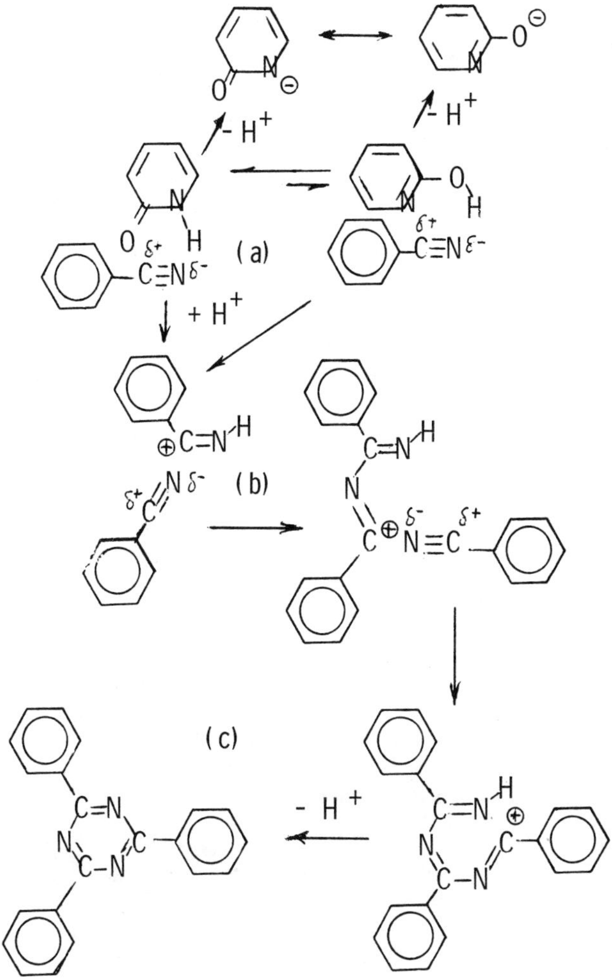

Figure 1. Proposed PUSH–PULL mechanism for catalytic trimerization of aromatic nitriles.

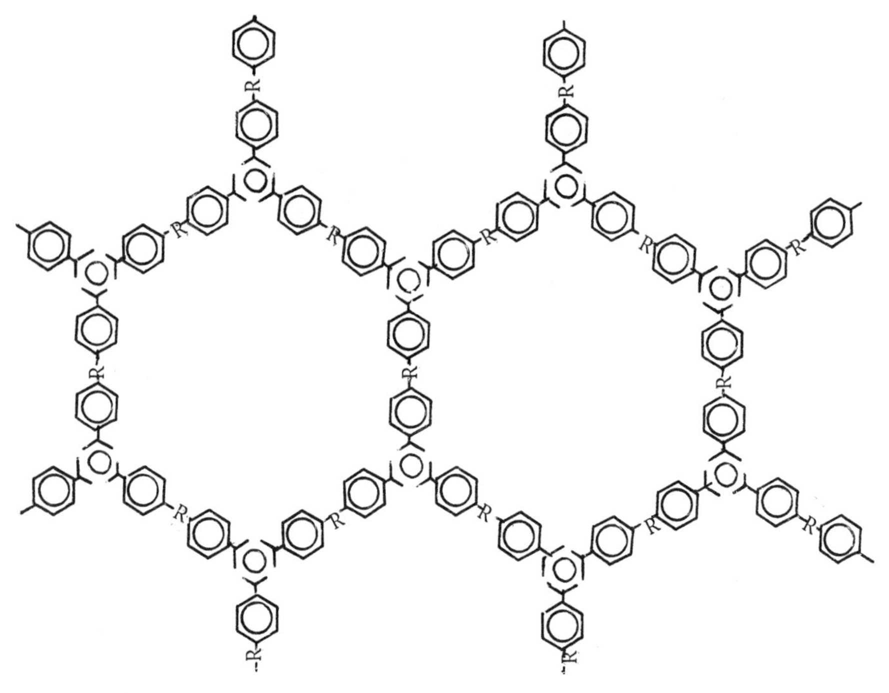

Figure 2. Triaryl-s-triazine ring polymer with ring–chain ring structure. Key: ⌬ *, triazine ring; R, imide oligomer.*

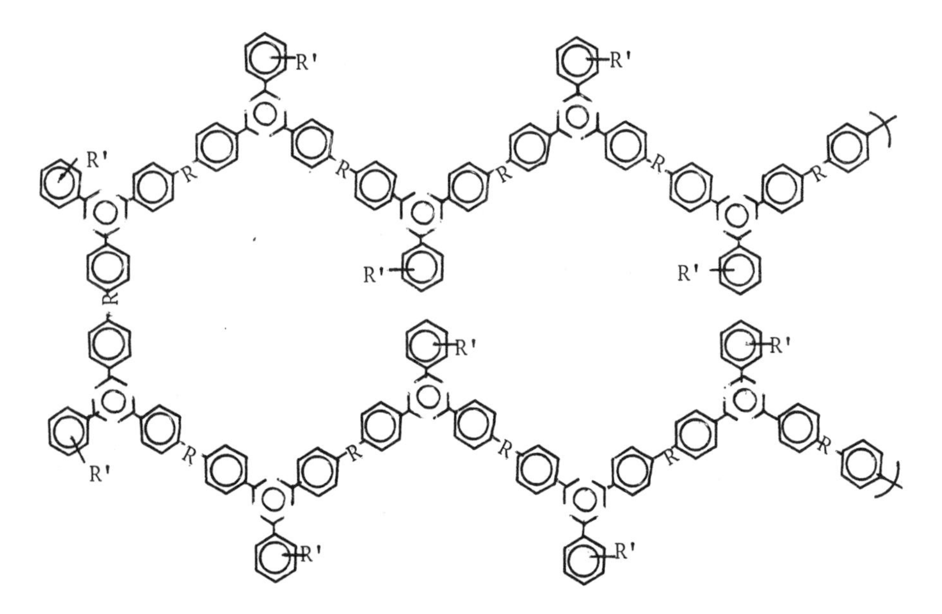

Figure 3. Triaryl-s-triazine ring polymer with ring–chain chain structure. Key: ⌬ *, triazine ring; R′, NH_2 etc.*

backbone, they are expected to be extremely stiff. Those having a predominant Ring-Chain Ring type structure would be rigid or brittle, whereas those having more Ring-Chain Chain type structures are expected to be more flexible.

Experimental

Materials. 3,3',4,4'-Benzophenonetetracarboxylic dianhydride (BTDA), 4,4'-methylenedianiline (MDA), p-aminobenzonitrile (PABN), and p-toluenesulfonic acid monohydrate (PTSA) were of commercial sources and used as received. Graphite fiber used as the reinforcing filler was Hercules HMS grade.

Synthesis of TSTR Cross-linked Polyimides (3). In a 100-milliliter glass flask, 9.27 grams (0.00302 mole) of BTDA was heated with 29 grams of anhydrous methanol until a clear solution was obtained. After cooling, 4.02 grams (0.00202 mole) of MDA, 2.36 grams (0.00200 mole) of PABN, and about 0.1 gram (0.25 mole percent on the basis of the nitrile content) of PTSA were introduced with stirring until all was dissolved. This clear solution of monomeric reactants was transferred to a 45-milliliter stainless steel pressure vessel equipped with a 1000 psi pressure gauge, a 1000 psi burst disk, and a needle valve. The vessel without the lid was heated gradually to about 100°C so that most of the methanol and some of the by-product (methanol and water) would evaporate off. The reaction vessel was then installed with the lid and heated under reduced pressure at a temperature up to about 200°C until no more methanol and/or water came off. The vessel was then filled with dry nitrogen gas and adjusted to a pressure of 750 psi and heated at 350-400°C for 8 hours. The polymeric product presumed to be TSTR cross-linked polyimide was found to be very hard and dark brown in color. It did not melt when heated up to 340°C. Its KBr infrared spectrum showed the absence of the nitrile band at 2240 cm^{-1}, but the broad bands at 1520 and 1380 cm^{-1}, which are characteristic s-triazine bands (7), in addition to imide bands at 1795, 1755, 1735 cm^{-1} appeared.

Another run was made with nearly the same procedure and composition except that 2.5 grams (0.00210 mole) of PABN was used. The resulting polymeric product was also dark brown in color, and did not melt up to 340°C; it appeared to be slightly softer than the product of the previous run. Comparison of the KBr infrared spectra of the two materials showed no significant difference.

Graphite Fiber Reinforced TSTR Polyimide Composites (3). The clear solutions of monomeric reactants prepared according to the method described in the preceding section were used to make prepregs by drum winding and impregnating Hercules HMS graphite fiber with a resin/fiber content of about 60/40 by weight. The prepregs were heated from 50 to 120°C for a couple of hours to reduce the solvent content to less than 10 percent by weight. Plies of partially imidized prepregs were then stacked between aluminum foil and heated in an oven at 200 to 250°C for several hours for complete imidization. The imidized prepregs were then placed in a mold and cured at 350°C under pressures in the range of 200 to about 1000 psi for several hours. The prepregs were further post cured in an oven at 316°C for an additional sixteen hours.

The flexural strengths (three-point method) of these two unidirectional HMS graphite fiber reinforced TSTR polyimide composites with or without post curing were measured with a universal Instron machine at room temperature and at 316°C. These results are shown in Table II.

Table II. FLEXURAL STRENGTH OF GRAPHITE FIBER/POLYIMIDE COMPOSITES[a]

Specimen	Flexural Strength, psi		
		316°C (600°F)	
	Room Temperature	No Post Cure	Post cure (16 hrs. 316°C)
TSTR-XL-PI/HMS			
1	155,000	148,000	167,500
2	130,200	145,700	157,500

[a]Resin/fiber = ~40/60 by weight.

Bunsen Burner Burning Test. A sample of graphite fiber reinforced TSTR polyimide composite was exposed to a Bunsen burner flame at about 1100 to 1250°C (Fig. 4). After 20 minutes in the flame, the composite still retained its integrity. No release of graphite fiber was observed (Fig. 5). A sample of graphite fiber reinforced composite made from PMR-15 polyimide (8) under the same conditions disintegrated within a couple of minutes with only graphite fibers remaining as also shown in Fig. 5.

Results and Discussion

Property-Structure Relationship. Table II shows the flexural strengths of the HMS graphite fiber reinforced TSTR polyimide composites at room temperature and at 316°C with or without post curing. The data clearly indicate that the graphite fiber reinforced composites prepared from aromatic nitrile-modified imide oligomers exhibited very good retention of flexural strength during short time exposure in air at 316°C. More important, the data also indicate that the flexural strength of the composites improved after a 16-hour post cure at 316°C. Apparently this resulted from an increase in TSTR formation during post cure.

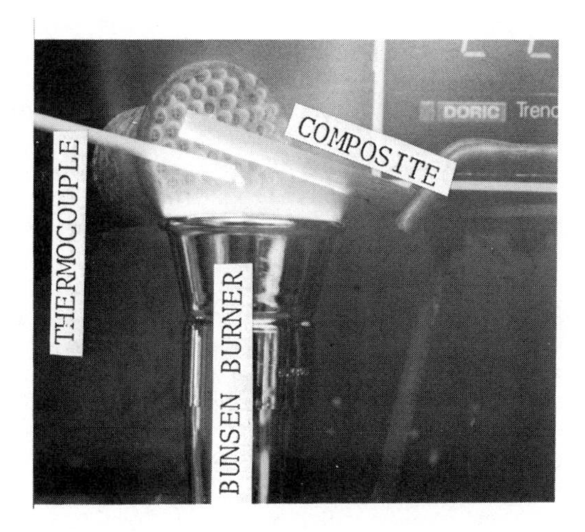

Figure 4. Exposure of triaryl-s-triazine ring polyimide composite to a bunsen burner flame (1100–1250°C).

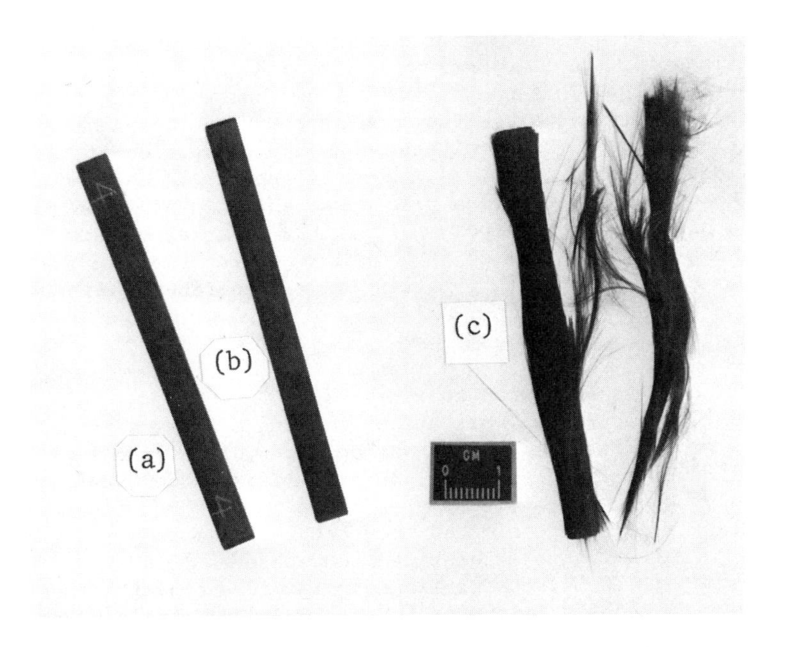

Figure 5. Bunsen burner burning test of graphite fiber reinforced polyimide composites. Key: a, TSTR-XL-PI/GrF specimen before test; b, TSTR-XL-PI/GrF specimen after test; and c, PMR-15-PI/GrF specimen after test.

The composites would reach maximum rigidity and thermal stability when all the TSTRs are completed in the cross-linking mode. Complete cross-linking via triazine ring formation results in a Ring-Chain Ring structure shown in Fig. 2. Excess of nitrile cross-linking agent above that required to endcap the imide oligomer takes part in polycyclotrimerization leading to chain termination. If the excess is sufficient just to block cross-linking, then only chain extension occurs resulting in the Ring-Chain Chain structure shown in Fig. 3. Thus the cross-linking density of the TSTR should decrease as the amount of excess nitrile is increased. A decrease in cross-linking density of a polymer should increase its flexibility. This is illustrated by a comparison of the flexural strengths of specimens 1 and 2 (Table II). Both specimens were derived from the same precursor except that in specimen 1, a 5% excess of PABN was used. As anticipated, the flexural strength of this specimen is about 10,000 psi higher than the more completely cross-linked specimen 2. Thus flexibility of an otherwise rigid material can be adjusted by optimizing cross-linking density which in turn can be controlled by the amount of excess nitrile added prior to curing.

Refractory Type Polymeric Materials. Pyrolysis of our TSTR cross-linked polyimide at about 1000°C leads to the formation of a refractory type material possibly of a graphitic ladder polymeric structure. These ladder polymers have previously been synthesized from aryl ketones with nitrogen containing groups on the ring (9). This phenomenon was not observed with conventional cross-linked polyimides such as polybisnorbornenylimide which disintegrated on pyrolysis; thus the conversion of TSTR polyimides to refractory type materials appears to be associated with the nitrile or triazine ring component. In addition to being a high temperature polymeric matrix material by itself, TSTR cross-linked polyimide is a promising precursor for the preparation of a refractory matrix for carbon fibers.

Conclusion

We note the following conclusions:

(1) Catalytic cyclopolymerization of imide oligomers containing nitrile groups is a convenient method for synthesizing TSTR cross-linked polyimides. The TSTR cross-linking density can be controlled by addition of excess nitrile.

(2) Graphite fiber reinforced TSTR cross-linked polyimides have good flexural strength at elevated temperatures (316°C).

(3) Pyrolysis of TSTR cross-linked polyimides converts them to refractory type materials which show promise as matrix materials for carbon fibers.

Literature Cited

1. Markle, R.A. in Carver, J.K.; Tess, R.W. Eds; " Applied Polymer Science "; American Chemical Society, Washington,D.C. 1975; p.560.
2. Blake, E.S.; Hammann, W.C.; Edwards, J.W.; Richard, T.E.; Ort, M.R. J. Chem. Eng. Data 1961, 16 , 87.
3. Hsu, Li-Chen. U.S.patent 4,159,262, June 26, 1979.
4. Hsu, Li-Chen. U.S.patent 4,061,856, December 6, 1977.
5. Hsu, Li-Chen. NASA TMX-71525; 1974.
6. Broadbent, H. Smith. Brigham Young University. Invited lecture on "Catalysis" at NASA-Lewis Research Center, Sep.5, 1969. Private communication.
7. The Aldrich Library of Infrared Spectra., "T4605-1 s-Triazine." Aldrich Company, p.1202.
8. Serafini, T.T. NASA TM-79039; 1979.
9. Cassidy, P.E.; "Thermally Stable Polymers--Syntheses and Properties," Marcel Dekker Inc., New York, N.Y. 1981, p.288.

RECEIVED December 27, 1981.

23

Heat and Water Resistant Polyphenyl-*as*-triazines from Terephthalamidrazone

FENG-CAI LU, LAN-MIN XING, and YU-LAN WANG

Institute of Chemistry, Academia Sinica, Beijing, China

Six new relatively high molecular weight polyphenyl-as-triazines PPT (I)-(VI) were synthesized from terephthalamidrazone and different aromatic dibenzils. These polymers were characterized by thermogravimetric analysis, differential thermal analysis, elemental analysis and infra-red spectroscopy. All properties were determined by using specimens in the form of films and varnished wire. Experimental results showed that these polymers exhibited good thermal oxidative stability, hydrolytic stability, chemical stability and tensile strength at room temperature with the exception of polymer (VI). The films of polymer (I)-(V) suffered 4.30-5.31% weight loss after isothermal aging in air at 300°C for 200 hours. There were no changes in weight when these films were treated in water at 200°C for 30 hours. Wire coated with polymer (I) at room temperature in water has a resistance of 10^{11} - 10^{12} ohms. After it was boiled in 40% sodium hydroxide solution for as long as 24 hours or submerged in water at 200°C for 100 hours, no change in insulation was detected.

Six new relatively high molecular weight polyphenyl-as-triazines PPT (I)-(VI) were prepared by solution cyclopolycondensation of terephthalamidrazone with various dibenzils as shown in Eq (I).

The polyphenyl-as-triazines in this paper are relatively high molecular weight polymers containing 1,2,4-triazines and aromatic nuclei in the main polymer chains. η_{inh} were between 1.39-2.16 dl/g (0.5% solution in cresol at 25°C) due to high purity of the two monomers concerned. A yellow, transparent, flexible film was prepared from each of these polymer solutions and varnished wire was obtained by coating the metallic wire

0097-6156/82/0195-0297$06.00/0
© 1982 American Chemical Society

with the polymer solution. With the exception of η_{inh}, all properties of these polymers were measured on such films and varnished wire.

$$n\, H_2N{-}HN{-}\overset{HN}{\overset{\|}{C}}{-}C_6H_4{-}\overset{NH}{\overset{\|}{C}}{-}NH{-}NH_2 + n\, C_6H_5{-}\overset{O}{\overset{\|}{C}}{-}\overset{O}{\overset{\|}{C}}{-}Ar{-}\overset{O}{\overset{\|}{C}}{-}\overset{O}{\overset{\|}{C}}{-}C_6H_5$$

$$\longrightarrow \left[\text{(phenyl-1,2,4-triazine)}{-}C_6H_4{-}\text{(phenyl-1,2,4-triazine)}{-}Ar{-} \right]_n + 4nH_2O \qquad (1)$$

Ar = I, II, III, IV, V, VI

Pure terephthalamidrazone which can be used as one of monomers in the synthesis of six polymers were prepared by an improved method. It was found that the oxidizing agent -- selenium oxide (SeO_2) commonly used in the synthesis of dibenzils was not satisfactory. In our previous work [1], selenium oxide was replaced by dimethyl sulfoxide (DMSO) in the presence of hydrobromic acid. This system was proved to be a simplified and less expensive procedure, which gave product of high purity.

Wrasidlo and Augl [2] reported that 1,4-bis(phenylglyoxalyl) benzene was prepared from 1,4-phenylene diacetic acid by reacting with thionyl chloride, followed by Friedel-Crafts reaction, and oxidation by SeO_2. However, we easily prepared 1,4-bis(phenylglyoxalyl)benzene by using 1,4-dicyanobenzene as the starting material which reacted with Grignard reagent of benzyl chloride and was then oxidized by dimethyl sulfoxide in the presence of hydrobromic acid. The other aromatic diacetyl compounds were also oxidized by dimethyl sulfoxide. In the literature [3], the solvent used for the reaction (4) was ether. In our previous work, mixed solvent (benzene and tetrahydrofuran) was used. As a result, this improvement offered the advantage for large scale preparation.

The routes used to prepare terephthalamidrazone and 1,4-bis-(phenylglyoxalyl)benzene are shown in Eq. (2)-(5):

$$NC{-}C_6H_4{-}CN + C_2H_5OH \longrightarrow C_2H_5O{-}\overset{HN}{\overset{\|}{C}}{-}C_6H_4{-}\overset{NH}{\overset{\|}{C}}{-}OC_2H_5 \qquad (2)$$

$$C_2H_5O{-}\overset{HN}{\overset{\|}{C}}{-}C_6H_4{-}\overset{NH}{\overset{\|}{C}}{-}OC_2H_5 + H_2NH_2N \longrightarrow H_2N{-}HN{-}\overset{HN}{\overset{\|}{C}}{-}C_6H_4{-}\overset{NH}{\overset{\|}{C}}{-}NH{-}NH_2 \qquad (3)$$

$$NC-C_6H_4-CN + C_6H_5CH_2MgCl \longrightarrow C_6H_5CH_2-\overset{O}{\overset{\|}{C}}-C_6H_4-\overset{O}{\overset{\|}{C}}-CH_2C_6H_5 \quad (4)$$

$$C_6H_5CH_2-\overset{O}{\overset{\|}{C}}-C_6H_4-\overset{O}{\overset{\|}{C}}-CH_2C_6H_5 \longrightarrow C_6H_5-\overset{O}{\overset{\|}{C}}-\overset{O}{\overset{\|}{C}}-C_6H_4-\overset{O}{\overset{\|}{C}}-\overset{O}{\overset{\|}{C}}-C_6H_5 \quad (5)$$

Experimental

Monomers. Diethyl terephthalimidate was prepared by mixing 7.68 g (0.06 mole) of 1,4-dicyanobenzene with 9 ml absolute ethyl alcohol and 330 ml anhydrous 1,4-dioxane and by passing dry hydrogen chloride gas into the solution at 0-5°C until saturation. Then the reaction vessel was closed, and maintained at this temperature for 5 days to insure completion of reaction. The resulting white solid was filtered, washed and air-dried to yield 15.5 g (90% yield) of diethyl terephthalimidate dihydrochloride. After neutralizing this dihydrochloride with 15-18% potassium hydroxide, the product was filtered, washed with water, and dried. The white solid (10.5 g) was then treated with petroleum ether, and diethyl terephthalimidate was formed as white crystalline plates, mp. 110-111°C (Lit. [4], mp. 110°C).

Analysis: $C_{12}H_{16}N_2O_2$
Calcd: C, 65.44; H, 7.32; N, 12.72%
Found: C, 65.45; H, 7.30; N, 12.69%

Terephthalamidrazone was prepared by reacting 6.6 g (0.03 mole) of diethyl terephthalimidate dissolved in 50 ml absolute ethyl alcohol with 5.01 g (90%, 0.09 mole) of hydrazine hydrate. This reaction took place in a corked vessel at room temperature, and light yellow solid gradually settled down. The yellow solid was then filtered, washed with anhydrous ether, and dried to yield 5.06 g (87.8% yield) of crude product. It was recrystallized from pyridine under N_2. After drying, 3.5 g (60% yield) of terephthalamidrazone which decomposited between 225-227°C (Lit. [5], Decomposition point: 223-225°C) was yielded.

Analysis: $C_8H_{12}N_6$
Calcd: C, 49.98; H, 6.29; N, 43.72%
Found: C, 50.36; H, 6.31; N, 43.36%

1,4-Bis(phenylglyoxalyl)benzene was prepared as previously reported [1].

1,4-Bis(phenyl acetyl)benzene was obtained from the reaction of 1,4-dicyanobenzene with Grignard reagent of benzyl chloride, mp. 175-177°C (54.4% yield), and recrystallized from ethyl acetate to yield white or light yellow crystals, mp. 178-180°C.

Analysis: $C_{22}H_{18}O_2$
Calcd: C, 84.00; H, 5.94%
Found: C, 83.90; H, 5.93%

Finally, 1,4-bis(phenyl acetyl)benzene was readily oxidized by dimethyl sulfoxide in the presence of hydrobromic acid to form 1,4-bis(phenylglyoxalyl)benzene, which was recrystallized from n-butyl alcohol, mp. 126-127°C (Lit. [6], mp. 125-126°C).

Analysis: $C_{22}H_{14}O_4$
Calcd: C, 77.18; H, 4.12%
Found: C, 77.06; H, 4.13%

In the synthesis of dibenzils (II,III,VI), the diacetyl compounds were prepared following a known procedure [3] and were then oxidized by dimethyl sulfoxide.

Ar =

II III IV

The dibenzils (IV, V) were prepared according to Relles et al. [7].

The properties of all the dibenzils are shown in Table I.

Model Compounds. Prior to polymer synthesis, the model compound I and II as shown in Eq. (6) were prepared seperately by refluxing (4 hours) 0.192 g (0.001 mole) of terephthalamidrazone with 0.42 g (0.002 mole) of benzil or 0.572 g (0.002 mole) of 4-phenylbenzil in the presence of anhydrous ethyl alcohol. The cooled solutions gave quantitative yield of pale yellow solids which were recrystallized from dimethyl formamide respectively, mp. 320-321°C; 329-332°C.

Analysis: $C_{36}H_{24}N_6$
Calcd: C, 79.98; H, 4.47; N, 15.54%
Found: C, 80.01; H, 4.44; N, 15.62%

Analysis: $C_{48}H_{32}N_6$
Calcd: C, 83.22; H, 4.65%
Found: C, 82.72; H, 4.55%

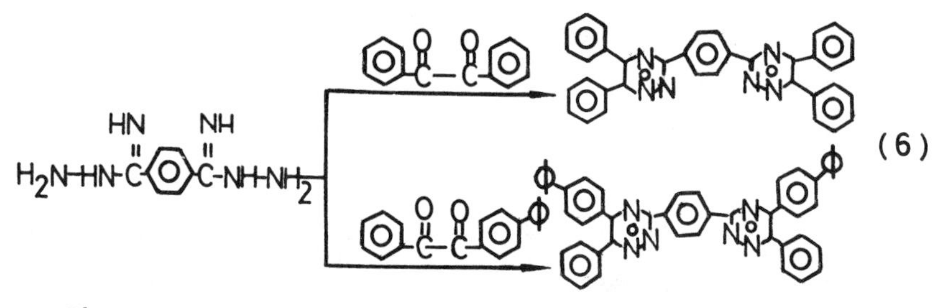

The IR spectra showed absorptions of 1,2,4-triazine at 1490, 1445, 1385-1390 and 1020 cm^{-1} [8] and are shown in Fig 1,2.

Polymers. The polymers were prepared by the following procedure. Stoichiometric quantities of the two monomers in fine powder forms were accurately weighed and mixed, and cresol (at 10-12% solid level w/w) was added immediately. The mixture was stirred vigorously at ambient temperature for 2 hours (PPT (I), PPT (II)) or at 70-80°C under N_2 for 4 hours (PPT (III)-PPT

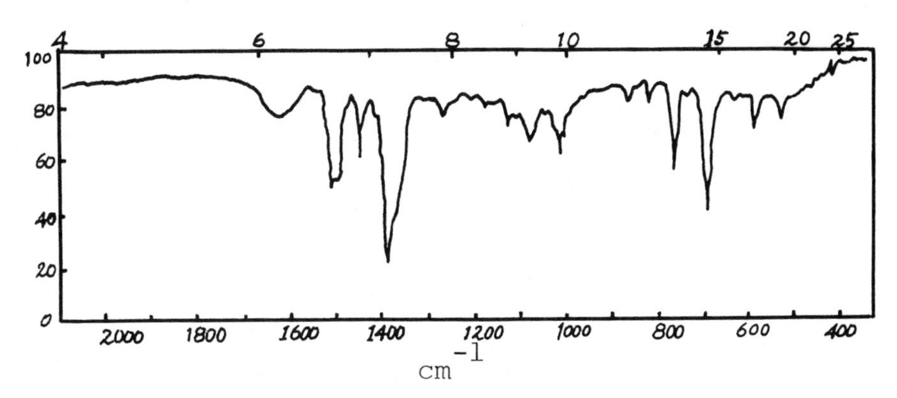

Figure 1. IR spectrum of Model Compound I.

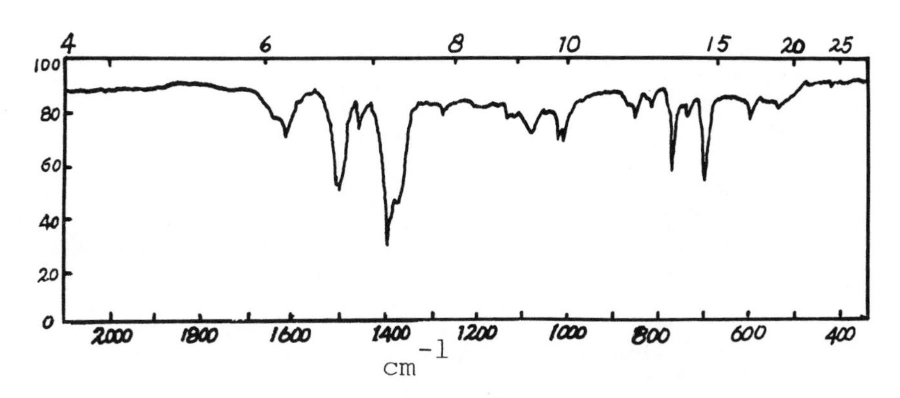

Figure 2. IR spectrum of Model Compound II.

Table I Characterizations of Dibenzils

No	Dibenzils Ar=	M.P °C	Formula	Elemental analysis C	H %*
I	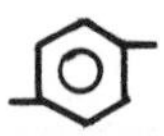	126–27(1) 125–26(6)	$C_{22}H_{14}O_4$	77.06 (77.18)	4.13 (4.12)
II		205-06 205-06(3)	$C_{28}H_{18}O_4$	80.83 (80.37)	4.37 (4.43)
III	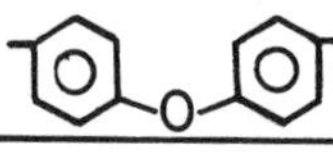 O	106-07 106-07(3)	$C_{28}H_{18}O_5$	77.57 (77.41)	4.21 (4.18)
IV	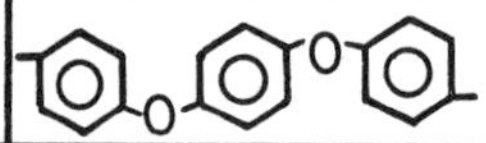 O O	159-60 159-60(7)	$C_{34}H_{22}O_6$	77.42 (77.60)	4.19 (4.20)
V	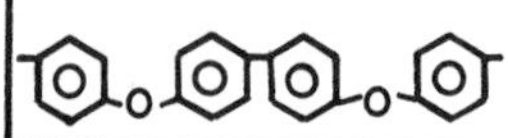 O O	168-69 168-69(7)	$C_{40}H_{26}O_6$	80.04 (79.70)	4.13 (4.30)
VI	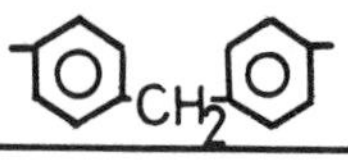 CH_2	145 -46 145-44.5(3)	$C_{29}H_{20}O_4$	80.74 (80.54)	4.67 (4.66)

* Theoretical values in parenthesis

(VI)). After 24 hours, η_{inh} were determined. A portion of solution was poured into ethyl alcohol in a Waring Blender to precipitate a yellow fiber or powder which was washed thoroughly with ethyl alcohol, and dried below 100°C in air for several hours. Films were prepared from the polymer solution by casting a portion of the solution onto a glass plate, followed by drying at 80-90°C. The films were cured at 300°C under N_2 for 1.5 hours. Varnished wire was obtained by coating the polymer solution on metallic wire and curing by a rolling machine in air.

Results and Discussion

Characterizations of the solution and isolated polymer are given in Table II. Elemental analysis agreed well with the theoretical values. Each of the solution was coated onto a sodium chloride plate as a film as described above. The IR spectra exhibited strong absorptions at the same wave numbers as those of model compounds and are shown in Fig. 3-8.

Thermal Stability. The thermal stabilities of polymers PPT (I)-PPT(VI) were determined by thermogravimetric analysis (TGA). A thermogram is shown in Fig. 9-11. All polymers underwent a two step decomposition in air. The first degradation occurred at about 400°C whereas the second degradation occurred at about 500°C. It was shown that the decomposition temperatures of these polymers determined under N_2 were generally lower than those in air.

Table II Characterization of PPT (I)-(VI)

Polymer No	Chemical structure	η_{inh} dl/g [1]	Tg °C [2]	Formula	Elemental analysis [3] C	H	N %
I	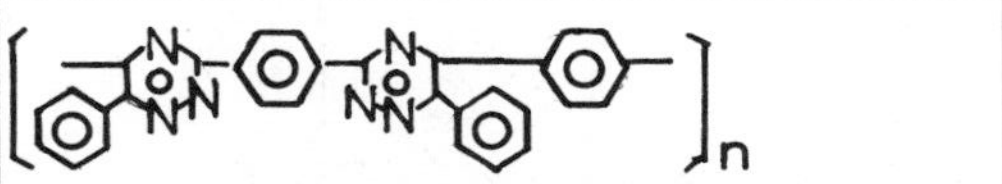	1.93	320	$C_{30}H_{18}N_6$	77.94 (77.94)	4.14 (3.93)	17.92 (18.16)
II	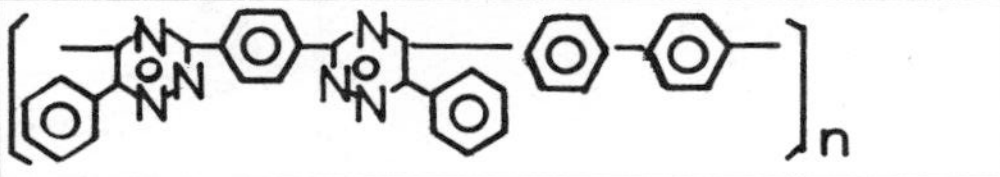	2.16	350	$C_{36}H_{22}N_6$	79.90 (80.27)	4.37 (4.12)	15.08 (15.60)
III	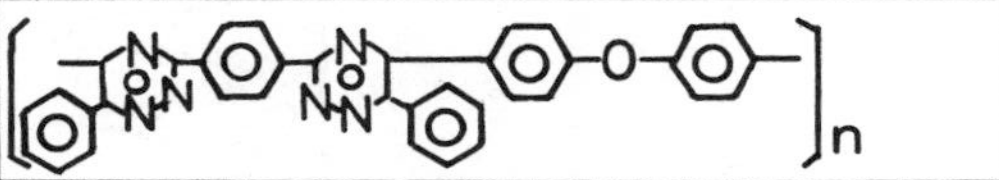	1.39	282	$C_{36}H_{22}N_6O$	77.77 (77.96)	4.13 (3.99)	14.70 (15.15)
IV	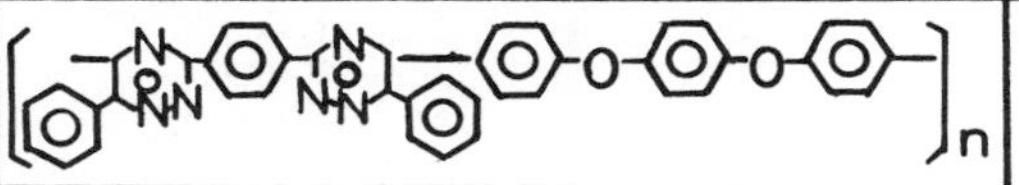	1.56	260	$C_{42}H_{26}N_6O_2$	78.33 (78.00)	3.97 (4.05)	12.75 (12.99)
V	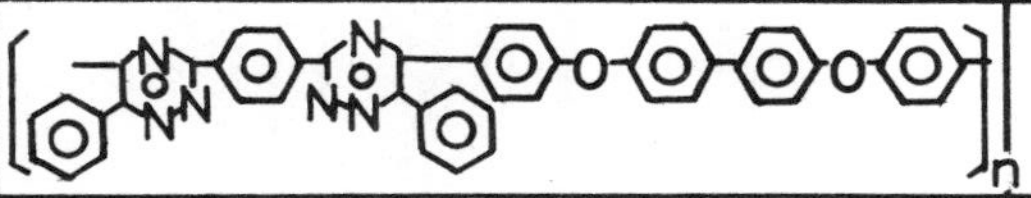	1.75	272	$C_{48}H_{30}N_6O_2$	79.43 (79.76)	4.09 (4.18)	11.51 (11.62)
VI	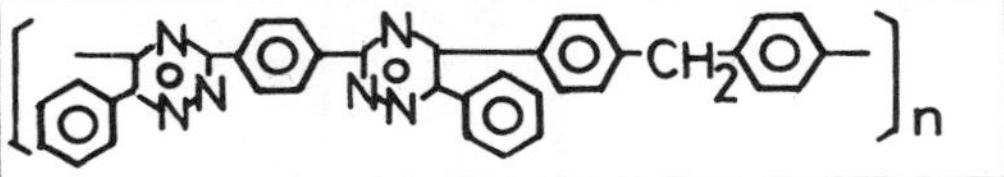	1.39	265	$C_{37}H_{24}N_6$	80.00 (80.41)	4.53 (4.37)	14.67 (15.21)

1) Inherent viscosity (0.5% in cresol at 25°C)
2) Tg was determined by DTA, the rate of heating: 10°C/min.
3) Theoretical values in parenthesis.

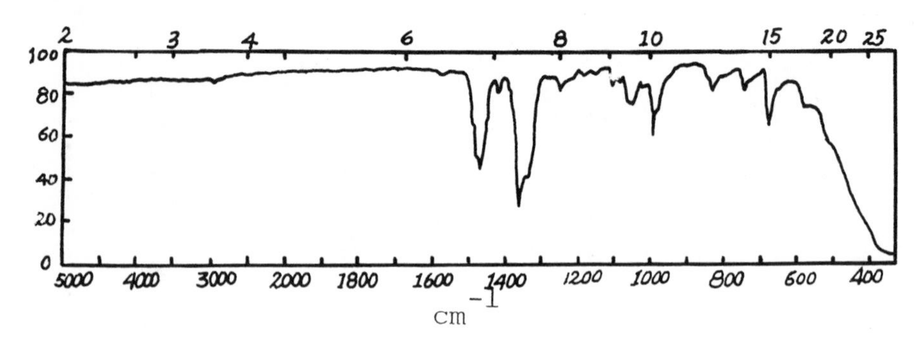

Figure 3. IR spectrum of PPT I.

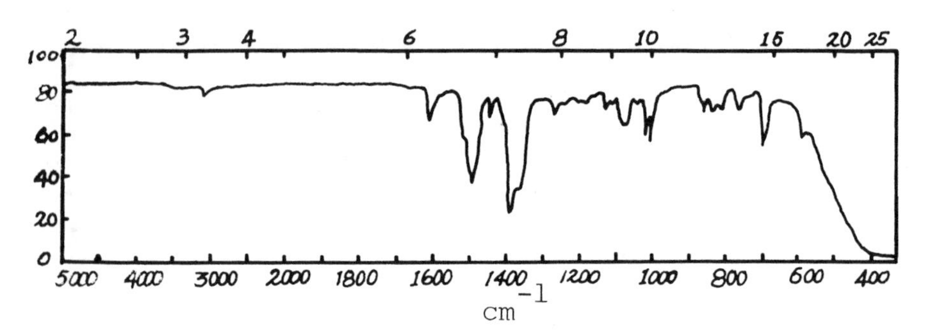

Figure 4. IR spectrum of PPT II.

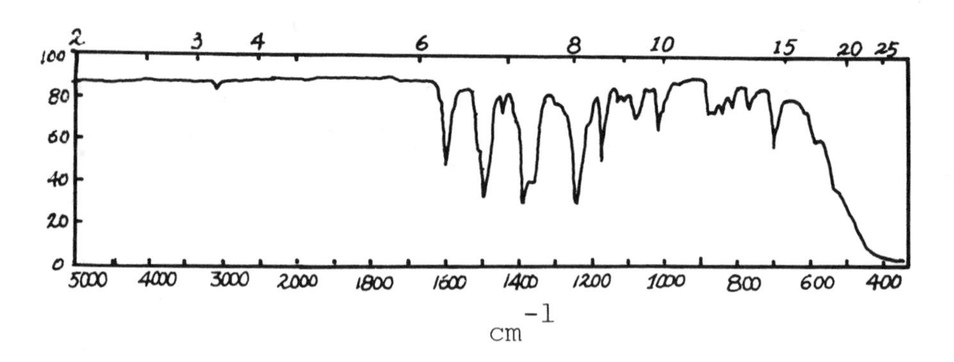

Figure 5. IR spectrum of PPT III.

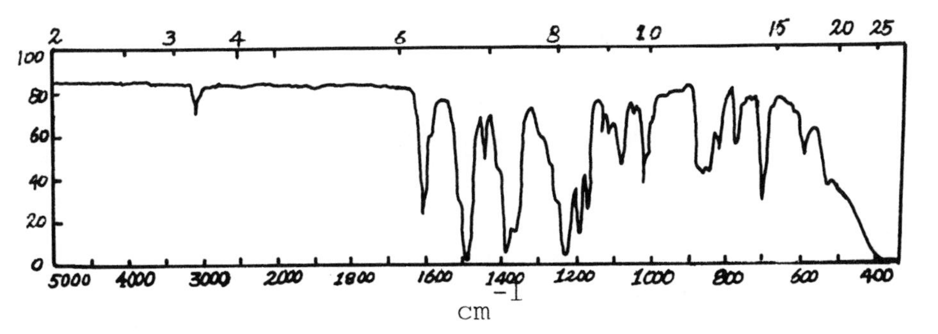

Figure 6. IR spectrum of PPT IV.

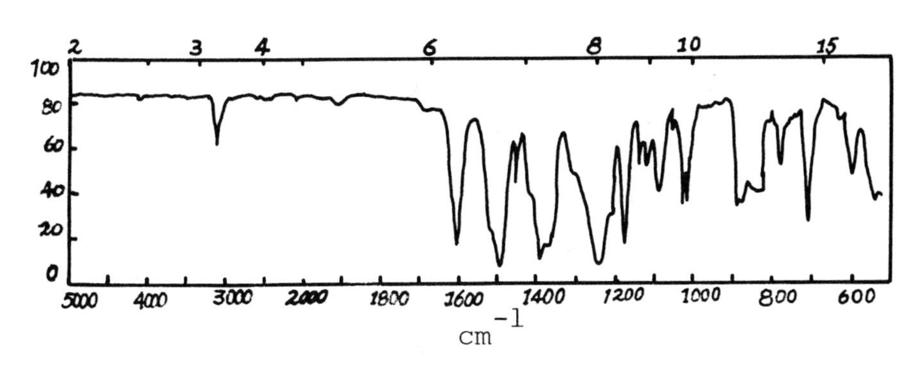

Figure 7. IR spectrum of PPT V.

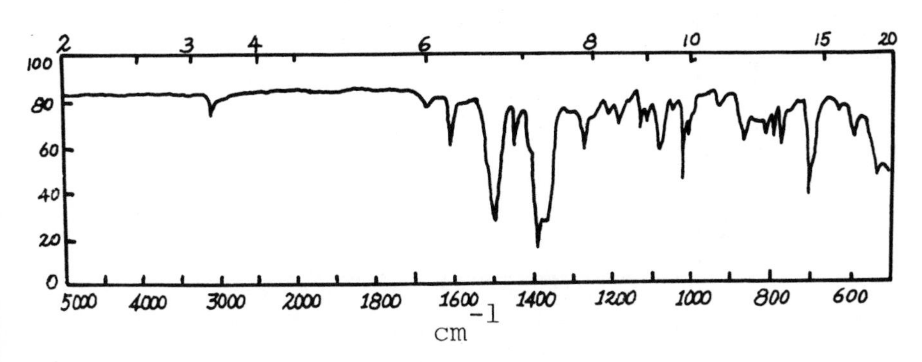

Figure 8. IR spectrum of PPT VI.

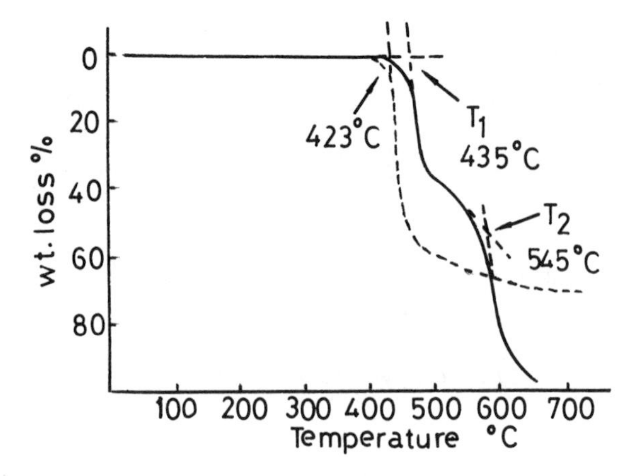

Figure 9. Thermogravimetric analysis of PPT I.

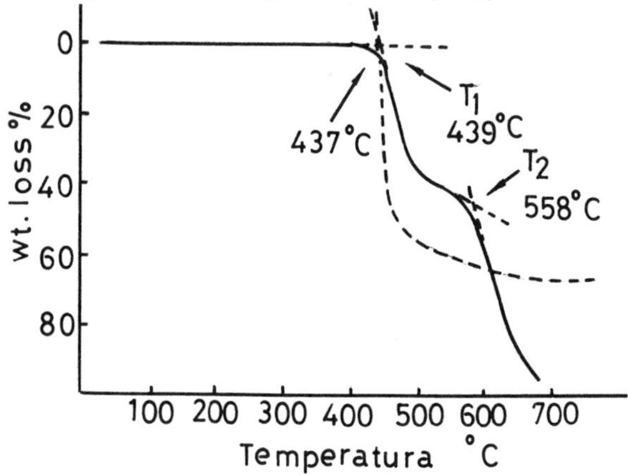

Figure 10. Thermogravimetric analysis of PPT II.

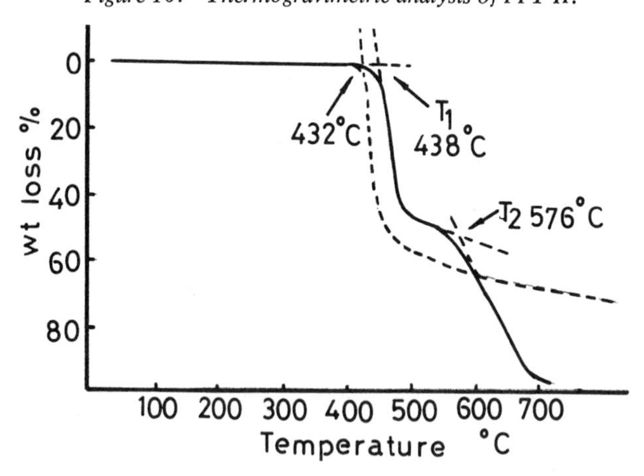

Figure 11. Thermogravimetric analysis of PPT III.

The thermal oxidative stabilities of these polymers were determined by isothermal aging of films at 300°C in air. These polymers containing a 1,4-phenylene group in the main chain structure exhibited higher thermal oxidative stabilities than those reported in literatures. The isothermal aging data of the polymers are presented in Table III. In addition, the order of thermal oxidative stability of these polymers is shown below:

PPT(I), PPT(II), PPT(III) > PPT(IV), PPT(V) > PPT(VI)

Table III Thermal Behavior of PPT (I) - PPT (VI)

Polymer No	PDT[1) °C]			Isothermal aging of PPT'S at 300°C in air, wt.loss %			
	Air T_1	Air T_2	N_2	50 hrs	100 hrs.	150 hrs.	200 hrs.
I	435	545	423	2.6	3.2	3.6	4.8
II	439	558	437	2.6	3.0	3.5	4.3
III	438	576	432	2.5	3.2	3.8	4.9
IV	415	500	406	2.4	3.3	4.1	5.1
V	430	500	410	2.6	3.5	4.1	5.1
VI	370	500	380	6.6	8.5	10.9	13.6

1) Polymer decomposition temperature (PDT) were determined by TGA, the rate of heating: 6°C/min.

The glass transition temperatures of these polymers were determined by differential thermal analysis (DTA). It seems that the presence of ether linkage in the backbone of polymers leads to lowering of Tg.

Hydrolytic Stability. The polymers exhibited not only thermal oxidative stability, but also good hydrolytic stability. The hydrolytic stabilities of these polymers in the form of films and varnished wire were evaluated in water and aqueous base. The electrical resistance of these films (Ø 1.5 mm, 0.035-0.040 mm thick) in boiling water was 10^{10} ohms. After treating in water at 200°C for 30 hours, the electrical resistances and weights of PPT (I)-PPT (V) remained essentially unchanged whereas the film of polyimide suffered a 9.43% weight loss and failed to be an insulating material under the same condition. In order to make further observation on hydrolytic stability, a varnished wire was made by coating PPT(I) onto stainless steel wire (1 meter in length, thickness of coating about 0.035-0.040 mm). Its electrical resistance was measured in water at 200°C for 100 hours. The value was about $4.6x10^9$ ohms as shown in Table IV. The breakdown voltage of a 10 cm long wire in water at room temperature was 7,000 volts. No electric breakdown can be detected when this wire was submerged in water under 2,000 volts for 5 minutes.

Table IV Insulation Resistance (Ω) Of PPT (I)-(VI) in water

Polymer No	100°C	200°C			
		6 hr[1)]	22 hr[1)]	30 hr[1)]	100 hr[2)]
I	$3x10^{10}$	$2x10^{10}$	$5x10^{10}$	$8x10^{10}$	$4.6x10^{9}$
II	$9x10^{10}$	—	$5x10^{9}$	$3x10^{9}$	—
III	$2.2x10^{10}$	$3x10^{10}$	$2.2x10^{9}$	$2x10^{9}$	—
IV	$9x10^{10}$	$9x10^{9}$	—	$9x10^{9}$	—
V	$9x10^{10}$	$9x10^{10}$	—	$2.6x10^{10}$	—
VI	$5x10^{10}$	$5x10^{7}$	—	film cracked	—
Polyimide	$1.4x10^{9}$	$1.3x10^{7}$	failed		

1) Measured in water at 100°C.
2) Measured in water at 200°C.

<u>Chemical Stability</u>. The wire made from PPT(I) exhibited good chemical resistance. For example, there was no change in electrical resistance after the wire was submerged in 5% hydrochloric acid solution or gasoline at room temperature for 24 hours or was boiled in 40% sodium hydroxide solution, lubricating oil, benzene or ethyl alcohol for 24 hours. These properties are shown in Table V.

<u>Mechanical Property</u>. The tensile strengths of these polymer films at room temperature were between 955-1427 Kg/cm^2.

All those properties show that these polymers can be preliminarily evaluated as good heat resistant, water resistant and/or chemical resistant materials.

Table V Chemical Stability of PPT (1)

Time \ Chemicals	HCl		NaOH		Lubri-cating oil	Gaso-line	Benzene	Ethyl alcohol
	5%	37%	5%	40%				
0 hr.	10^{12}	10^{11}	10^{12}	10^{12}	10^{12}	10^{12}	10^{12}	10^{12}
R.T. for 24 hr.	10^{12}	10^{8}	10^{12}	10^{12}	10^{12}	10^{12}	10^{12}	10^{12}
boiled for 24hr.	—	—	10^{11}	10^{11}	10^{12}	—	$10^{10}-10^{11}$	$10^{10}-10^{11}$

Literature Cited

[1]. Lu, F.C.; Xu, X.M.; Wang, B.G.; Chang, J.B.; Xing, L.M.; Wang, Y.L.; Gao, Y.M.; Bai, L.N., Gaofenzi Tongxun (Beijing, China).

[2]. Wrasidlo, W.; Augl, J.M., J. Polym. Sci. 1969, A-1, 7(12), 3393.

[3]. Ogliaruso, M.A., J. Org. Chem.1963, 28, 2725

[4]. Greth, E.; Elias, H.G., Makromol. Chem.1969, 125, 24.

[5]. Evers, R.C., J. Polym. Sci., Polym. Chem. Ed. 1973, 11(7), 1449.

[6]. Schmitt, P.J.; Comoy, P.; Boitard, J.; Suguet, M., Bull. Soc. Chim. 1956, (4), 636.

[7]. Relles, H.M.; Orlando, C.M.; Heath, D.R.; Schluenz, R.W.; Manello, J.S.; Hoff, S., J.Polym. Sci., Polym. Chem. Ed. 1977, 15(10), 2441.

[8]. Wahl, B.; Wöhrle, D., Makromol. Chem.1975, 176, 849.

RECEIVED November 23, 1981.

24

Polycocyclotrimerization of Isocyanates

K. H. HSIEH and J. E. KRESTA

University of Detroit, Polymer Institute, Detroit, MI 48221

The cyclotrimerization of monofunctional isocyanates, preparation and properties of polycocyclotrimers based on isocyanates were investigated.

It was found that the reactivity of isocyanates in the cyclotrimerization reactions increased with the presence of the electron withdrawing groups in the vicinity of the isocyanate group, with the increased nucleophilicity of a catalyst and relative permitivity of the solvent system.

Polycocyclotrimers were prepared by polycocyclotrimerization of difunctional isocyanates of variable chain length or difunctional isocyanate with monofunctional isocyanate. The stress-strain and viscoelastic properties of resulting polymers were determined. It was found that copolycyclotrimers prepared from diisocyanates of the variable chain length had typical properties of phase separated block copolymers.

It is well established that under certain reaction conditions isocyanates can form cyclic trimers or linear polymers. Shashoua et al. (1) have found that catalysis and reaction temperature are key factors which determine the composition of the resulting reaction products. They observed that the formation of linear polymers proceeded only at low temperatures ($< -20^{o}C$) and the formation of cyclic trimers at ambient or higher temperatures. Under certain steric conditions formation of polycyclic structures were observed by Butler et al. (2,3) and Iwakura (4).

The polycyclotrimerization of polyfunctional isocyanates (or NCO-terminated prepolymers) produces polymer networks containing heterocyclic perhydro-1,3,5-triazine-2,4,6-trione (isocyanurate) rings as crosslinks:

0097-6156/82/0195-0311$06.00/0
© 1982 American Chemical Society

$$3 \sim NCO \xrightarrow{cat.} \text{(isocyanurate ring: } (C=O)_3 N_3 \text{ with } \sim \text{ on each N)}$$

The isocyanurate rings are thermally stable (decomposition started at temperatures above 400°C (5)) and therefore cyclotrimerization of difunctional isocyanates can be utilized for preparation of thermally stable polymers. (7,8)

The properties of polycyclotrimers depend on the crosslink density, on the flexibility of the isocyanate chain, and on the degree of the completion of the cyclotrimerization reaction.

The crosslink density can be varied by the polycocyclotrimerization of difunctional isocyanates of variable chain length or by polycocyclotrimerization of difunctional isocyanate with monofunctional isocyanate:

$$2 \sim NCO + R\text{-}NCO \xrightarrow{cat.} \text{(isocyanurate ring: } (C=O)_3 N_3 \text{ with } \sim \text{ on two N and R on the third N)}$$

In this paper, conditions for successful polycocyclotrimerization of isocyanates and properties of resulting polycocyclotrimers will be discussed.

Experimental

All chemicals used in this study were reagent grade. Butyl isocyanate (BuNCO, 99% from the Upjohn Chemical Co.), hexamethylene diisocyanate (HDI, 99% from the Mobay Chemical Co.), phenyl isocyanate (PhNCO, 99%, from the Upjohn Chemical Co.), p-tolyl isocyanate (MePhNCO, 99% from the Aldrich Chemical Co.), p-chlorophenyl isocyanate (ClPhNCO, 99%, from the Aldrich Chemical Co.) and cyclohexyl isocyanate (CHI, 98%, from the Aldrich Chemical Co.) were purified by vacuum distillation. Methylene diphenyl diisocyanate (MDI, 99%+, from the Mobay Chemical Co.) was used without further purification. N,N-Dimethylformamide (DMF, reagent grade, from the Mallinckrodt) was dried by molecular sieves 4A. The NCO-terminated prepolymers were prepared from poly(oxytetramethylenediol) (POTMD, mol. wt. 650, 1000, 2000, Quaker Oats Chem. Co.) and MDI.

Sodium ethoxide, trimethyl-2-hydroxypropyl ammonium 2-ethylhexanoate (TMHAE) and N,N',N"-tris(dimethylaminopropyl) sym. hexahydrotriazine (TDAPH) were used as catalysts.

Measurement of Kinetics

The isocyanate solution (50 ml, 1.0N, DMF) was transferred into the thermostated three-necked flask and stirred magnetically under dry nitrogen. Then 50 ml of a catalyst in DMF was placed into the dropping funnel which was heated by means of an electric jacket. When the constant temperature was reached in both the flask and the funnel, the catalyst solution was poured into the flask and the whole mixture was thoroughly mixed. At various time intervals samples were taken and analyzed by the dibutylamine method (12).

Preparation of Polycyclotrimers and Polycocyclotrimers

HDI-BuNCO Polycocyclotrimers. The mixtures of HDI with BuNCO (of various equivalent ratios) were prepared in the three-necked flask under the nitrogen blanket. Then the catalyst was added at room temperature and after intensive mixing the mixture was poured into a mold before gelation. The curing conditions were the following: 60°C/20 hrs., 100°C/20 hrs., and 120°C/20 hrs.

Prepolymer-MDI Polycocyclotrimers. The NCO terminated prepolymers were prepared by reacting MDI with poly(oxytetramethylenediol) (MW=1000, 2000) in the reaction kettle, equipped with the nitrogen inlet and outlet, mechanical stirrer, heating mantle and a thermometer. The reaction was carried out at 70°C without any catalyst and was terminated when the NCO content reached the theoretical values. (11)

The polycocyclotrimers were prepared by casting of the mixture of prepolymer, MDI (at various molar ratios) and catalyst in a mold at 110°C for 17 hrs.

Characterization of Polycocyclotrimers

Infrared spectra of polycocyclotrimers were recorded using a Sargent-Welch Infrared Spectrophotometer, Model 3-300. The dynamic viscoelastic properties of copolymers were measured by using Rheovibron Model DDV-II (Toyo Measuring Instruments Co.). The measurements were carried out at intervals of 1-2°C in the transition region (otherwise at 3-5°C) at a frequency of 110 Hz. The thermal analysis of the copolymers were carried out using a Perkin-Elmer Differential Scanning Calorimeter, Model DSC-2; the tensile strength, elongation of polycocyclotrimers were measured at different temperatures by employing an Instron Tester, Model 1113. The crosshead speed was 2 in/min.

Results and Discussion

Cyclotrimerization of Isocyanates. The previous investigations of the kinetics of cyclotrimerization of isocyanates revealed that the cyclotrimerization reaction proceeded through initiation, propagation, transfer and termination steps (6). The reactivity of isocyanates in the cyclotrimerization reaction depends on the structure of isocyanate, nucleophilicity of a catalyst and the relative permitivity of the solvent system. Isocyanates with electron withdrawing substituents adjacent to the NCO groups are more reactive than isocyanates with electron donating groups. Therefore, aromatic isocyanates cyclotrimerize significantly faster than aliphatic or cycloaliphatic ones. (Table I) The kinetic data for the cyclotrimerization of substituted phenyl isocyanates correlated with the Hammett equation gave the ρ value of +1.57 (Fig. 1). The large positive ρ value supports the

TABLE I

REACTIVITY OF ISOCYANATES IN CYCLOTRIMERIZATION REACTION

Catalyst = Sodium ethoxide
T = 40°C
Solvent = DMF

Isocyanate	$\frac{(k_{isoc})cat}{(k_{BuNCO})cat}$
Prepolymer (MDI-POTMD 650)	974.5
ClPhNCO	370.5
PhNCO	177.3
MDI	166.5
MePhNCO	86.5
HDI	3.85
BuNCO	1
CHI	0.41

idea that the cyclotrimerization reaction proceeds via attack of the nucleophile (a catalyst or a propagating chain) on the electrophilic carbon atom of the isocyanate group. It is interesting to note that the NCO-terminated prepolymers cyclotrimerize faster than the original isocyanates. The phenomenon is due to the cocatalytic effect of urethane groups which participate in the induced polarization of the isocyanate groups. (10)

It was previously reported by us that the reaction rate constants for cyclotrimerization increased with the increase of a nucleophilicity of the catalyst (catalytic center). (8,9)

The solvents used in the cyclotrimerization of isocyanate had a very strong effect on the reaction rate: with the increase of the relative permitivity of the solvent system, the cyclotrimerization rate constants increased. The experimental data correlate well with the Kirkwood equation:

$$\ln k_{cat} = \ln k_{cat_o} - \frac{1}{kT}\left(\frac{D-1}{2D+1}\right)\left[\frac{\mu^2_C}{r^3_C} + \frac{\mu^2_M}{r^3_M} - \frac{\mu^2_{CM^{\ddagger}}}{r^3_{CM^{\ddagger}}}\right]$$

where μ_c, r_C, μ_M, r_M, $\mu_{CM^{\ddagger}}$, $r_{CM^{\ddagger}}$ are dipole moments and radii for catalyst, isocyanate and activated complex $CM^{\ddagger}$, respectively.

The observed positive value for the second term (see slope, Fig. 2) indicated that the activated complex (created between monomer and catalyst, or propagating chain) had a larger separation of charges than the reactants in the initial stage.

Besides the relative permitivity, the specific solvation of reactants by solvents affected the cyclotrimerization reaction. It was observed that in the solvent systems with the constant relative permitivity, the reaction rate decreased with the increasing solvation of reactive groups. (8)

Preparation and Properties of Polycocyclotrimers of Isocyanates. The formation of isocyanurate rings in the polycyclotrimerization or polycocyclotrimerization reaction of isocyanates is characterized by the disappearance of the NCO groups (ν(N=C=O) 2280 cm^{-1}) and by the appearance of a new IR absorption band at 1700 cm^{-1}, assigned for the stretching vibration ν(C=O) of the isocyanurate ring (12).

The polycocyclotrimers of isocyanates of variable crosslink density were prepared by polycocyclotrimerization of difunctional isocyanate (HDI) with monofunctional isocyanate (BuNCO) and by polycocyclotrimerization of difunctional isocyanates of variable chain length using casting and molding techniques.

It was found, that a comparable reactivity and mobility of comonomers and good solubility of a trimerization catalyst in monomers (or monomer-solvent system) were desirable for the successful preparation of polycocyclotrimers.

The properties of HDI-BuNCO random polycocyclotrimers are summarized in Table II. The polycocyclotrimers had the appearance of elastoplastics and with the decrease of the HDI/BuNCO eqv. ratio (with the decrease of the crosslink density) elongation increased and the ultimate tensile strength decreased. However at the higher testing temperature (100°C) there was a significant drop in mechanical properties. This decrease of the mechanical properties with temperature was probably a result of the incompletion of the cocyclotrimerization reaction and the formation

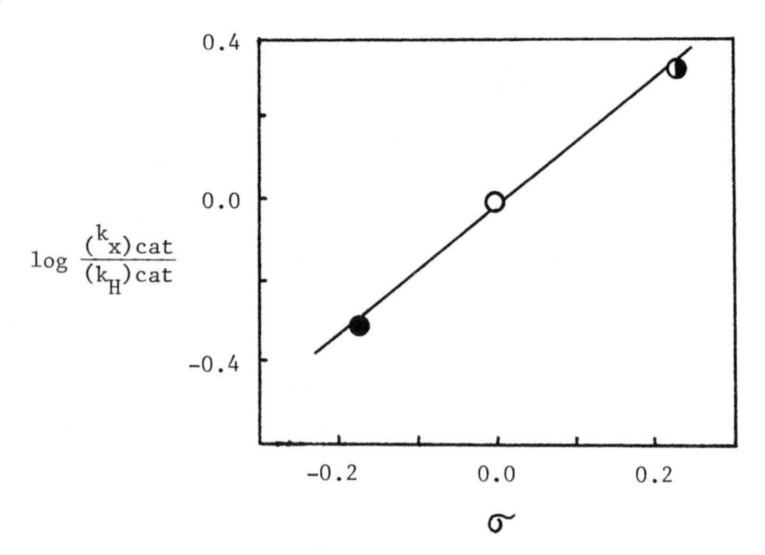

Figure 1. The Hammett correlation for cyclotrimerization of substituted phenyl isocyanates. Key: ●, p-*MePhNCO;* ○, *PhNCO;* ◑, p-*ClPhNCO; T, 40°C; catalyst, sodium ethoxide; solvent, DMF.*

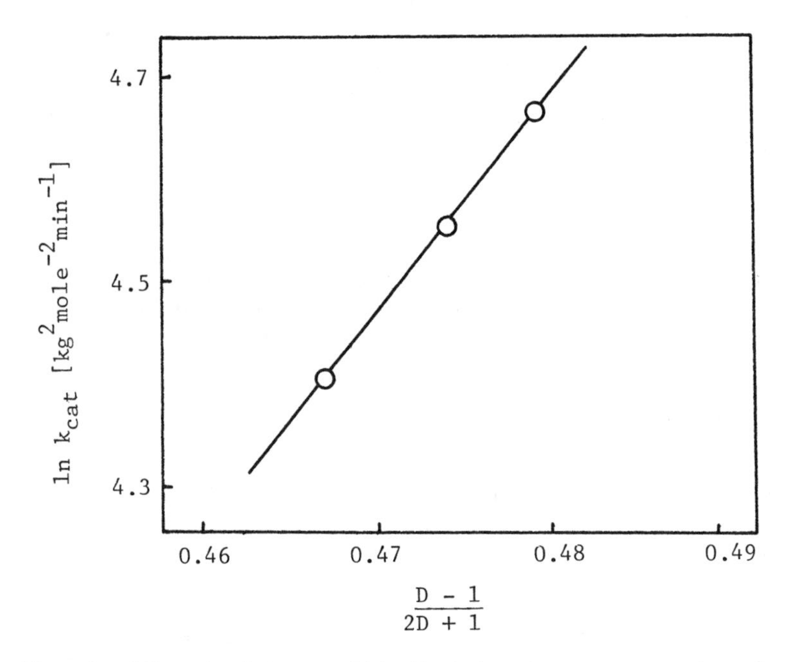

Figure 2. Effect of relative permitivity D of the solvent system on the cyclotrimerization of phenyl isocyanate. Conditions: solvent system, acetonitrile–ethyl acetate; catalyst, TMHAE; and temperature, 25°C.

TABLE II

PROPERTIES OF HDI-BuNCO POLYCOCYCLOTRIMERS

HDI/BuNCO Eq. Ratio	1.5/1	2.0/1	2.5/1	3/1	3.5/1	4/1	10/1	1.0/0
Tensile Strength (MPa)	2.57	10.98	37.31	51.48	54.66	59.60	66.09	59.65
Elongation %	1247	635	130	30	30	30	30	30
Rigid Ring Content (wt. %)	46.7	47.2	47.6	47.9	48.1	48.3	49.2	50

Catalyst: TMHAE

of the highly branched low-molecular weight fractions dispersed in the polymer network. The incompletion of the cocyclotrimerization reaction is probably associated with the limited mobility of the chain attached NCO groups at high conversion and the formation of the highly branched low-molecular weight fractions is connected with the termination reaction of growing chains by the highly mobile monofunctional isocyanate:

$$\sim\!\sim NCO + 2\ BuNCO \xrightarrow{cat.} \sim\!\sim N\!\!<\!\!\begin{matrix} C(=O) - N(Bu) - C(=O) \\ C(=O) - N(Bu) - \end{matrix}$$

Better results were obtained by utilizing the second method for the preparation of polycocyclotrimers. This polycocyclotrimers were prepared from the NCO-terminated prepolymers (MDI-POTMD, MW = 1000, 2000) and difunctional isocyanate (MDI) using trimerization catalyst TDAPH.

The polycyclotrimers and polycocyclotrimers had typical properties of block copolymers. The dynamic mechanical measurements revealed that the phase separation occurred in those polymers forming the soft segment (consisted mainly of polyether chains) and hard segment (isocyanurate rings and urethane groups) domains. The driving force for this phase separation came from incompatibility (different solubility parameters) of segmental units. The α-relaxation transition on the loss modulus E" curve (maximum) is due to the microbrownian segmental motion associated with the glass transition of the soft segment phase. This Tg transition temperature (~-30°C) is higher than Tg of the pure soft phase (Tg of the polyoxytetramethylene is -79°C (13)) which indicates that some phase mixing occurred. (See Fig. 3)

In the case, when POTMD of mol. wt. 2000 was used in the preparation of polycocyclotrimer, Tg value decreased to -67°C, which indicates a better phase separation. In general, a better phase separation is achieved when longer chain diol is used or when polycocyclotrimerization is performed in solution. Similar correlations were reported previously for urethane block copolymers. (14,15)

The properties of polycocyclotrimers significantly changed with the concentration of hard segments. The values of the storage modulus E' increased (and were less sensitive to temperature) with the increase of the concentration of hard segments. (See Fig. 5, Table III) The maximum of the loss modulus E" steadily decreased and finally completely disappeared with the increase of the content of the hard segment. Similarly, the maximum of the loss tangent (tan δ) shifted to the higher temperature under those conditions. (Fig. 4) All mentioned changes

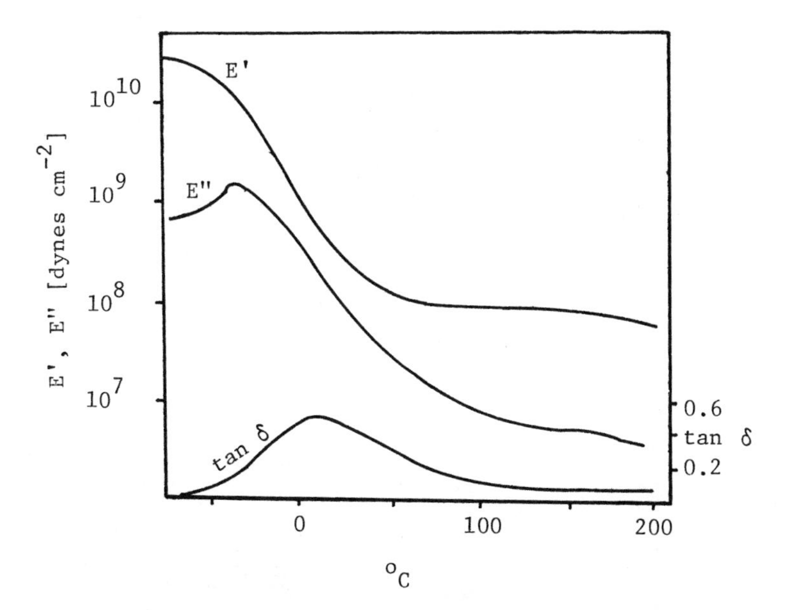

Figure 3. Dependence of the storage (E′), loss (E″) moduli and tan δ of polycyclotrimer with 33.4% hard segments on temperature.

TABLE III

THE PROPERTIES OF POLYCOCYLOTRIMERS PREPARED FROM MDI/MDI-POTMD PREPOLYMER

Prepolymer	MDI/Prepolymer Mol. Ratio	Catalyst (% wt)	Hard Segment Content (wt %)	Tensile Strength MPa		Elongation %		Endotherm (°C) (DSC)	E" Viscoelastic Transition Tg (°C)
				25°C	100°C	25°C	100°C		
MDI-POTMD 1000	0/1	TDAPH (0.21)	33.4	4.33	1.23	103	23	196	-32
MDI-POTMD 1000	1/1	TDAPH (0.14)	42.9	22.46	6.01	115	48	195.2	-22
MDI-POTMD 1000	2/1	TDAPH (0.21)	50.0	31.48	14.43	69	56	199.5	-30
MDI-POTMD 1000	4/1	TDAPH (0.35)	60.0	50.98	30.69	41	28	206.6	-
MDI-POTMD 1000	6/1	TDAPH (0.28)	66.7	60.16	41.92	34	23	220.5	-
MDI-POTMD 1000	8/1	TDAPH (0.3)	71.4	68.70	50.91	19	12	--	-
MDI-POTMD 2000	2/1	TDAPH (0.24)	33.4	7.62	3.68	34	15	204.9	-67

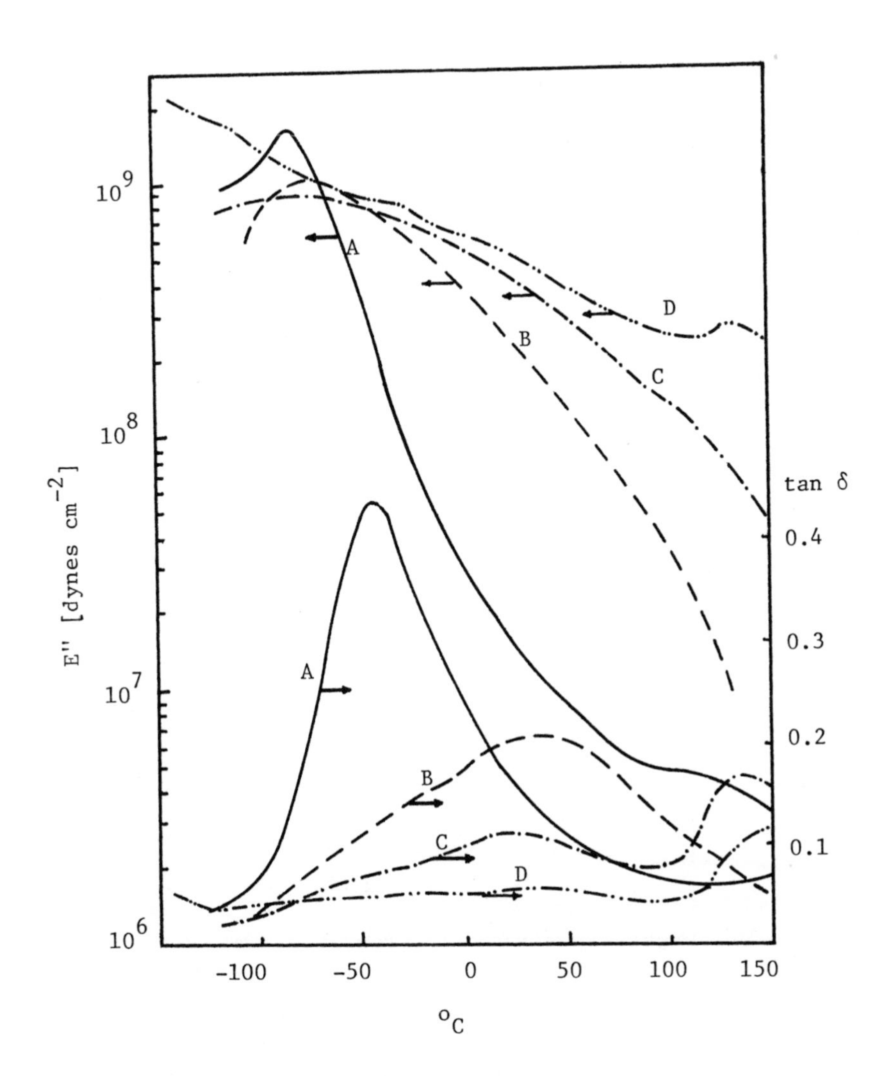

Figure 4. Temperature dependence of the loss modulus, E'', and tan δ for MDI/MDI–POTMD 1000 polycocyclotrimers containing 33.4 (A), 42.9 (B), 50 (C), and 60 (D) wt % of hard segments.

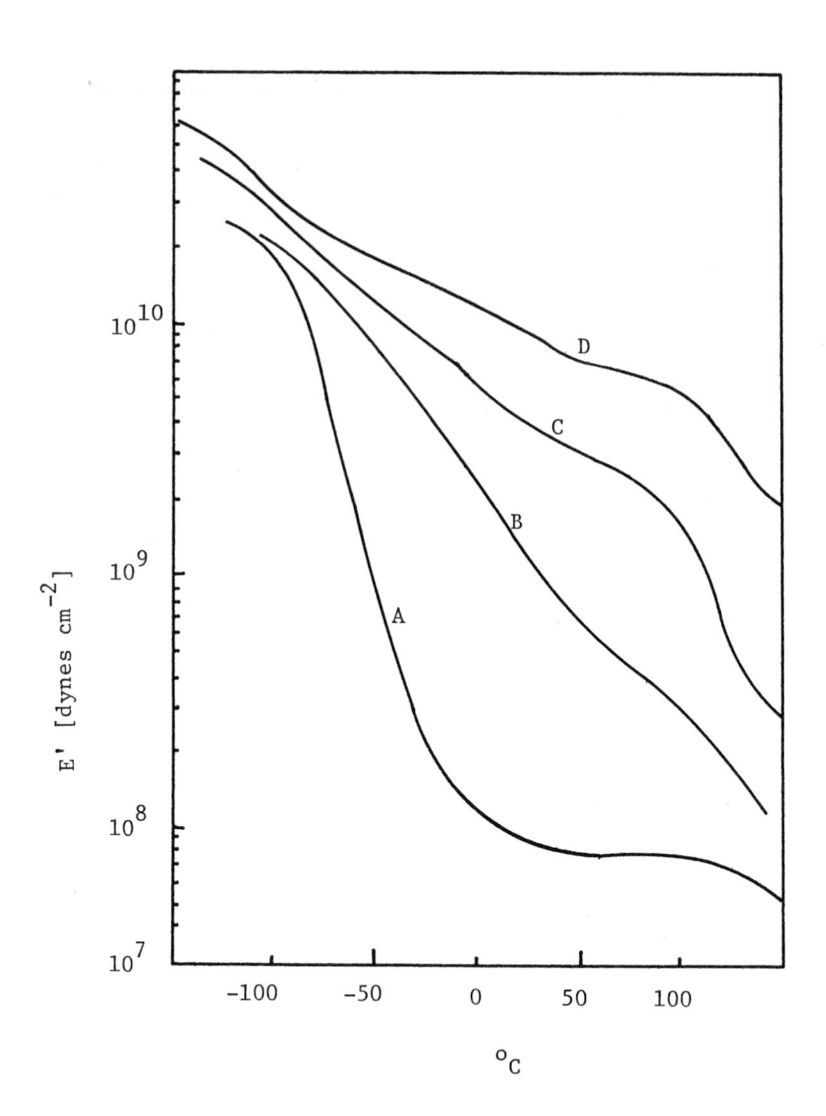

Figure 5. Temperature dependence of the storage modulus, E′, for MDI/MDI–POTMD 1000 polycocyclotrimers containing 33.4 (A), 42.9 (B), 50 (C), and 60 (D) wt % of hard segments.

in E', E" and tan δ are characteristic for the advancing crosslinking process (increase of the crosslink density) in the hard segment. (16) At the same time, the relaxation transition temperature of the crosslinked hard segment (as measured by DSC) increased with the increase of the molar ratio MDI/prepolymer. (Table III) This phenomenon is consistant with the decrease of the segmental motion within hard segment due to the crosslinking reaction. (16)

The effect of the hard segment content on the stress-strain behavior of polycocyclotrimers was also investigated. The results are summarized in Table III. It was found that with the increase of the content of the hard segment, the tensile strength at ambient and 100°C increased and elongation decreased and at the same time polycocyclotrimer changed from the elastomeric material to the brittle, high modulus plastic. This change is associated with the inversion of the continuous and dispersed phases which occur in polycocylotrimer at approximately 40-50% wt. of hard segment. Similar behaviors were described in other block copolymers containing soft and hard segments. (14)

Acknowledgement

The authors wish to thank the NSF-Engineering Program Office for the financial support of this work.

Literature Cited

1. Shashoua, V.E.; Sweeny, W.; Tietz, R.F. J. Am. Chem. Soc. 1960, 82, 866.
2. Butler, G.B.; Cornfield, G.C. J. Macromol. Sci. Chem. 1971, 5, 1889.
3. Butler, B.G.; Cornfield, G.C.; Aso, C. "Progress in Polymer Science"; Jenkins, A.D., Ed. Pergamon Press; Oxford, 1975; 4, p. 112, 185.
4. Iwakura, Y.; Uno, K.; Ichikawa, K. J. Polym. Sci. 1964, A2, 3387.
5. Kordomenos, P.; Kresta, J.E. Macromolecules 1981, 14, 1434.
6. Kresta, J.E.; Shen, C.S.; Frisch, K.C. ACS Org. Coatings and Plastics Preprints 1976, 36, (2), 674.
7. Sasaki, N.; Yokoyama, T.; Tanaka, T. J. Polym. Sci. 1970, A-1, 11, 248.
8. Kresta, J.E.; Hsieh, K.H.; Shen, S.C.; Lin, I.S. SPE Technical Papers 1980, 26, 462.
9. Kresta, J.E.; Lin, I.S.; Hsieh, K.H.; Frisch, K.C. ACS Org. Coatings and Plastics Preprints 1979, 40, 910.
10. Kresta, J.E.; Hsieh, K.H. Makromol. Chem. 1978, 179, 2779.
11. Saunders, J.H.; Frisch, K.C. "Polyurethanes: Chemistry and Technology"; Wiley-Interscience: N.Y., 1962; Part I.

12. David, D.J.; Staley, H.B. "Analytical Chemistry of Polyurethanes"; Wiley-Interscience: N.Y., 1969; p. 86.
13. Brandrur, J.; Immergut, E.H. "Polymer Handbook"; Wiley-Interscience: N.Y., 1967; p. III-80.
14. Van Bogart, J.W.C.; Lilaonitkul, A.; Cooper, S.L. "Multiphase Polymers"; Cooper, S.L.; Estes, G.M., Ed. Advances in Chemistry Series, 176, ACS, Washington, DC, 1979; p. 3.
15. Seefried, C.G.; Koleske, J.V.; Critchfield, F.E. J. Appl. Polym. Sci. 1975, 19, 2493.
16. Murayama, T. "Dynamic Mechanical Analysis of Polymeric Materials"; Elsevier: Amsterdam, 1978; p. 81-93.

RECEIVED April 28, 1982.

25

Polymerization of Bisphthalonitriles: Metal Free Phthalocyanine Formation

N. P. MARULLO
Clemson University, Clemson, SC 29631

ARTHUR W. SNOW
Naval Research Laboratory, Washington, DC 20375

Previous reports on the formation of network polyphthalocyanines by thermal polymerization of bisphthalonitriles have in fact been found to give materials containing a low percentage of phthalocyanine ring structures. Reactions carried out either under nitrogen or in air result instead in extensive formation of triazines via trimerization of cyano groups, and in the formation of imide or amide structures. Formation of network polymers having a high phthalocyanine ring content can be achieved by carrying out the polymerization of bis-phthalonitriles in the presence of suitable coreactants. The use of these coreactants also markedly reduce the time and temperature required for polymer formation.

The pioneering attempt by Marvel[1] to polymerize a bisphthalonitrile to a metal free polymeric phthalocyanine has prompted a continuing interest in devising starting materials and procedures for the preparation of this and related classes of polymeric macrocycles. Motivation for effort in this area is the expectation that polymeric phthalocyanines could have a wide range of interesting properties among which are electrical conductivity, electrochromism and high thermal stability. The synthetic approach taken by Marvel for the preparation of polymeric metal free phthalocyanines was analogous to the classical Linstead procedure which involves reaction of a phthalonitrile with lithium amyloxide to initially generate the dilithio phthalocyanine which can then be easily hydrolyzed in situ to give the dihydrophthalocyanine.

0097-6156/82/0195-0325$06.00/0
© 1982 American Chemical Society

$$\text{(phthalonitrile: benzene-1,2-dicarbonitrile)} \xrightarrow{\text{Li OAm}} \xrightarrow{H_2O} \text{DIHYDROPHTHALOCYANINE}$$

Several attempts to employ this same route for the preparation of polymeric metal free phthalocyanines employing a bis-phthalonitrile were unsuccessful (1).

$$\text{bis(3,4-dicyanophenyl) ether} \not\rightarrow \text{POLYPHTHALOCYANINE}$$

In recent years, Griffith and co-workers have successfully synthesized and polymerized a series of ether and amide linked bis-phthalonitriles to intensely colored network materials having high thermal stability[2]. These materials were generated by heating the appropriate bis-phthalonitrile in an open aluminum container until the melt cured to a rigid solid. On the basis of the nature of the starting materials and the development of an intense green color, the polymerization and concomitant cross linking reaction was believed to result mainly in phthalocyanine formation. However, cognizance was taken of the possibility of alternate structures. One particular example is the polymerization of neat bis(3,4-dicyanophenyl) ether of bisphenol A, (I), to a material with excellent thermal stability[2].

$$\underset{\text{I}}{(NC)_2C_6H_3\text{–R–}C_6H_3(CN)_2} \xrightarrow[\text{7 days}]{280^\circ} \underset{\text{II}}{\text{POLYPHTHALOCYANINE}}$$

$$R = \text{–O–}C_6H_4\text{–}C(CH_3)_2\text{–}C_6H_4\text{–O–}$$

Discussion and Results

Since the polymerization of these bis-phthalonitriles were carried out without benefit of those conditions which are known to promote phthalocyanine ring formation, some question exists as to the chemical nature of these materials. In addition to the phthalocyanine ring, isoindoline (III) or triazine (IV) structures are also possible. Structure III is well known and, indeed, is an isolable intermediate in the alkoxide[3] or NH_3[4] catalyzed conversion of simple phthalonitrile to phthalocyanine. Furthermore, the triazine (R=H) is the major product of the high temperature reaction of phthalonitrile containing trace amounts of H_2O[5]. Although the literature (6) contains an assertion to the preparation of dehydrophthalocyanine (V where R=H) in an impure form, no evidence was given to support such a structure. Furthermore, if the macrocyclic ring has a planar geometry then the cycle is a 16 π anti-aromatic system. In view of the anticipated anti-aromaticity of V as well as the lack of authenticity to its preparation, structure V is not considered by us to be a viable structure for consideration in this current work.

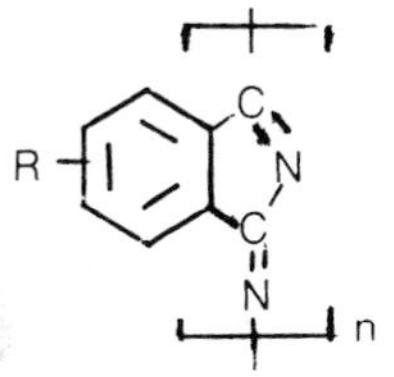

III

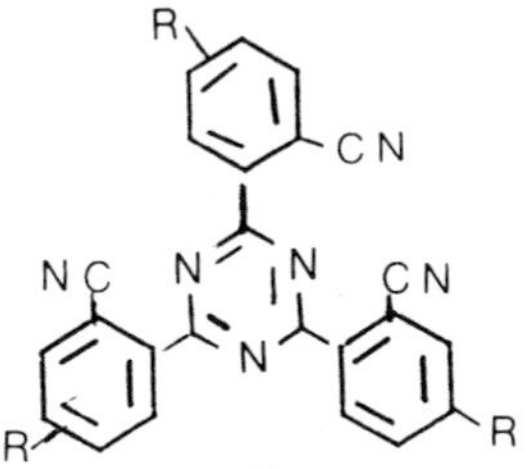

IV

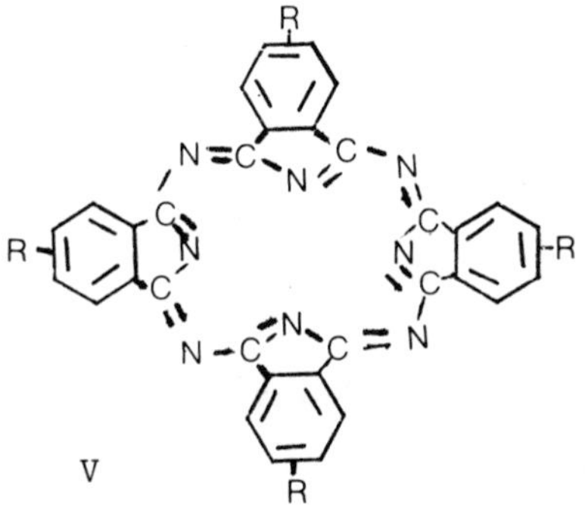

V

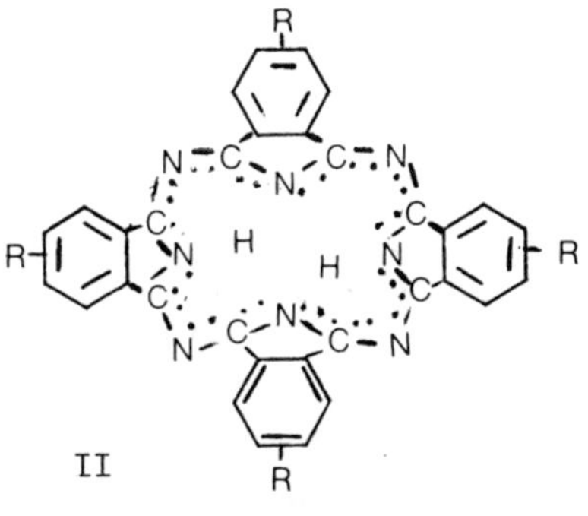

II

In order to determine the extent to which the procedure employed by Griffith and co-workers generates phthalocyanine structures, both the amount of unreacted starting material and of phthalocyanine present in the reaction mixture need to be analytically determined. To this end, compounds VI and VII, which serve as models for I and II, were prepared by scheme 1.

$R{-}OH + O_2N{-}C_6H_3(CN)_2 \xrightarrow[DMSO]{K_2CO_3} RO{-}C_6H_3(CN)_2$

VI

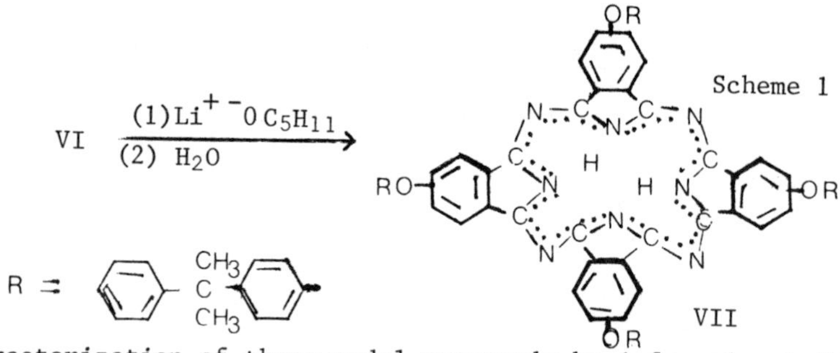

Scheme 1

$$\text{VI} \xrightarrow[\text{(2) } H_2O]{\text{(1) } Li^{+}\ {}^{-}OC_5H_{11}} \text{VII}$$

R = $C_6H_5-C(CH_3)_2-C_6H_4-$

Characterization of these model compounds by infrared, visible-ultraviolet, NMR (3), and mass spectroscopy and elemental analysis is consistent with the assigned structures. Phthalocyanines, both metal free or those containing metals, are generally characterized as being extremely insoluble materials. However, the four bulky cumylphenoxy substituents of compound VII apparently interfere with the close packing of phthalocyanine rings and thereby makes VII very soluble in a wide variety of organic solvents. Furthermore, it was also found that samples of the bis-phthalonitrile (I) which were allowed to cure only to an extent of about 30% (as measured by the decrease in the cyano absorbance in the infrared spectrum) were also dioxane soluble. The ready solubility of VII and of partially cured samples of I therefore permitted easy and accurate quantitative determination of the extent of phthalocyanine ring formation by the use of visible spectroscopy. As can be seen from Fig. 1, the visible spectrum of VII is very similar to that of phthalocyanine throughout the 500 to 750 mm region, and that of partially cured I is identical to that of VII. This confirms the validity of VII as an analytical model for the polyphthalocyanine II, and, coupled with their solubility characteristics, permits easy and accurate quantitative determination of the phthalocyanine ring content of partially cured samples of I and VI.

Table 1 summarizes the analytical results obtained for the thermal reaction of I and of VI. With the exception of the experiment in which methanol was added to VI, the extent of phthalocyanine formation was very small. Moreover, the infrared spectra of these reacted samples exhibited absorptions characteristic of amide or imide carbonyl as well as those for triazine. Under these conditions, phthalocyanine formation is not favored. For example, the yield of phthalocyanine ring formation, based on the consumption of starting material, was less than 5% for the model phthalonitrile VI, and less than 10% for the bis-phthalonitrile I. In a separate experiment a thin film of I deposited on a salt plate was allowed to cure while exposed to the atmosphere. Periodic infrared examination of the curing film showed that as the cyano absorbance diminished the absorbance in the carbonyl region increased until the latter was the predominant infrared band. These results demonstrate that

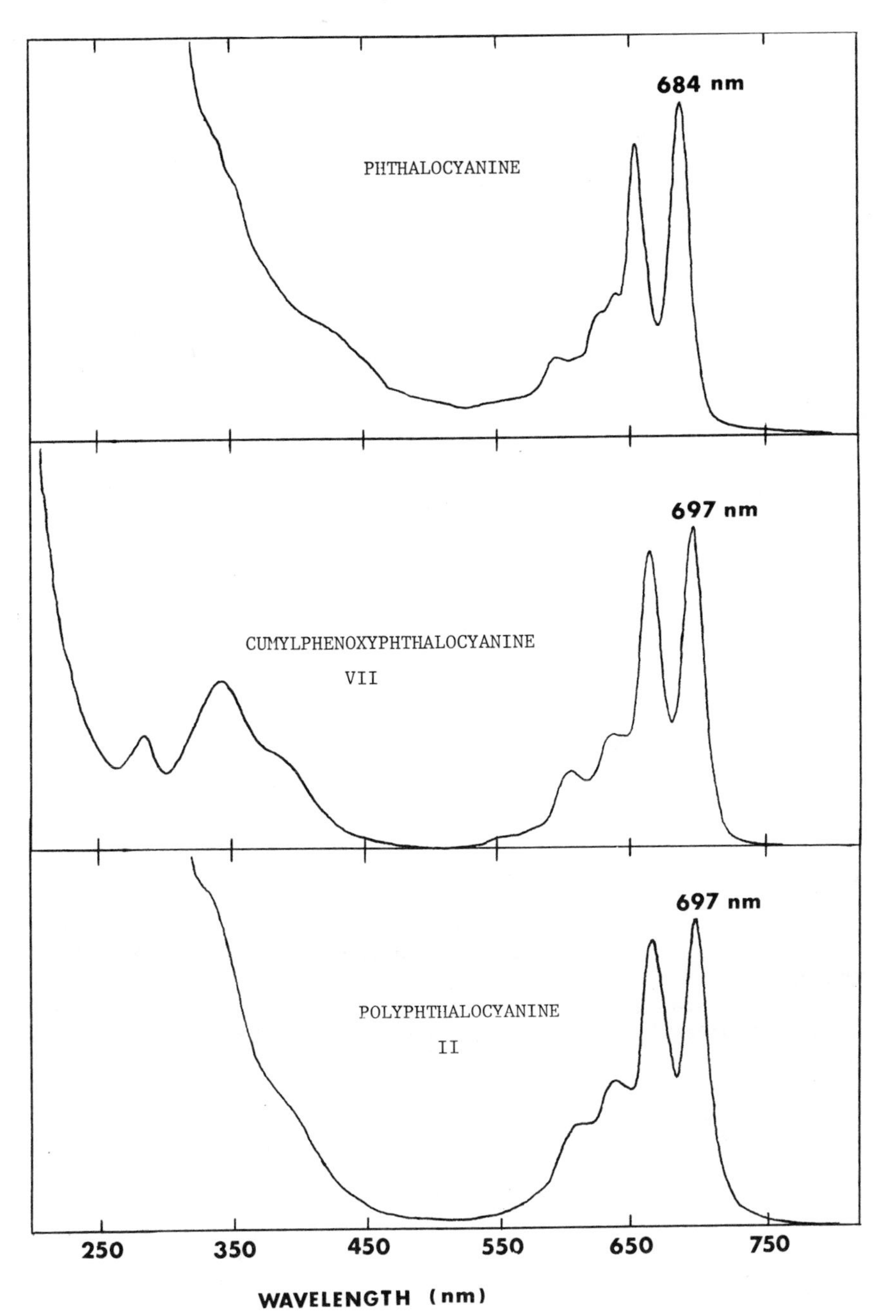

Figure 1. Similarity of the electronic spectra of phthalocyanine, with cumylphenoxyphthalocyanine (VII) and the polyphthalocyanine (II).

thermal cure of phthalonitriles preferentially give triazines (IV) and perhaps linear isoindolines (III) when reacted in sealed tubes or under nitrogen. If the cure is carried out in the open then extensive oxidation and or hydrolysis occurs.

TABLE I
THERMAL CURE OF PHTHALONITRILES I AND VI
ANALYTICAL RESULTS

Sample (Conditions)	Reaction Time (hr)	% Decrease in C ≡ N	% Rx. Pc in Mixture
Cpd. VI 300°C N_2 stream	189	9.6	0.12
	333	67	2.3
Cpd. VI 300°C open tube	50	46	0.06
Cpd. VI 300°C sealed tube	50	<3	0.2
Cpd. VI 300°C +2.4% CH_3OH sealed tube	60	92	40
Cpd. I 280°C N_2 stream	14.5	3	0.07
	25	6	0.13
	39.5	19	0.35
Cpd. I 310°C sealed tube	14	24	2.0

The effect of methanol in increasing both the rate of reaction as well as the relative amount of phthalocyanine demonstrates the susceptibility of the reaction to catalysis. Our approach to catalysis and promotion of phthalocyanine formation is based upon employing a co-reactant which would function to meet the following mechanistic requirements:

(1) sufficient nucleophilicity to promote cyclization of the o-dicyano groups to an isoindoline intermediate,

(2) ready displacement of the nucleophilic moiety during the macrocycle forming step,

(3) facile redox.

The last requirement is critical for formation of metal free phthalocyanines directly from phthalonitriles. The phthalocyanine ring is an aromatic 18 π electron octaazaannulene.

Cyclotetramerization of eight cyano functions must be accompanied with transfer of two additional electrons in order to give a phthalocyanine. If the co-reactant is not able to undergo ready oxidation, then alternate products such as triazene or linear isoindoline would be the expected products since their formation does not require a redox reaction to occur.

Table 2 summarizes some of our initial efforts with regard to catalysis and to the promotion of phthalocyanine formation. The tabulated reaction times are those required to cause a 90% or more decrease in C≡N infrared absorbance at a reaction temperature of 165°C and demonstrate the marked catalytic effect of each additive. The powerful base, benzyl trimethylammonium hydroxide was employed as the solvent free solid and is by far the most powerful catalyst, however, it gave no detectable quantity (less than one part per million) of phthalocyanine. This is consistent with the proposed mechanistic criteria and dramatically demonstrates that in the absence of a viable redox pathway, reactions other than phthalocyanine formation can occur very rapidly. All other entries gave substantial amounts of phthalocyanine; hydroquinone being superior in this regard.

TABLE 2
REACTION OF VI WITH VARIOUS ADDITIVES
(4:1 MOLAR RATIO, 165°C)

Additive	Reaction Time (hr)	% Pc
Benzyl trimethyl ammonium hydroxide	10 min. (100°C)	None
Hydroquinone	21	63
Catechol	1.5	36
Resorcinol	3.5	32
Benzoin	21	36

Hydroquinone, catechol and benzoin were also found to drastically reduce the temperature and time required for the polymerization of I, increase the phthalocyanine content of the polymer (II), and improve its thermal stability. For example, without additive, I required 80 hours of heating at 300°C to effect an 87% reaction of CN groups. However, a stoichiometric mixture of I and hydroquinone underwent 88% reaction when heated at 165°C for 2 hours followed by 250°C for 1 hour. With added catechol, DSC (Fig. 2) carried out isothermally at 202°C showed a well defined exotherm indicating complete cure after 27 minutes. Bulk samples of the polymer (II) cured with these additives have an intensely blue or blue-green appearance with a purple reflex indicating high phthalocyanine ring content. These color characteristics are similar to those of the model phthalocyanine (VII) as well as unsubstituted phthalocyanine. Furthermore, the infrared spectrum of a sample of I cured as a thin film with added hydroquinone is very similar to that of the model phthalocyanine including the distinctly weak but sharp N-H band at 3290 cm^{-1}. Notably absent are bands corresponding to carbonyl or triazine functional groups.

Thermal gravimetric analysis shown in Fig. 3, indicates some improvement in the thermal stability of the polymer cured with the appropriate additive.

In conclusion, the bis-phthalonitrile (I) can be polymerized at moderate temperatures and short reaction times to a network polymer having a high metal free phthalocyanine content by employing a suitable co-reactant.

Experimental

Preparation of 4-Cumylphenoxy Phthalonitrile VI. Compound VI was prepared according to the general procedure reported for ether substituted phthalonitriles (2). Finely ground anhydrous K_2CO_3, 19.6 g (0.14 mol) was added in small portions to a solution of 19.1 g (0.09 mol) of 4-cumylphenol and 15.6 g (0.09 mol) of 4-nitrophthalonitrile dissolved in 150 ml of dry DMSO in a nitrogen atmosphere. The K_2CO_3 was added in 1 to 2 g portions at 1/2 to 1 hour intervals over an eight hour period and stirred at room temperature for 24 hours. The reaction mixture was filtered and the filtrate slowly added with stirring to 400 ml of water, neutralized with HCl and the crude product taken up in 100 ml of CH_2Cl_2. The aqueous layer was extracted twice with 50 ml portions of CH_2Cl_2 and the combined CH_2Cl_2 extracts washed first with 100 ml of 5% Na_2CO_3 , then with water and dried. Removal of solvent at reduced pressure gave 21.3 g (70%) of VI. Two recrystallizations from methanol gave large platelets, m p 90 ^{o}; i r (KBr) 3080-3020 (=CH), 2238 (CN), 1580 (c=c), 1500-1430 (aromatic): nmr ($CDCl_3$)δ 1.70 (singlet, 6H), 7.27 (multiplet 12H): mass spectrum parent ion 338.

Anal. Calcd. for $C_{23}H_{18}N_2O$: C, 81.65; H, 5.32; N, 8.28 Found: C, 81.86; H, 5.28; N 8.23.

Preparation of Tetracumylphenoxyphthalocyanine (VII). To a solution of lithium amyloxide in amyl alcohol prepared by careful addition of 0.103 g (0.01 5 g atom) of lithium to 20 ml of dry amyl alcohol, was added 3 g (0.089 mol) of VI and then heated to reflux for 3 hrs. The solid product was precipitated by addition of 200 ml of methanol, filtered and then soxlet extracted with methanol. The solid product is soluble in ordinary organic solvents such as benzene, CH_2Cl_2 or dioxane: ir(film, evaporated soln. from dioxane) 3290 (NH), 3090 - 3030 (=CH), 1608 (c = c), 1510- 1470 (aromatic); nmr (CH_2Cl_2) 1.72 (singlet,

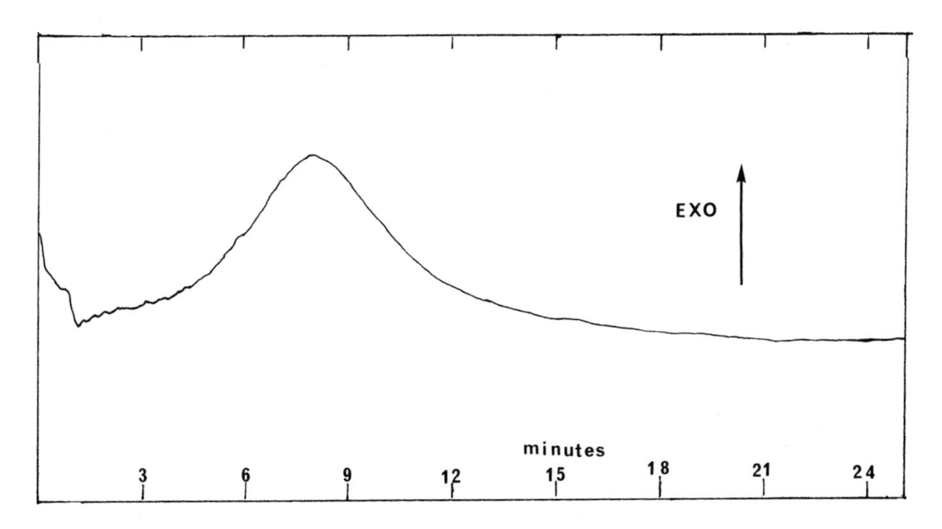

Figure 2. Differential scanning calorimetry of a 4:1 molar mixture of I to catechol, carried out isothermally at 202°C.

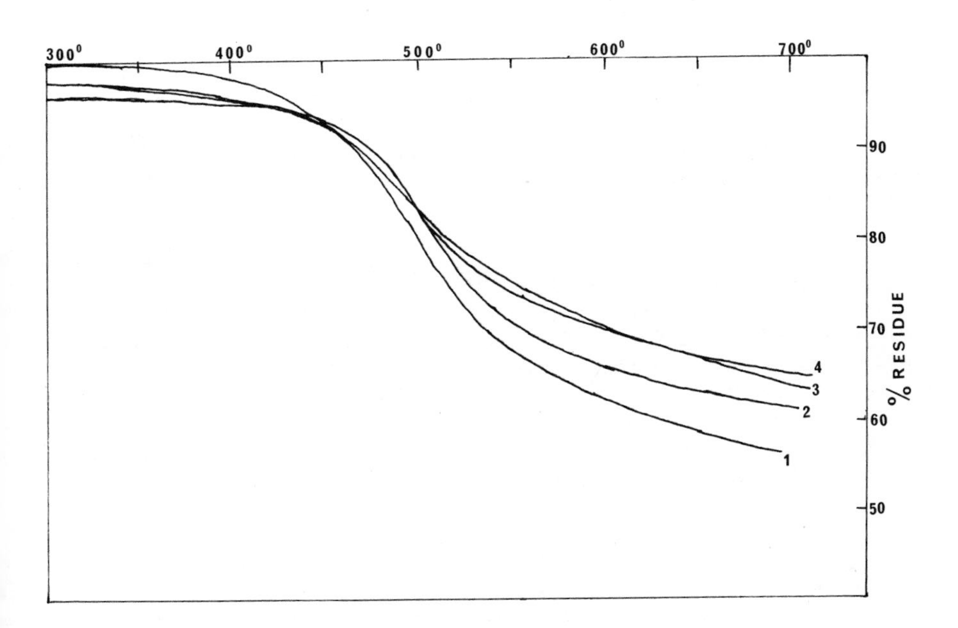

Figure 3. Thermogravimetric analysis (nitrogen) of I cured for 7 days at 280° (curve 1), and of samples of I with hydroquinone (2), catechol (3), and benzoin (4), cured for 2 h at 165° and 1 h at 250°.

12H) 7.32 (multiplet, 24 H); vis.-uv(dioxane) 697, 663, 635, 602, 390, 241, 286.

Anal. Calcd. for $C_{92}H_{74}N_8O_4$: C, 81.51; H, 5.50; N 8.27. Found: C, 81.61; H, 5.62; N 8.43.

Thermal Cure Reactions. Approximately 150 mg of VI or I were placed in 5 mm diameter tubes which subsequently were either sealed, kept open to the atmosphere or appropriately fitted so as to maintain a very slow stream of nitrogen over the reaction mixture. The tubes were placed in an electrically heated air bath and samples withdrawn periodically and monitored for disappearance of starting material and for formation of phthalocyanine rings as described in the Analytical section. The results are tabulated in Table I.

Thermal cure reactions were also carried out as thin films sandwiched between salt plates and the edges sealed with Teflon, or on a single plate with the surface exposed to the atmosphere. This permitted an easy method of monitoring the extent of cure beyond the gelation point of I and determination of the i r characteristics of highly cured samples.

Reaction of VI with Benzyltrimethylammonium Hydroxide. The hydroxide, 0.125 mmol, (60 mg of a 35% methanol solution) was placed in a tube and the methanol evaporated by warming at reduced pressure. Subsequent addition of 175 mg (0.5 mmol) of VI immediately resulted in the formation of a green color at the contact surface of VI and the solid hydroxide. Warming of the solid mixture to 100°C over a 10 minute period appeared to give complete reaction and heating was discontined. The i r spectrum showed greater than 95% decrease in cyano absorbtion and the visible spectrum showed no absorbance characteristic of phthalocyanine chromophore.

Reaction of VI with Various Additives (Table II). Compound VI (0.5 mmol) and 0.125 mmol of the additive were placed in sealed tubes and heated at 165°. The tubes were initially monitored in terms of change from a fluid to solid and those which solidified were opened and the i r spectra taken to determine the extent of reaction. Those samples which had not undergone sufficient reaction were resealed and reheated. When the i r showed that sufficient reaction had occurred, the contents were analyzed for phthalocyanine chromophore content.

Analytical Measurements. The change in concentration of I or of VI was monitored by determing the change in absorbance of the cyano group at 2238 cm^{-1}. This was done by melting and subsequent cooling of either I or VI between salt plates which resulted in transparent glassy films. Quantitative changes in the cyano absorbance were easily determined using the peak ratio method employing the sharp signal at 1020 cm^{-1} as an internal reference.

Phthalocyanine content was determined by electronic absorbtion at 697 nm (dioxane). Analytically pure samples of VII were used to generate a calibration curve which was then used to determine the amount of phthalocyanine chromophore generated from reactions of VI and also of I.

Acknowledgement

NPM gratefully acknowledges the financial support of the NRL and the use of its facilities. The authors thank Mr. T. Price for samples of the monomer I and for the TGA, Dr. P.Peyser for the DSC, Dr. C. Poranski for the carbon 13 NMR, and Drs. Ting, Keller and Griffith for their support and helpful discussions.

References

1. C. S. Marvel and Michael M. Martin, J. Amer. Chem. Soc. 1958, 80, 6600.
2. T. M. Keller and J. R. Griffith, "Resins for Aerospace," ACS Symposium Series 132, 1980, p. 25 - 34.
3. F. Borodkin, Zhur. Priklad. Khim., 1958, 31, 813.
4. P. J. Brach, S. J. Grammatica, O. A. Ossanna and L. Weinberger J. Hetero. Chem., 7, 1970, 1403.
5. S. D. Ross and M. Fineman, J. Amer. Chem. Soc., 1950, 72, 3302.
6. K. Venkataraman , "Chemistry of Synthetic Dyes", Vol. 5; Academic Press,N.Y., 1971; p 285.

RECEIVED December 14, 1981.

26

Characterization of the Cure of Diether-Linked Phthalonitrile Resins

R. Y. TING
Naval Research Laboratory, Orlando, FL 32856

T. M. KELLER, T. R. PRICE, and C. F. PORANSKI, JR.
Naval Research Laboratory, Washington, DC 20375

New organic resin systems are in constant demand because of the increasing use of fiber-reinforced organic matrix composites to replace metallic components in weight-critical aerospace and advanced marine applications. High temperature capability, low moisture absorption, and improved storage and processing properties are emphasized for new composite matrix materials. Diether-linked phthalonitrile resins recently developed at NRL promise to meet these requirements and are being evaluated as potential candidates for such applications. The linking structure R in the monomers include bisphenol-A, bisphenol-S, bisphenol-A6F, resorcinol, hydroquinone and dihydroxybiphenol. The results of the characterization of cured phthalonitrile resins are presented. This includes their thermal properties from thermal gravimetric analysis and differential scanning calorimetry, spectroscopic properties from the infra-red and nuclear magnetic resonance techniques, and mechanical properties from torsional pendulum analysis and fracture mechanics evaluation.

Compared to conventional metallic structural materials, fiber-reinforced organic matrix composites have a very high specific modulus and tensile strength. For this reason, they are being used in many aerospace and marine systems in which weight reduction is an important consideration. This reduction is especially critical for the successful development of new vertical and short take-off and landing (V/STOL) aircraft in order to compensate for the large engines required for vertical lift.

Composite materials have been used successfully, and conventional matrix materials such as epoxies are being considered for application in V/STOL systems. Epoxy polymers, however, suffer many shortcomings. They are extremely sensitive to moisture absorption, which causes a large reduction in their glass

This chapter not subject to U.S. copyright.
Published, 1982, American Chemical Society

transition temperature, thus limiting their maximum use to temperatures less than approximately 135°C in a wet environment. Furthermore, epoxy prepregs require cold storage at all times and are therefore difficult to handle, costly to store, and susceptible to chemical degradation. Because of the complex chemistry and variations in formulation of epoxy resins, providing quality analysis and control for these resins presents a serious problem for large quantity, multi-batch procurements, especially those involving several manufacturers. As a result of these problems, new organic resins are being sought to meet operational requirements at higher temperatures, to be relatively insensitive to moisture, and to provide improved room-temperature storage properties and processability equivalent to the state-of-the-art expoxies.

In recent years, research efforts at the Naval Research Laboratory have led to the development of a new class of resins called phthalonitriles (1), which seem to provide the desired properties for advanced composite applications. The resin that has a diamide structure with an eight-carbon aliphatic chain linkage, called the C-10 monomer, has been studied extensively and successfully demonstrated to be a potential matrix material (2). The C-10 polymer exhibited long-term stability at temperatures up to 245°C (3), and its moisture uptake is less than that of epoxies or polyimides (4). The resin monomer also represents a chemically simple and pure system for easy quality control (5), and is virtually inert at room temperature for easy storage and handling (2). Unfortunately, the synthetic procedure for the C-10 phthalonitrile resin involves a reaction that has relatively low yield, and requires expensive starting materials. Continued efforts to overcome these difficulties have been successful, however, and a series of second-generation phthalonitrile resins with diether-linking structure have been developed (6). In this report, the results of the characterization and the cure of these new materials will be presented.

Experimental

Material. The detail of the synthesis of diether-linked phthalonitrile resins has been given in an earlier report (6). Briefly, a mixture of bisphenol, 4-nitrophthalonitrile and an excess amount of anhydrous potassium carbonate stirred in dry dimethyl sulfoxide resulted in phthalonitrile monomers:

$$\mathrm{HOROH} + 2\ O_2N\text{-}C_6H_3(CN)_2 \xrightarrow[\mathrm{DMSO}]{K_2CO_3} (NC)_2C_6H_3\text{-}ORO\text{-}C_6H_3(CN)_2$$

HEAT

POLYMER

The HOROH component may be bisphenol-A, bisphenol-S, dihydroxy-biphenyl, resorcinol, hydroquinone or bisphenol-A6F:

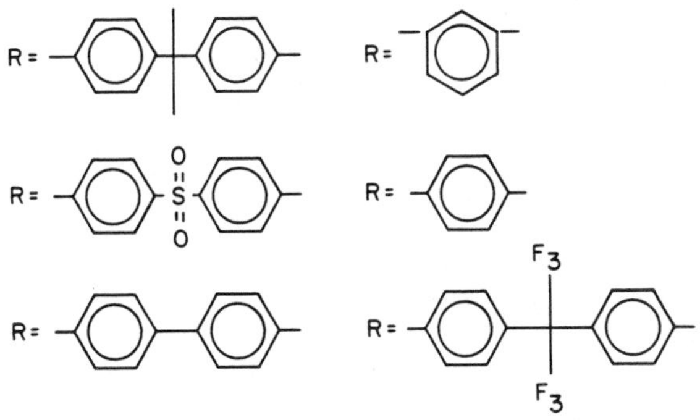

These resins have been shown to exhibit even better properties than the first-generation phthalonitriles in terms of thermal and oxidative stability and moisture sensitivity. But, most important of all, this new reaction pathway is short and simple, and takes advantage of inexpensive starting materials. Thus, it seems to have opened up possibilities for designing polymers with a wide variety of structural variations, and for lowering the material cost to make the resins competitive with other existing composite matrix systems.

Thermal Analysis. Differential scanning calorimetry (DSC) was carried out for all resins using a Perkin-Elmer DSC-2 system. The scanning was performed over the temperature range of 27°C to 267°C at a heating rate of 20°C/min under a dry nitrogen atmosphere. Thermogravimetric analysis (TGA) was also carried out for cured resins by using a DuPont 900 Series unit. Samples were analyzed over the temperature interval of 20°C to 700°C by determining sample weight loss as a function of temperature at a heating rate of 10°C/min while continuously purged with dry nitrogen.

Torsional Pendulum Analysis (TPA). A freely oscillating torsional pendulum (7) operating at ca. 1 Hz was used for the determination of dynamic shear modulus of all cured samples as a function of temperature. The procedure recommended in ASTM-D-2236-70 was followed.

Fracture Energy. The fracture energy of cured phthalonitrile samples was measured by using standard compact tension specimens (8). A precrack was always introduced at the end of the saw-cut with a razor blade. Specimens were fractured in an INSTRON at a crosshead speed of 0.125 cm/min. The critical failure load was measured and polymer fracture energy determined by using the equation of Schultz (9).

Results and Discussion

DSC results of uncured diether-linked phthalonitrile resins, shown in Figure 1A, indicated a single melting peak for an initial scan. The melting temperatures so determined ranged from 199°C for the bisphenol-A linked resin to 234°C for the bisphenol-A6F resin. Each repeated DSC scan always showed an exotherm at about 127°C (see Figure 1B) or a lower temperature, indicative of the formation of a meta-stable crystalline phase upon heating and cooling. Furthermore, multiple, broadened melting was evident during the second scan, a situation similar to that observed in aliphatic-linked phthalonitrile resins that was "B"-staged in prepregging operations (2). This thermal behavior may be used to develop proper degree of B-staging for prepregs of these resins. The "B"-staged prepolymer is non-reacting at room temperature, an appealing feature of these materials for the preparation of prepregs.

Neat phthalonitrile monomers normally were polymerized by melting the resin and then continuously heating them in air at 280°C for 6 days. The relative thermal stabilities in air of the cured resins were studied by heating the sample isothermally and observing the weight loss. No observable weight loss was found after heating at 250°C for 1000 hours. At 280°C, the biphenyl linked polymer exhibited only 1.6% weight loss after 3200 hours. The next most stable polymer was that formed from resorcinol, followed by bisphenol-S, bisphenol-A and bisphenol A6F in that order. TGA results, shown in Figure 2, also indicated that no weight loss could be detected for any polymer up to 300°C. As the temperature was further increased, respective weight loss data showed that the biphenyl-linked polymer was again the most stable material, whereas the bisphenol-S linked polymer was the least stable of the four systems studied.

The water absorptivity of cured resins was found to depend greatly on the linking structure R. Polar groups in the linking structure are expected to attract water. The results of gravimetric measurements, shown in Figure 3, indicated that the bisphenol-A and bisphenol-A6F linked polymers had very low affinity for water, absorbing less than 1.5% water on immersion. On the other hand, the bisphenol-S linked polymer absorbed more water (up to 3.5%), which may be attributed to the presence of the polar sulfone group.

Figure 4 shows the TPA results for the bisphenol-A and the bisphenol-S linked polymers cured at 280°C for six days. Both the dynamic shear modulus and the mechanical loss factor are given as a function of temperature from -150°C to about +300°C. During a TPA run, a temperature scan covering the complete glass-to-rubber transition could not be achieved because the sample softened as the glass transition temperature, T_g, was approached.

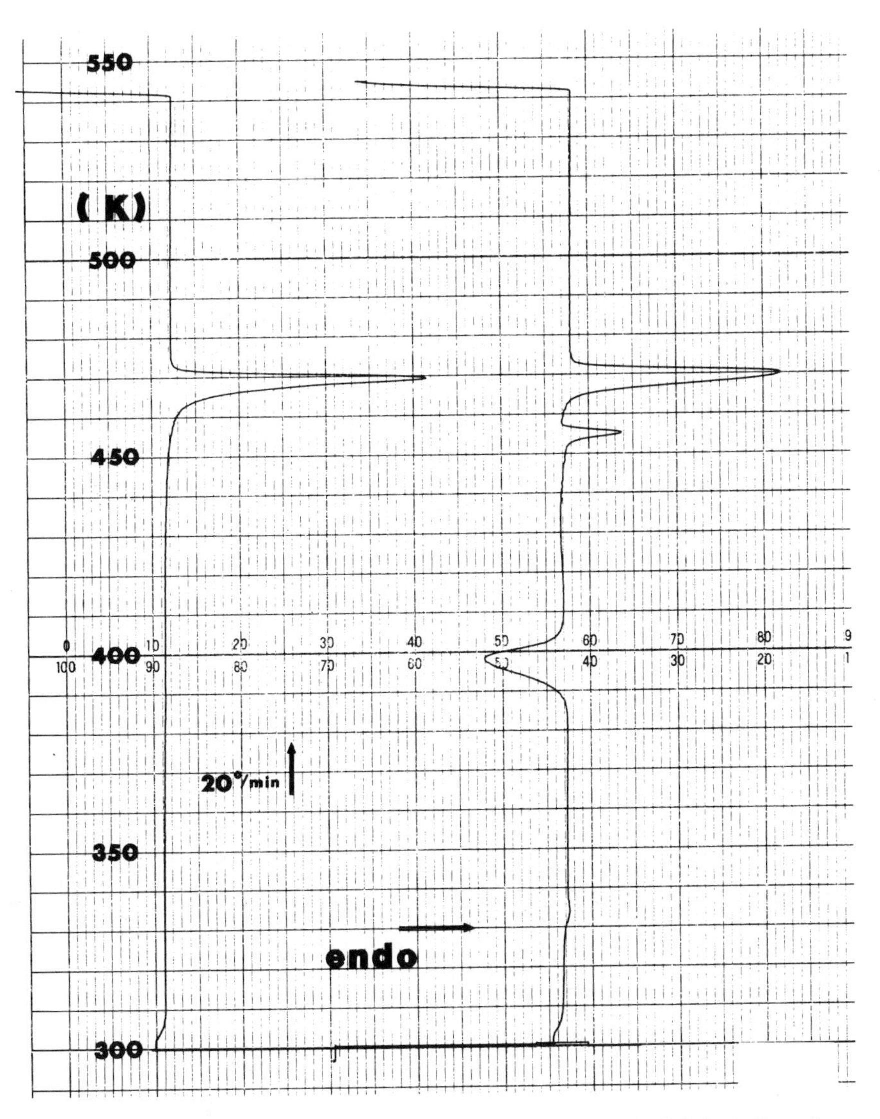

Figure 1. Differential scanning calorimetry of a bisphenol-A phthalonitrile resin.

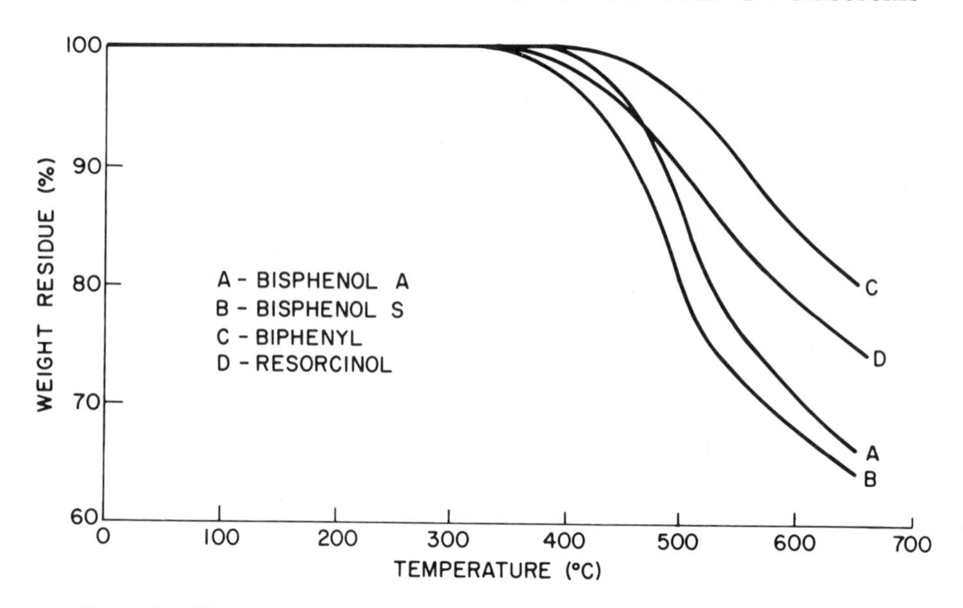

Figure 2. Thermogravimetric analysis results of various diether-linked phthalonitrile resins cured for 6 days at 280°C.

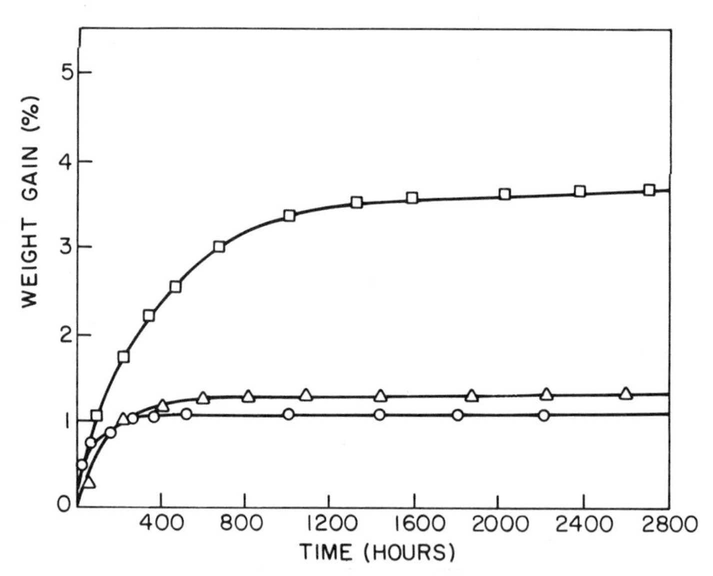

Figure 3. Water absorption of various phthalonitrile polymers. Key: □, bisphenol S; △, bisphenol A; and ○, bisphenol A6F.

Therefore, in this study T_g was taken as the temperature at which the modulus decreased rapidly with a corresponding increase in the value of the loss factor. This was usually the point where the TPA experiment had to be terminated due to sample softening. In addition to the glass transition process, two secondary relaxation processes in the glassy state were observed at approximately -50°C and +150°C, respectively. The peak at -50°C may be attributed to the effect of absorbed water (10). The peak intensity is only 0.04, as opposed to 0.1 for the cured C-10 phthalonitrile (11). This is another indication that the diether-linked phthalonitrile resins are less sensitive to moisture than the diamide-linked systems. The dynamic loss curve for the C-10 polymer (11) also showed a transition at -130°C, related to the molecular motions of the amide linkages and the alkyl chain between them. This peak was not observed in the diether-linked polymers, presumably because the aromatic linkages are more rigid and would not allow intra-chain segmental motions to take place at such a low temperature. The relaxation peak at 150°C is similar to that found in the C-10 polymer cured at 240°C for five days (11). The origin of this peak is still unknown at the present time. The mechanical properties including the fracture energy, bending modulus, shear modulus and the glass transition temperature of the polymers are shown in Table I. The properties of a tetrafunctional epoxy (Narmco 5208), cured according to a manufacturer's suggested procedure, are also included in Table I for comparison. Although the cured phthalonitrile resins are only marginally tougher than the epoxy and comparable in stiffness, they offer higher use temperature (or T_g). The exceptional case was the bisphenol-A6F linked resin, which is extemely brittle and has a very low glass transition temperature.

Based on these results, some phthalonitrile resins were eliminated from further study. The bisphenol-S phthalonitrile resin was not considered because of its high water absorptivity. The melting point of hydroquinone-linked monomers was higher than that of the other resins. The texture, color and solubility of this compound also varied from batch to batch although the melting temperature and IR and proton nmr spectra were nearly identical. The bisphenol-A6F compound was quite attractive based on its thermal stability and water up-take. However, the precursor material for the synthesis of this monomer is no longer available. The mechanical properties of dihydroxybiphenyl phthalonitrile did not represent any improvement over those of the bisphenol-A polymer. The TGA result showed that the biphenyl polymer was more stable, as expected from its linking structure, but the precursor was much more expensive than bisphenol-A. Based on the considerations of modulus, glass transition temperature, thermal stability, water absorptivity as well as material cost, it was decided to examine further the properties of bisphenol-A linked phthalonitrile resin.

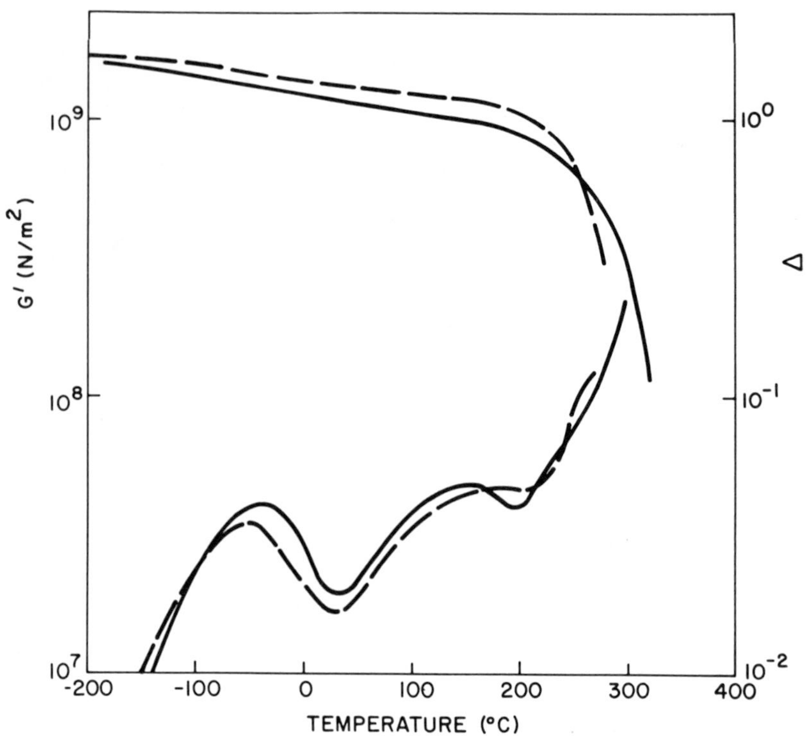

Figure 4. Dynamic mechanical properties of bisphenol-linked phthalonitrile polymers. Key: ——, bisphenol A; and – – –, bisphenol S.

Table I. Room Temperature Mechanical Properties of Cured Phthalonitrile Resins

Linking Structure	Fracture Energy (J/m^2)	Bending Modulus (GN/m^2)	Shear Modulus (GN/m^2)	T_g (C)
Bis-A	99	3.92	1.20	320
Bis-S	202	2.25	1.32	300
Bi-phenyl	140	4.35	1.22	300
Bis-A6F	53	4.02	1.25	225
5208 Epoxy	76	3.89	1.25	260
90:10 mixture*	60	5.18	1.74	246

*90% bisphenol-A phthalonitrile and 10% bisphenol-A.

The polymerization of any phthalonitrile resin could be initiated by simply heating the resin monomer at a temperature higher than its melting temperature. However, as was indicated earlier, the neat polymerization of bisphenol-A linked phthalonitrile resin required several days of continuous heating at 260-290°C before a viscosity increase became detectable in the melt. Extended post-cure was also necessary for the development of adequate mechanical strength, because extremely pure phthalonitrile resin was stubbornly resistant to polymerization. In order to reduce the cure time, bisphenol compounds were used as coreactants in the polymerization reactions. The results indicated that as long as the initiating agent and the monomer were compatible, both the temperature and the time for gelation could be greatly diminished. The mechanisms involved in this accelerated cure were discussed in a recent report (12). It is believed that bisphenol probably attacks the nitrile groups of the phthalonitrile monomer to afford initially an 1-aryloxyisoindolenine unit, which reacts with other nitrile groups to form polymeric materials.

ROH + CN CN → OR N NH + CN CN (excess)
↓
RO N N H n

Mixtures of the bisphenol-A linked resin monomer and bisphenol-A coreactant in various molar ratios were prepared and DSC analyses carried out. The results suggested that a 90:10 mixture sample was most acceptable, and the DSC scan for such a sample is shown in Figure 5. A weak endotherm at 137°C indicates the presence of bisphenol-A coreactant, whereas the position of the melting peak of the monomer is practically unchanged from that of the neat resin (Figure 1A). After staging the sample at 277°C for 30 minutes, an isothermal time-scan on DSC showed the appearance of an exotherm, indicating the initiation of polymerization after an induction period of about 15 minutes. This mixture sample was heated at 210°C for 48 hours with gelation occurring in less than 24 hours (i.e. overnight). TGA results indicated that a post-cure of the sample at 250°C for 46 hours produced a polymer which had a lower rate of decomposition than that from the neat polymerization of 6 days at 280°C. An infra-red spectroscopic technique, developed by Marullo and Snow (13), was used to monitor the time required for the disappearance of the C≡N group as an indication of the advancement of the cure. This result is given in Table II, where the effect of added bisphenol-A in accelerating the cure of bisphenol-A phthalonitrile is evident. Carbon-13 nmr spectra were also obtained for this mixture in perdeutero-dimethyl sulfoxide (DMSO-d_6). Samples of the mixture were heated for 3.5 and 7 hours at 200°C, cooled and ground to a fine powder form. 200-mg portions of the powder samples were then dissolved in 2 ml of DMSO-d_6. The resin mixture was then heated at 200°C for 3, 5 and 7 hours while stirred in 2 ml of DMSO-d_6. Not all of the reacted material dissolved, however. The aromatic carbon region of the carbon-13 nmr spectrum of an unheated sample is shown in Figure 6. The four lines marked with arrows are from the bisphenol-A resin. The relative intensities do not necessarily reflect the relative concentration of the two components. But the intensities of these four lines did decrease with the heating time of the mixture until after 7 hours they were barely observable. The DMSO-d_6 solutions of the heated mixtures were intensely green, indicative of the presence of some type of reaction product. However, no new lines were evident in the spectra of the heated samples. This fact was attributed to very low concentrations of soluble reaction products, below the level required for carbon-13 nmr detection.

Mechanical test data for the 90:10 mixture sample cured at 210°C for 48 hours followed by a two day post cure at 250°C were included in Table I. Although the TGA result seemed to indicate this to be a desirable cure schedule, the resulting polymer was definitely too brittle. Post-cure at 250°C was also insufficient in providing a glass transition temperature higher than that of the epoxy sample. Furthermore, bisphenol-A was proven to be too volatile at the required cure temperatures of higher than 200°C. To alleviate these problems, new polyphenols have been synthesized and are presently being evaluated as potential curing agents:

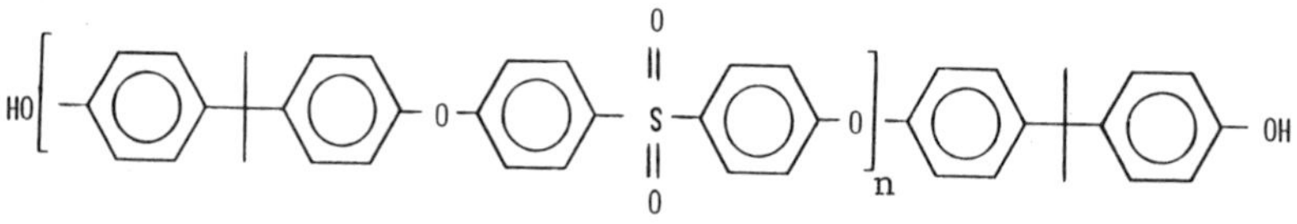

HO[–C₆H₄–C(CH₃)₂–C₆H₄–O–C₆H₄–SO₂–C₆H₄–O]$_n$–C₆H₄–C(CH₃)₂–C₆H₄–OH

where n=3 or 9. TGA results of samples of bisphenol-A phthalonitrile cured with the polyphenols at 250°C for 21-48 hours, followed by approximately one day post curing at 280°C, are shown in Figure 7 (curves B and C). Compared to neat bisphenol-A phthalonitrile cured for 7 days at 280°C (curve A), they volatilized at a greater rate at lower temperatures. However, the rate decreased as temperature increased and at 700°C there was about 12% more weight residue. Furthermore, it should also be noted that high molecular weight elastomeric polymeric materials containing polysulfone-ether linkages and terminated by ether units are commercially available. The approach

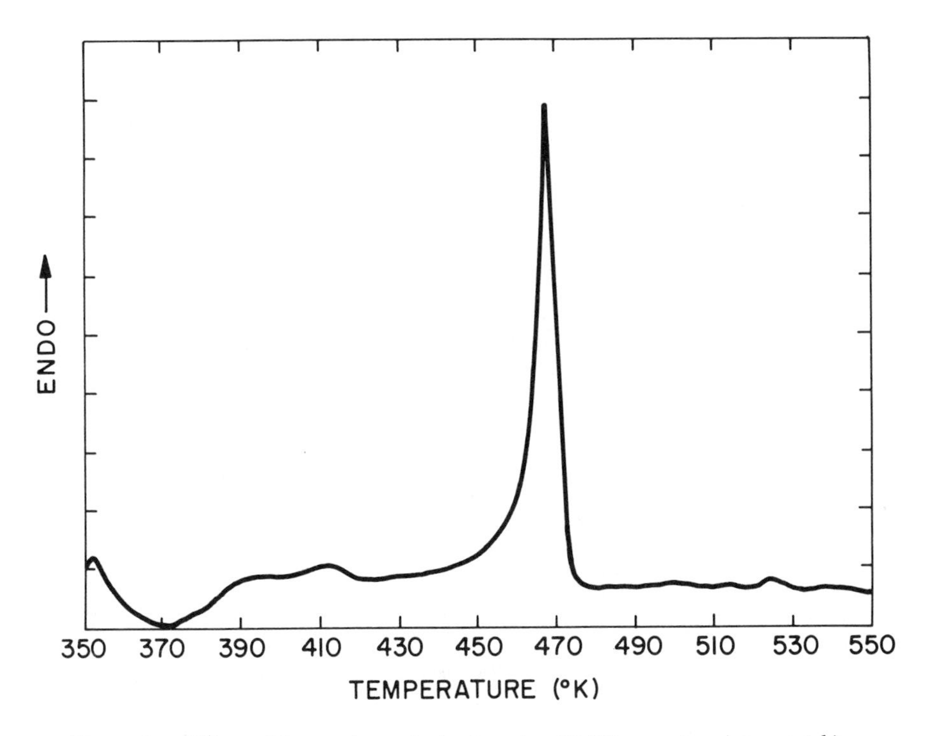

Figure 5. Differential scanning calorimetry of a 90/10 sample mixture of bisphenol-A phthalonitrile and bisphenol-A.

Table II. IR Monitoring of the Cure of Bisphenol-A Phthalonitrile Resins

	Time (hours) required for	
Sample	50% loss of C≡N	75% loss of C≡N
Neat resin cured at 300°C	27	43
Neat resin cured at 280°C	62	240
90:10 mixture* sample cured at 210°C	7	12

*90% bisphenol-A phthalonitrile and 10% bisphenol-A.

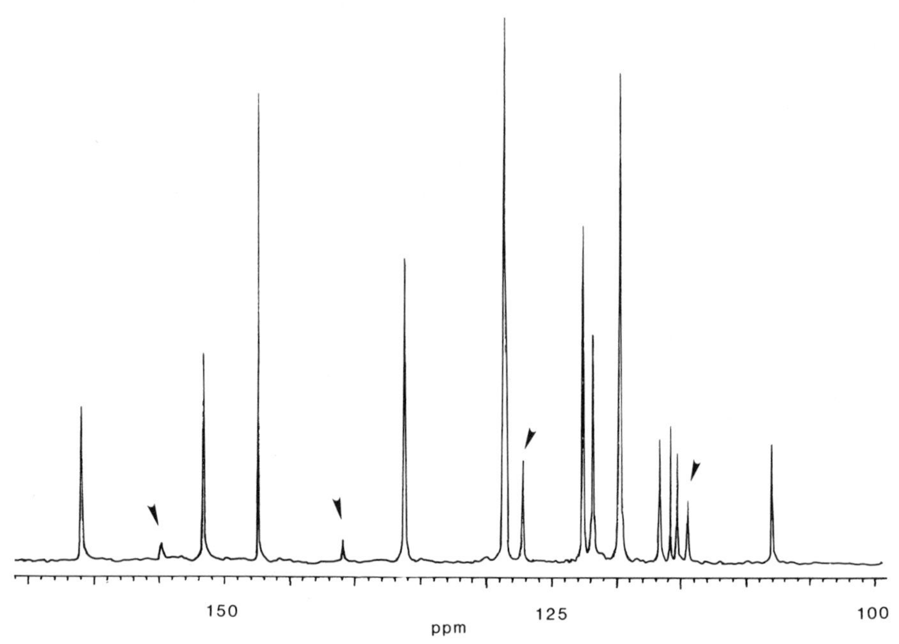

Figure 6. Aromatic portion of the NMR spectrum of an unheated mixture sample.

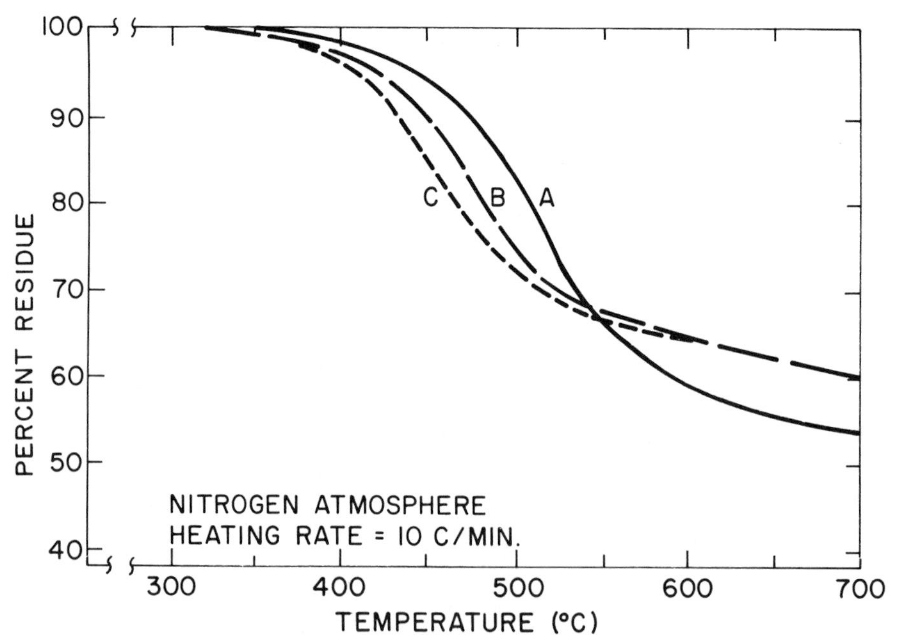

Figure 7. Thermogravimetric analysis results of bisphenol-A phthalonitrile resins cured for 7 days at 280°C (A), 22 h at 250°C and 22 h at 280°C with 15% polyphenol (B), and 48 h at 250°C and 24 h at 280°C with 20% polyphenol (C).

presented here therefore suggests that it would be possible to economically produce a corresponding rubbery, high temperature stable polyphenol. Such a polyphenol, in addition to expediting the cure of bisphenol-A linked phthalonitrile monomers, could also generate small elastomeric domains for the toughening of the base resin. Similar approaches involving the reaction of either CTBN or ATBN rubbers with epoxy resins have been used to toughen epoxy polymers (14). When used as composite matrix materials, the rubber-modified resins also show distinct advantages in improving the processability and bondability and in increasing the fatigue design limit of the composite (15).

Acknowledgement

This work was sponsored by the Naval Air Systems Command. The authors gratefully acknowledge the assistance of Paul Peyser, Robert L. Cottington and Harry C. Nash in portions of the experimental work.

Literature Cited

1. Griffith, J. R., O'Rear, J. G., and Walton, T. R. in "Copolymers, Polyblends and Composites", ed. N. A. Platzer, Advances in Chem. Ser. of ACS, 1975, 142, 458.
2. Ting, R. Y. and Nash, H. C., Polym. Eng. Sci., 1981, 21, 441.
3. Walton, T. R., Griffith, J. R. and O'Rear, J. G. in "Adhesion Science and Technology", Polym. Sci. Tech., ed. L. H. Lee, Plenum Press, N. Y., 1975, Vol. 9, p. 655.
4. Bascom, W. D., Bitner, J. L., and Cottington, R. L. in "High Performance Composites and Adhesives for V/STOL Aircraft", ed. L. B. Lockhart, Jr., NRL Memo Rept. 4005, Washington, DC (May 1979).
5. Poranski, C. F. and Moniz, W. B., in "Resins for Aerospace," ed. C. A. May, Am. Chem. Soc. Symp. Ser., 1980, 132, 337.
6. Keller, T. M. and Griffith, J. R., ACS Organic Coat. Plast. Chem. Preprint, 1979, 40, 781.
7. Nelson, L. E., Mechanical Properties of Polymers and Composites, Dekker, N. Y., 1974.
8. Knott, J. F., Fundamentals of Fracture Mechanics, Butterworths, England, 1973.
9. Schultz, W., in Fracture Mechanics of Aircraft Structures, ed. H. Liebowitz, AGARDograph, AGARD Rept. AG-176, NTIS, Springfield, VA, 1974.
10. Gillham, J. K., ACS Organic Coat-Plast. Chem. Preprint, 1978, 38, 598.
11. Bascom, W. D., Cottington, R. L., and Ting, R. Y., J. Materials Sci., 1980, 15, 2097.
12. Keller, T. M., Price, T. R., and Griffith, J. R., ACS Org. Coat. Plast. Chem. Preprint, 1980, 43, 804.

13. Marullo, N. P. and Snow, A., ACS Polymer Preprint, 1981, 22, 46.
14. Riew, C. K., Rowe, E. H., and Siebert, A. R., in Toughness and Brittleness of Plastics, ed. Dennin and Crugnola, ACS Advances in Chemistry Series, 1976, 154, 326.
15. Moulton, R. J. and Ting, R. Y., to appear in the Proc. Int'l Conf. Composite Structutes, Sept. 16-18, 1981, Paisley, Scotland.

RECEIVED December 15, 1981.

27

Polyamides Via the Phosphorylation Reaction: Poly-*p*-benzamide (PBA) and *N*-Methyl PBA Copolymers

J. PRESTON[1] and W. R. KRIGBAUM
Duke University, Department of Chemistry, Durham, NC 27706

J. ASRAR[2]
Monsanto Triangle Park Development Center, Inc.,
Research Triangle Park, NC 27709

The monomers 4-N-(4'-aminobenzamido) benzoic acid I and 4-N-methyl-4-N-(4'-aminobenzamido) benzoic acid II were polymerized via the Yamazaki phosphorylation reaction to yield, respectively, poly-p-benzamide (PBA) and a 1:1 copolymer of PBA and poly-p-N-methylbenzamide (PMBA). The PBA obtained from I was found to have a higher inherent viscosity value, 2.4, than any previously reported value for PBA obtained from p-aminobenzoic acid polycondensed via the Yamazaki reaction. The PMBA obtained from II was found to be extraordinarily soluble but to have a relatively low inherent viscosity value, 0.4; only weak and brittle films were obtained from II whereas all of the other polymers could be cast to tough films. Random copolymers prepared from mixtures of I and II or from mixtures of II and p-aminobenzoic acid were found to have inherent viscosity values that decreased monotonically with N-methyl substitution. Nevertheless, all of the polymers had approximately the same molecular weight, i.e., $M_w = \sim 8,000$. Thermal stability of the copolymers was not greatly effected at mole percent substitutions up to and including 25 percent.

Several routes for the synthesis of poly-p-benzamide (PBA) have been reported. The routes[1-2] that yield high molecular weight PBA (i.e., $\eta_{inh} \geq 3.5$ determined in conc. sulfuric acid) have employed acid chloride monomers, such as I or II (1-2).

[1] Current address: Monsanto Textiles Company, Pensacola, FL.
[2] Current address: University of Lowell, Lowell, MA.

0097-6156/82/0195-0351$06.00/0
© 1982 American Chemical Society

$$HCl\cdot NH_2-C_6H_4-CO-Cl \longrightarrow \left[-NH-C_6H_4-CO-\right]_x \qquad [1]$$

I PBA

$$HCl\cdot NH_2-C_6H_4-CO-NH-C_6H_4-CO-Cl \longrightarrow PBA \qquad [2]$$

II

The properties of the PBA prepared from I or from II appear to be the same despite a report (3) that the properties are not comparable for PBA prepared [2-3] *via* the solid state polymerization of monomer (III) and monomer (IV) with preformed amide linkages. Probably the differences in properties between the PBA prepared by [3] and [4]

$$C_6H_5-O-CO-NH-C_6H_4-CO-OH \xrightarrow{\Delta} PBA \qquad [3]$$

III

$$C_6H_5-O-CO-NH-C_6H_4-CO-NH-C_6H_4-CO-OH \xrightarrow{\Delta} PBA \qquad [4]$$

IV

can best be ascribed to differences in molecular weight (η_{inh} = 0.17 and 1.5, respectively).

A highly convenient synthesis [5] for the preparation of PBA from *p*-aminobenzoic acid (PAB) has been reported by Yamazaki (4).

$$NH_2-C_6H_4-CO-OH \xrightarrow[\text{pyridine}]{(\phi O)_3P} PBA \qquad [5]$$

PAB

However, only moderate molecular weight (i.e., η_{inh} = 1.7 $\pm$ 0.2) PBA has been prepared by this route.

In the present work, monomers (e.g., V) having preformed amide linkages (as in II and IV) have been used in the Yamazaki reaction

$$NH_2-\langle\bigcirc\rangle-CO-NH-\langle\bigcirc\rangle-COOH \longrightarrow PBA \qquad [6]$$

V

in an effort to obtain higher molecular weight PBA than that using p-aminobenzoic acid as monomer. Another object of this work is to enhance the solubility of PBA copolymers containing N-methyl substituted amide linkages. Thus, monomers (e.g., VI) having preformed N-methyl amide linkages have been used in the Yamazaki reaction to make PBA copolymers of varying degrees of N-methyl substitution [7-9].

$$NH_2-\langle\bigcirc\rangle-CO-\underset{}{N}(CH_3)-\langle\bigcirc\rangle-COOH \longrightarrow$$

VI

$$\left[-NH-\langle\bigcirc\rangle-CO-N(CH_3)-\langle\bigcirc\rangle-CO-\right]_x \qquad [7]$$

$$V + VI \longrightarrow \left[\left(-NH-\langle\bigcirc\rangle-CO-\right)_3-N(CH_3)-\langle\bigcirc\rangle-CO-\right]_x \qquad [8]$$

$$VI + n\ PAB \longrightarrow \left[\left(-NH-\langle\bigcirc\rangle-CO-\right)_{n+1}-N(CH_3)-\langle\bigcirc\rangle-CO-\right]_x \qquad [9]$$

The synthesis of some aromatic polyamides (e.g., VII)

$$\left[-N(CH_3)-\langle\bigcirc\rangle-N(CH_3)-CO-\langle\bigcirc\rangle-CO-\right]_x$$

VII

of moderately high molecular weight ([η] up to 1.3) containing N-methyl amide linkages have been reported previously, but these polymers have been produced by methods employing acid chlorides (5,6). Copolymers having varying degrees of N-methyl groups apparently have not been reported previously. A polymer corresponding to VII, but having N-phenyl groups, has been reported to be fiber forming (7). Also, the conversion of poly-p-phenyleneterephthalamide to an N-alkylated (C_{3-18}) product has been described (8); however, the inherent viscosity values for these polymers are extremely low (< 0.15), suggesting that considerable degradation of the polymers probably occurs during their synthesis.

Discussion and Results

Preparation of Monomers. Monomers V and VI were prepared by reduction of the corresponding nitro intermediates which were synthesized by reaction of p-nitrobenzoyl chloride with, respectively, p-aminobenzoic acid and N-methyl-p-aminobenzoic acid. The properties of the monomer and corresponding intermediates are given in Table I.

Table I

Melting Points of Monomer and Intermediates

	M.P., °C[a]
V	274-276[b] (340-342)[c]
VI	247-248 (196-198)

a. Values in parentheses are for the corresponding dinitro intermediate.

b. Determined by heating sample in capillary m.p. tube; at the indicated temperature, the sample bubbles and resolidifies (polymerizes?) to a white solid that turns brown ∿ 380°C.

c. Determined by heating sample in capillary m.p. tube; the sample "gatherers" at ∿ 330°C, melts at the indicated temperature and resolidifies.

Polymerization. Monomers V and VI were polycondensed at 100°C using standard Yamazaki reaction conditions. A copolymer was prepared in a similar manner by polymerization of equimolar amounts of V and VI [8]. VI also was copolymerized [9] with different amounts of p-aminobenzoic acid (PAB) to yield PBA copolymers having varying degrees of N-methyl amide substitution. The several polymers, with their mole percent N-methyl PBA contents and inherent viscosities, are given in Table II.

Table II

Molecular Weights of PBA and Copolymers

No.	Monomers(s):	% $N-CH_3$ groups	Sulfuric Acid η_{inh} [a]	Sulfuric Acid M_w	Organic Solvents[b] M_n	Organic Solvents[b] M_w	Organic Solvents[b] M_w/M_n
1.	V	0.0	2.35	8,400	51,600	182,000	3.53
2.	PAB	0.0	1.85	--	--	--	--
3.	PAB + VI	5.0	1.28	--	18,200	65,800	3.62
4.	PAB + VI	11.1	1.10	--	12,600	44,100	3.50
5.	II + VI	25.0	0.65	--	--	--	--
6.	PAB + VI	25.0	0.58	7,600	3,700	10,700	2.91
7.	VI	50.0	0.35	7,600	3,300	7,000	2.10
8.	$CH_3-NH-C_6H_4-COOH$	100.0	0.1[c]	--	--	--	--

a. Determined in conc. sulfuric acid on a 0.1% solution at 30°C.
b. Determined in N-methylpyrrolidone containing 5% LiCl and 1% H_2O.
c. Ref. 9.

The PBA produced from monomer V has the highest inherent viscosity reported to date for any PBA prepared via the Yamazaki reaction. The reason for this probably lies in the fact that fewer by-products are produced in the polycondensation of V compared to that of PAB. Hence, there is less chance for degradation of the PBA obtained. An extrapolation of a plot of the number of preformed amide linkages in monomers that may be used to produce PBA shows that this approach probably could yield higher molecular weight polymer, but that a large number of preformed linkages would be required to produce PBA having an inherent viscosity > = 3.5. The latter value is needed for PBA to be used for the preparation of commercial quality fibers. It can be seen (Figure 1) that the inherent viscosity values for the polymers of Table II decrease monotonically with N-methyl substitution. However, from these data it cannot be concluded whether the drop in the inherent viscosity values is due to a decrease in molecular weight or whether the N-methylated polymers are more flexible and have a more highly coiled conformation. It was necessary, therefore, to determine the molecular weights of the several polymers to answer this question (see next section).

The polymerization of N-methyl-p-aminobenzoic acid has been reported to give polymer having a very low inherent viscosity, $\eta_{inh} = 0.1$ (9). One of the reasons for this probably is related to the low reactivity of the substituted amino group under the Yamazaki reaction conditions. Preforming the N-methyl linkages removes the obstacle of having to form this linkage during polycondensation. The N-methyl amide linkage appears to be stable during the course of the formation of polymers via the Yamazaki reaction because all of the polymers produced here, both substituted and unsubstituted, have approximately the same molecular weight.

Molecular Weight Determinations. An attempt was made to determine M_w and M_n from light scattering and GPC data, respectively, collected in N-methylpyrrolidone (NMP) containing 5% LiCl and 1% water. The results appear in columns 6 and 7 of Table II. Clearly there was aggregation in this solvent, although the extent of aggregation decreases progressively with the degree of N-methylation. As shown in column 8, only for the copolymer having 50% N-methylation (number 7) does the M_w/M_n ratio approximate the value, 2.0, expected for a most probable molecular weight distribution. Light scattering data were therefore collected in 96% sulfuric acid. These M_w values, shown in column 5 of Table II, indicate that the molecular weight is nearly independent of the degree of methylation. Two conclusions can be drawn from these results. First, one does not obtain molecularly dispersed solutions of PBA in a solvent such as NMP. Although the extent of aggregation is progressively reduced by N-methylation, if does not vanish until 50% of the amide

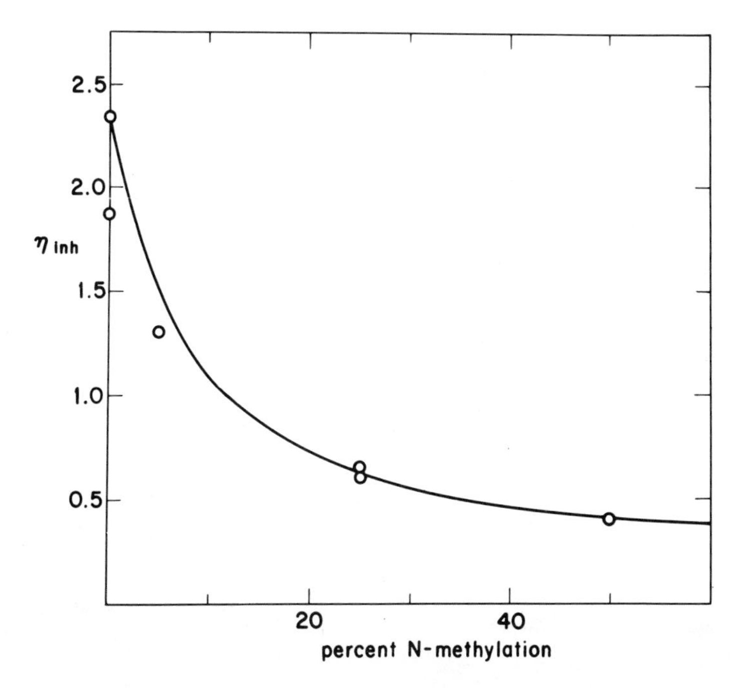

Figure 1. Relationship of inherent viscosity to the degree of N-*methyl substitution.*

nitrogen atoms are methylated. Secondly, the inherent viscosity measured in sulfuric acid decreases by a very large factor, 6.7, as the percentage of N-methyl substitution is increased from zero to 50%. Since M_w remains nearly constant, this decrease must be due to a progressive increase in chain coiling. Tsvetkov and coworkers (6) compared the hydrodynamic properties of poly-*p*-phenyleneterephthalamide (PPD-T) and the fully N-methylated polymer, VII. They found the Kuhn length of the latter is only 1/20th that of the former, and interpreted this as an indication that methylation tends to reduce the fraction of amide bonds having a planar *trans* conformation:

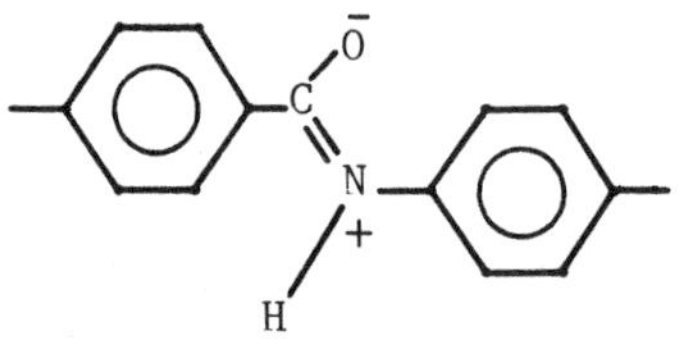

PBA (rodlike) 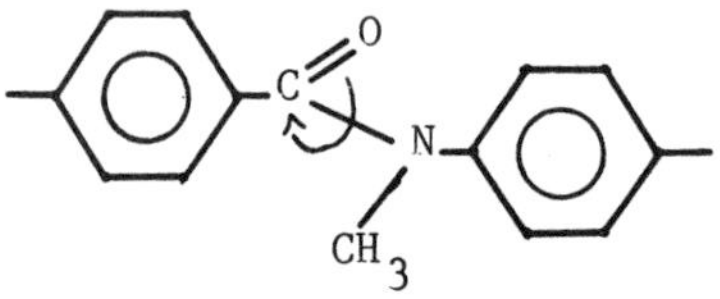 N-CH_3 PBA (non-rodlike)

The results obtained here confirm their findings, and extend them by showing that even a small percent of N-methylation is sufficient to change the hydrodynamic properties of the *para*-linked aromatic polyamide molecule. The agreement between the molecular weights measured for polymer 7 (50% N-methyl substitutive) in 96% sulfuric acid and the N-methyl pyrrolidone with 5% LiCl and 1% water implies that this polymer forms a molecularly disperse solution in the latter solvent. This observation differs significantly from the report by Tsvetkov *et al*. (6) and Koton and Nozova (5) that the fully methylated polymer only formed molecularly disperse solutions in sulfuric acid.

The M_w values obtained by us for PBA in sulfuric acid are in reasonably good agreement with those reported by Schaefgen *et al*. (10) for PBA having comparable inherent viscosity values.

Film Casting. Solutions of the various copolymers were prepared in NMP containing 5% dissolved lithium chloride. The solubility of the copolymer containing 5 mole % N-methylamide groups was similar to that of PBA, although slightly larger. Rather thick solutions were formed at 3-5% solids, and very thick solutions were formed at 9% solids, from the copolymer containing 11 mole % N-methylamide groups; films could be cast readily from these solutions. The same copolymer produced a rubber-like mass at 16% solids. The copolymer containing 25 mole % N-methylamide groups formed solutions of moderate viscosity at 9-10% solids and clear, tough films could be cast. Single filaments could be produced from these solutions by withdrawing a stirring rod from then and coagulating the polymer in water. Solutions of 30-50% solids could be prepared in DMAc and NMP (with or without added salt) from the copolymer containing 50 mole % N-methyl amide groups but these solutions were thin. Glassy films could be cast on glass, but these fragmented when placed in water.

Thermal Properties. Some wholly N-methylated aromatic polyamides have been reported previously and found to be considerably lower in thermal stability than the corresponding unsubstituted polymers. Thus, Koton reported that the softening point of VII was ~ 247°C, a temperature much below that of PPD-T, which shows no softening point below 400°C.

For our series of copolymers, no strong influence by N-methyl substitution on the thermal properties is observed until the degree of substitution exceeds 25 mole percent. This can be seen in Figure 2, which shows the thermogravimetric analysis (TGA) curves of the several polymers. The differential of the TGA losses (DTG) show (Table III) no differences within experimental error, except for the copolymer containing 50 mole percent N-methyl substitution. The DTG value for the latter polymer is approximately 55°C lower than that of the unsubstituted PBA and the other copolymers; the total amount of weight loss at 600°C is also greater. It might be noted that this weight loss is not in proportion to the weight of the methyl groups initially present. The small break in the DTG of the 50% methylated polymer at 215°C may reflect the start of some thermal cracking of the N-methyl group. Although Koton did not show a TGA curve for VII, he did comment on a dual break in the TGA.

Table III

Thermal Properties of PBA and PBA Copolymers

% Me Substituted	Transitions: DSC[a]	DTA[b]	DTG
0	295	528	526
5	-	509	511
11	-	512	514
25	-	521	521
50	225[c], 320	522(320)[d]	429(215)[e]

a. Exotherm, b. Endotherm, c. Weak endotherm, d. Exotherm, e. Second major break in TG curve.

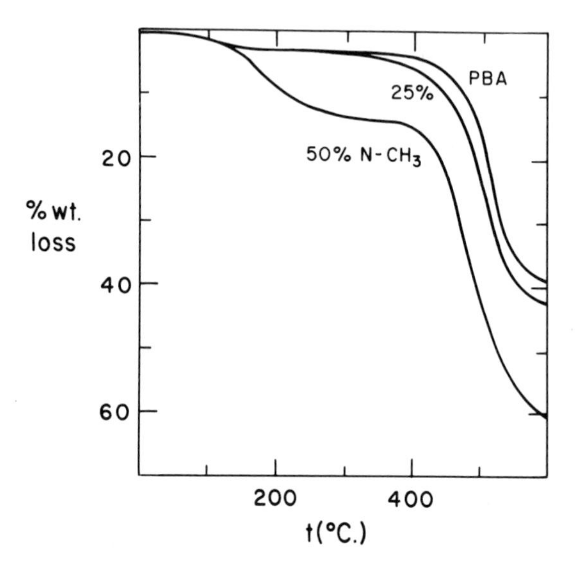

*Figure 2. Weight loss curves for poly-*p*-benzamide and* N*-methyl poly-*p*-benzamide copolymers.*

The differential thermal analysis (DTA) values (Table III) also reflect relatively little effect (i.e., within experimental error) of the presence of the methyl groups until the level of N-methyl substitution reaches 50 mole percent. This result is somewhat surprising, because it might have been predicted that elimination of hydrogen bonding capability might exert a strong influence on the melting behavior, as in the well-known case of nylon 11:

% methyl substitution	m.p., °C
0	186
25	160
50	104
75	66
100	60

It would appear that the inherently stable PBA structure (present in all of the copolymers) controls the DTA melting point, which is about 530°C. Nevertheless, an exothermic transition, also seen in the differential scanning calormetric (DSC) analysis, occurs at 320°C, and is probably associated with thermal fragmentation of the methyl groups.

Experimental

Synthesis of Monomers. 4-N-(4'-Nitrobenzamide)benzoic acid was prepared by reacting in a blender jar 18.6 g (0.1 mole) of *p*-nitrobenzoyl chloride in 30 ml of tetrahydrofuran with 13.7 g of *p*-aminobenzoic acid in 100 ml of water containing 4 g (0.1 mole) of sodium hydroxide. The crude product was collected on a filter, washed, and crystallized from dimethylacetamide (DMAc)/water for a 98% yield of pure product. 4-N-Methyl-4-N-(4'-nitrobenzamido)benzoic acid was prepared in similar manner using sodium carbonate as a acid acceptor. Recrystallization of the crude product from 95% ethanol gave a 68% yield of pure product.

The nitro precursors above were dissolved in DMAc and hydrogenated to the desired monomers in a Parr bomb at 50-55°C and at an initial pressure of about 60 psi. After evaporation of DMAc on a rotary evaporator, the products were recrystallized from ethanol in 95% yield.

Polymerizations. In a typical preparation, 0.006 mole of monomer was dissolved in 30 ml of N-methylpyrrolidone containing 4% lithium chloride and 10 ml of pyridine. The reaction mixture was heated to 100°C while the flask was swept with a flow of nitrogen and 0.006 moles of triphenylphosphite was added. After a 6 hour reaction time, the viscous mass was poured into a blender jar, collected on a filter and dried under vacuum at 80°C for 12 hours.

Acknowledgment

The authors wish to express their appreciation to Dr. A. S. Kenyon of Monsanto Company for the molecular weight determinations by light scattering and GPC. This work was supported by National Science Foundation Grant DMR 78-8451 to W. R. K.

Literature Cited

1. Pikl, J., U.S. Pat. 3,541,056 (1970), assigned to the DuPont Company.
2. Johnson, T. A., U.S. Pat. 3,719,642 (1973), assigned to the DuPont Company.
3. Iwakura, Y.; Uno, K., Chau, N., J. Polym. Sci., 1973, 11, 2391.
4. Yamazaki, N.; Matsumoto, M.; Higashi, F., J. Polym. Sci. 1975, 13, 1373.
5. Koton, M. M.; Nosova, G. E., Vysokomol. Soed., 1978, B20, 711.
6. Tsvetkov, V. N.; Koton, M. M.; Shtennikova, I. N.; Lavrenko, P. N.; Peker, T. V.; Okatova, O. V.; Novakowskii, V. B.; Nosova, G. I., Vysokomol. Soed., 1979, A21, No. 8, 1711; in translation, Polymer Sciences USSR, 1980, 21, 1883.
7. Stephens, C. W., U.S. Pat. 3,296,201 (1967), assigned to the DuPont Company.
8. Takayanagi, M.; Katayose, T., J. Polym. Sci., Polym. Chem. Ed., 1981, 19, 1133.
9. Preston, J.; Hofferbert, W. L., J. Polymer Sci., Polymer Symposia, 1978, 65, 13 (1978).
10. Schaefgen, J. R.; Foldi, V. S.; Logullo, F. M.; Good, V. H.; Gulrich, L. W.; Killiam, F. L., Polymer Preprints, 1976, 17(1), 69.

RECEIVED January 7, 1982.

28

Polymers and Fibers of *para*- and *meta*-Phenylene Oxadiazole–*N*-Methyl Hydrazide Copolymers: High Strength–High Modulus Materials of Increased Molecular Flexibility

HARTWIG C. BACH and HELMUTH E. HINDERER

Monsanto Company, Pensacola Technical Center, Pensacola, FL 32575

All-p-phenylene oxadiazole/N-methyl hydrazide copolymers have yielded high performance fibers with tenacities of 18-21 gpd and moduli above 300 gpd through an unconventional polycondensation reaction of terephthalic acid, dimethyl terephthalate and hydrazine sulfate in fuming sulfuric acid and subsequent "reaction spinning" into aqueous acid. Fibers spun had been demonstrated to be useful as tire-reinforcing materials of good fatigue characteristics.

Partial replacement of para- with meta-phenylene rings in these oxadiazole/N-methyl hydrazide copolymers yielded fibers with good tenacities (12-15 gpd). Yarn elongation-to-break and toughness were found to be increased substantially with increasing meta/para ratios. The Goodrich Disk Fatigue Test showed tire cord of partially meta-oriented oxadiazole/N-methyl hydrazide copolymers to be superior in fatigue properties to their all-p-phenylene homologues.

As expected, inherent viscosities of the copolymers prepared under identical reaction conditions decreased with increasing m-phenylene content while the rate of methyl hydrazide formation was similar for polymers of 0 and 50% meta substitution.

Recently, we reported (1, 2) a new class of high strength/high modulus fibers prepared from all-p-phenylene oxadiazole/N-methyl hydrazide copolymers. Oxadiazole homopolymers and fibers had been described earlier by Iwakura (3), Imai (4), and Frazer (5). All-p-phenylene oxadiazole/N-methyl hydrazide copolymer fibers were shown by us to give tire cord with excellent end use performance including good fatigue characteristics. In concurrent

0097-6156/82/0195-0363$06.00/0
© 1982 American Chemical Society

investigations, we studied compositions in which molecular flexibility and, hopefully, thereby yarn elongation-to-break and wear resistance would be further increased by replacement of some of the para- with meta-phenylene units in the polymer backbone. These hopes were fulfilled as reported in this paper.

Fibers (6) of p-phenylene oxadiazole/N-methyl hydrazide copolymers were prepared by polymerization of terephthalic acid (TA), dimethyl terephthalate (DMT) and hydrazine sulfate in fuming sulfuric acid (Oleum) and "reaction spinning" of the resulting polymer solution into aqueous sulfuric acid. Replacement of some of the dimethyl terephthalate used with dimethyl isophthalate (DMI) allowed the preparation of p-/m-phenylene oxadiazole/N-methyl hydrazide copolymers. In the polymerization, a p-/m-phenylene oxadiazole polymer is formed which is methylated during the reaction:

$$HOOC-C_6H_4-COOH + CH_3OOC-C_6H_4-COOCH_3 + H_2N-NH_2 \cdot H_2SO_4$$

$$\xrightarrow[\text{IN OLEUM}]{120-150^\circ C} \left[\left(C_6H_4-\text{oxadiazole}\right)\left(C_6H_4-\text{N-}CH_3\text{-oxadiazolium}^{\oplus}\right)\right] \; HSO_4^{\ominus}$$

During the "reaction spinning" process, the methylated oxadiazole units are hydrolyzed to mono-methylated hydrazide linkages yielding fibers of the p-/m-phenylene oxadiazole/N-methyl hydrazide copolymers:

$$\left[\left(C_6H_4-\text{oxadiazole}\right)\left(C_6H_4-\text{N-}CH_3\text{-oxadiazolium}^{\oplus}\right)\right] \; HSO_4^{\ominus}$$

$$\downarrow \; +H_2O, \; -H_2SO_4$$

$$\left[\left(C_6H_4-\text{oxadiazole}\right)\left(C_6H_4-\overset{O}{\overset{\|}{C}}-\overset{CH_3}{\overset{|}{N}}-NH-\overset{O}{\overset{\|}{C}}\right)\right]$$

By proper combination of acid-ester/para-meta monomer ratios, a variety of oxadiazole/N-methyl hydrazide copolymers could be prepared and evaluated in fiber form. High molecular weight polymers were easily prepared with inherent viscosities of 3-6. The sulfuric acid dopes were spun to fiber by dry-jet or wet-jet solution spinning into dilute sulfuric acid at ambient temperature in a conventional solution spinning process including a fiber neutralization step as well as a hot-drawing operation @350-400°C. Representative fiber properties are given in Table I.

Table I

Fiber Properties of Aromatic Oxadiazole/N-methyl Hydrazide Copolymers with Partial m-Phenylene Orientation

MONOMER CHARGE Mole %		TENACITY[a] gpd	ELONGATION %	MODULUS gpd	TOUGHNESS g cm/den cm
ACID/ESTER	PARA/META				
100/0	75/25	5.9	2.9	282	.10
90/10	"	6.1	4.1	249	.15
75/25	"	14.4	7.2	303	.56
60/40	"	10.9	6.6	252	.39
50/50	"	9.8	6.2	242	.35
75/25	100/0	9.5	4.0	301	.21
"	90/10	8.4	4.8	276	.23
"	75/25	14.4	7.2	303	.56
"	50/50	12.2	9.4	240	.66
50/50	100/0	14.6	6.3	347	.49
"	90/10	12.8	6.3	291	.44
"	75/25	9.8	6.2	242	.35
"	50/50	9.9	10.9	179	.61
"	25/75	6.8	10.8	133	.49
"	0/100	3.3	15.9	76	.39

[a]Properties given are average of five bundle breaks, 10 cm gauge length, 10% extension/min.

As the data in Table I show, yarn elongation-to-break increases substantially as the polymer backbone is made more flexible by substituting m-phenylene for some of the p-phenylene units. To a lesser degree, elongation is also increased by increasing chain mobility through conversion of oxadiazole to N-methyl hydrazide units. As will be shown later, increased yarn elongation through greater backbone flexibility of these p-/m-phenylene oxadiazole/N-methyl hydrazide copolymers coincides with improved fatigue performance of tire cord made from such yarn.

Inherent chain flexibility also manifested itself quite strongly in a decrease of polymer inherent viscosity with increased m-phenylene content for a series of copolymers prepared under identical polymerization conditions. Data obtained are shown in Table II.

Table II

Effect of Meta-Orientation on Inherent Viscosity

Polymerization: 33% oleum, 140°C, 6 Hrs.

PARA/META (MOLE RATIO)	INHERENT VISCOSITY
100/0	8.4
90/10	6.9
80/20	6.0
75/25	5.6
70/30	5.4
60/40	4.8
50/50	4.5

During the polymerization of aromatic diacids and diesters with hydrazine in fuming sulfuric acid, aromatic oxadiazole units are formed first which are subsequently methylated by monomethyl sulfate or its homologues (derived from reaction of the diester with H_2SO_4) forming an aromatic N-methyl oxadiazolium polymer. This polymer is hydrolyzed during the spinning operation to the fiber polymer, i.e., p-/m-phenylene oxadiazole/N-methyl hydrazide copolymer.

In order to clarify whether the presence of meta-oriented units would change the rate of oxadiazole methylation and thus would change the ultimate oxadiazole/hydrazide ratio in the fiber, identical polymerizations of all-para- and 50/50 para/meta-oriented monomer mixtures were sampled at various intervals, the polymer dope samples were coagulated and hydrolyzed in dioxane/water and the polymer compositions determined. As data in Table III show, essentially identical oxadiazole/N-methyl hydrazide ratios were obtained for the all-para and the mixed para/meta compositions.

One of the main reasons for investigation of the partially _meta_-oriented aromatic oxadiazole/N-methyl hydrazide copolymers was the expectation that their fibers would show even greater fatigue resistance than the all-_para_ compositions. To test this hypothesis, tire cord of a 75/25 para-/meta-phenylene 75/25 oxadiazole/N-methyl hydrazide copolymer was tested against a comparable (in fiber properties, see Table I) 100/0 50/50 all-para-phenylene control. The test was the Goodrich laboratory disk test in which the tire cord sample, encased in a rubber matrix, is being alternated between 7.5% compression and 2.5% elongation and the retention of tensile strength is being measured. As data in Table IV show, fatigue resistance of the partially meta-oriented composition was found to be increased substantially over the all-

Table III

Effect of Meta-Orientation on Polymer Composition

Polymerization: 22% oleum, 137°C

REACTION TIME (Hrs.)	COMPOSITION			
	ALL PARA (50/50, TA/DMT)		50% META (50/50, TA/DMI)	
	INHERENT VISCOSITY	COMPOSITION Ox/Hy	INHERENT VISCOSITY	COMPOSITION Ox/Hy
2	3.4	88/12	1.5	88/12
4	4.5	71/29	2.0	73/27
6	4.9	61/39	2.3	63/37
8	4.9	50/50	2.4	53/47

Table IV

Goodrich Disk Fatigue Testing

Strength Retention, %

TIME, Min.	CYCLES	75/25 PARA/META	100/0 PARA/META
10	18,000	95	83
100	180,000	92	79
1,000	1,800,000	94	77
10,000	18,000,000	80	72

para 50/50 oxadiazole/N-methyl hydrazide analogue. If a 75/25 all-para composition had been used, the comparison would have been even more favorable for the partially meta-oriented yarn due to the inherent stiffness (2) of oxadiazole-rich compositions.

As shown in this paper, aromatic oxadiazole/N-methyl hydrazide copolymers with partial m-phenylene orientation give yarn with increased elongation-to-break and tire cord with improved fatigue resistance when compared to the all-para analogue. Whether ultimately the same high strength (18-21 gpd) as demonstrated for the latter (1) could be obtained with aromatic oxadiazole/N-methyl hydrazide copolymers of partial meta-orientation remains to be proven.

Experimental

A typical preparation of an aromatic oxadiazole/N-methyl hydrazide copolymer and fiber was conducted as follows:

> 1807 g. (10.88 moles) of terephthalic acid, 704 g. (3.63 moles) of dimethyl isophthalate, and 1915 g. (14.72 moles) of hydrazine sulfate were added to 17.5 kg of 30% oleum in a 10 gallon glass-lined Pfaudler reactor. The mixture was stirred overnight at 70°C to effect dissolution of the monomers, and then was heated to 145°C for 6 hrs. The polymer solution was diluted with 9.5 kg of 24% oleum, transferred to a spinning container, pressured to 100 psi with inert gas and dry-jet wet-spun to fiber by extruding it downward through a 25 hole 4-mil spinneret into an aqueous coagulation bath maintained at 25°C. The spinneret was heated to 162°C and kept 1/2" above the spin bath liquid. After passage under a guide at the bottom of the spin bath, the fiber was passed over a water wash roll, a neutralizing roll rotating in 4% aqueous sodium carbonate solution, through a water bath at 95°C, over another wash roll, a steam-heated drying roll and was finally hot-stretched at 357°C over a hot shoe. The fiber obtained at 1010 fpm had a tenacity of 14.4 gpd, an elongation of 7.2% and an initial modulus of 303 gpd.

Acknowledgment

We would like to express our deep appreciation to W. J. Boles, Jr. and W. E. Waters for their dedicated assistance, to D. A. Zaukelies for conducting the fatigue test and to J. H. Saunders and K. R. Lea for their support.

Literature Cited

1. Bach, H. C., Dobinson, F., Lea, K. R., and Saunders, J. H., J. Appl. Pol. Sci., 1979, 23, 2125.
2. Bach, H. C., Hinderer, H. E., and Silver, F. M., Am. Chem. Soc. Polym. Preprints, 1978, 19 (2), 671.
3. Iwakura, Y., Uno, K., and Hara, S., J. Polym. Sci., 1965, A-3, 45.
4. Imai, Y., J. Appl. Polym. Sci., 1970, 14, 225.
5. Frazer, A. H. and Wallenberger, F. T., J. Polym. Sci., 1964, A-2, 1157.
6. Bach, H. C., U.S. Patent 4,115,503, 1978, assigned to Monsanto Company.
7. Bach, H. C., U.S. Patent 4,202,962, 1980, assigned to Monsanto Company.

RECEIVED November 30, 1981.

29

Free Radical Copolymerization of 4,7-Dihydro-1,3-dioxepins with Maleic Anhydride and Maleimides: Preparation of Head-to-Head Poly(4-hydroxycrotonic acid) and Poly(γ-crotonolactone)

B. M. CULBERTSON and A. E. AULABAUGH[1]

Ashland Chemical Company, Columbus, OH 43216

Acting as electron donor monomers, 4,7-dihydro-1,3-dioxepins undergo free-radical initiated copolymerization with maleic anhydride (MA) to give good yields of low to moderate molecular weight, 1:1 alternating copolymers. Under similar conditions, the same 1,3-dioxepins copolymerize in a nonequimolar fashion with maleimides to give good yields of soluble copolymers. Total hydrolysis of the anhydride copolymers produce previously unreported head-to-head poly (4-hydroxycrotonic acid). The hydrolyzed copolymers readily cyclize to previously unreported head-to-head poly (γ-crotonolactone). In the presence of strong acids the 1,3-dioxepin-alt-MA copolymers rearrange to poly (γ-crotonolactone). At elevated temperatures (>200°C), the anhydride copolymers also thermally rearrange to poly (γ-crotonolactone), eliminating aldehyde or ketone. The synthesis and some of the properties of these new polymers are briefly explored.

Water soluble polymers are of special interest today for biomedical, oil recovery, water purification, boiler scale prevention, coatings and many other applications. Many of the polymers under study contain hydroxyl and/or carboxylic acid functionalities, which contributed needed hydrophilicity, reactivity and other properties to the base polymer.

As part of an ongoing program to prepare and study new, water soluble polymers, attention was focused on the possible radical copolymerization of 4,7-dihydro-1,3-dioxepins with maleic anhydride (MA) and maleimides.

A wide variety of 4,7-dihydro-1,3-dioxepins or 1,3-dioxep-5-enes are known (1,2). These cyclic acetal or ketal compounds are conveniently prepared by the acid catalyzed reaction of cis-2-

[1] Current address: University of Wisconsin, Madison, WI 53706.

0097-6156/82/0195-0371$06.00/0
© 1982 American Chemical Society

butene-1,4-diol with aldehydes, ketones and acetals (1). For this study, monomers I and II were prepared. The unsymmetrical isomer of II, i.e., III was also prepared from II.

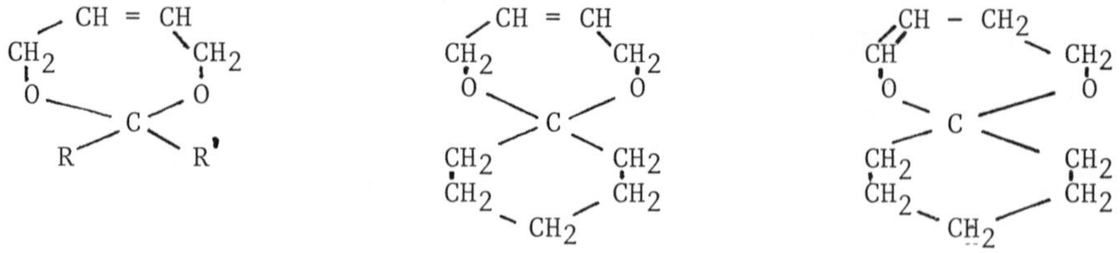

```
   CH = CH              CH = CH               CH - CH2
CH2       CH2        CH2       CH2         CH          CH2
O          O         O          O          O            O
     C                    C                      C
R          R'        CH2       CH2         CH2          CH2
                     CH2       CH2         CH2          CH2
                          CH2                    CH2

      I                   II                    III
```

IA. R=H, R'=H
IB. R=H, R'=CH $(CH_3)_2$
IC. R=H, R'=C_6H_5
ID. R=CH_3, R'=CH_3
IE. R=H, R'= $-CH=CH_2$

This study was undertaken to find out if 1,3-dioxepins, functioning as electron donor monomers, would copolymerize with the strong electron acceptor monomer MA, to give a new family of heretofore unreported 1:1 alternating copolymers. It was hoped that tying the hydroxyl groups of the cis-2-butene-1,4- diol into a cyclic planar or chair (3,16) structure would decrease steric hindrance and enhance double bond polymerization reactivity. Sterling (4) had previously shown that 2-substituted 1,3-dioxepins would copolymerize with vinylidene chloride, butadiene and methyl isopropenyl ketone and it is well known that unsaturated cyclic olefins (5), including unsaturated cyclic ethers (6-8) copolymerize in a 1:1 alternating fashion with MA. A second goal was to find out if this synthetic route could provide a path to heretofore unreported head-to-head poly (4-hydroxycrotonic acid) and poly (γ-crotonolactone) via hydrolysis and rearrangement reactions of the 1,3-dioxepin-alt-MA copolymers.

Since our initial report (9) on this subject, Yokoyama and Hall (10) published their results on the free-radical initiated copolymerization of IA with MA to obtain a 1:1 alternating copolymer. Yokoyama and Hall (10) also confirmed that the IA-MA copolymer is a source of head-to-head poly (4-hydroxycrotonic acid).

Measurements - HNMR spectra were determined on a Bruker WP200 NMR spectrometer at 200 MH_z, using tetramethylsilane as a reference. Infrared (IR) spectra (KBr) were recorded with a Perkin-Elmer 580 spectrophotometer. Molecular weights ($\overline{M}_w$) and dispersivity ($\overline{M}_w/\overline{M}_n$) values were estimated from data collected on the Waters GPC 301 equipment, using tetrahydrofuran solvent. Inherent viscosities, η_{inh}, were determined in Cannon-Ubbelohde viscometers (0.4g/100ml solvent at 30°C). Differential thermal and thermogravimetric analyses profiles were obtained on the DuPont 900 DTA, equipped with a DSC cell, 950 TGA and 990 thermal analyzer units. A Hewlett-Packard 5610A Gas Chromatograph (gc), with 10% FFAP on Chromosorb W (AW-DMCS) column using N_2 carrier gas was used for monomer purity studies. Acid numbers were determined by nonaqueous titration of the copolymers in acetone, using 0.1N sodium hydroxide titrant, phenolphthalein indicator and 0.1N sulfuric acid back titrant. Elemental analyses were performed by Huffman Laboratories, Inc., Wheatridge, Colorado.

Solvents and Initiators. All polymerization solvents, ethyl acetate (EA),1,2-dichloroethane (DCE), methyl ethyl ketone (MEK), cyclohexanone (CH), toluene and tetrahydrofuran (THF) were purified by standard procedures (12) and stored under N_2 or over molecular sieves. All other solvents used, N,N-dimethylformamide (DMF), dimethylsulfoxide (DMSO), γ-butyrolactone, hexane, diethyl ether, acetone, etc., were AR grade materials. Initiators, azobisisobutyronitrile (AIBN), di-t-butylperoxide (DTBP), lauryl peroxide (LP) and benzoyl peroxide (BPO) were recrystallized (AIBN and BPO) or used as received from suppliers.

Monomers. MA was purified by multiple recrystallizations and/or subliniation. The monomer N-phenylmaleimide was prepared by a known procedure (13) and recrystallized several times to obtain high purity. Treatment of cis-2-butene-1,4-diol with a two-fold molar excess of acetic anhydride for 6 hr. at 100°C was the method of choice to prepare cis-2-butene-1,4-diol diacetate, bp 55-60°C/0.1mm [lit.: 120-1°C/18mm (28)]). Dimethyl maleate and di-ti-butyl fumarate were dried over sodium sulfate and distilled. The cis-2-butene-1,4-diol starting material contained ca. 8% trans isomer, as shown by NMR.

Synthesis of 4,7-Dihydro-1,3-dioxepins (IA). Paraformaldehyde (45.0g, 1,5 mol) was combined with cis-2-butene-1,4-diol (134.0g, 1.52 mol), 0.37g p-toluenesulfonic acid and 40 ml toluene. The mixture was heated at reflux with a nitrogen purge until approximately 16 ml of water was collected in a Dean-Stark trap. Distillation of the remaining reaction mixture gave 130g (80% yield) of crude IA, bp 120-125°C. Redistillation over potassium carbonate gave polymerization grade monomer, with gc showing >99% purity: bp 125-127°C/760mm [lit.: 124°C/760mm (14), 127.8-128.2°C/734mm (11), 127°C/760mm (15)]. The NMR spectrum ($CDCl_3$) gave peaks at δ5.68 (-CH=CH-), δ4.8 ($-OCH_2O-$), δ4.2ppm ($-CH_2O-$), and no exchangeable protons.

Synthesis of 4,7-Dihydro-2-isopropyl-1,3-dioxepin (IB). Isobutyraldehyde (79.4g, 1.10 mol) was combined slowly with cis-2-butene-1,4-diol (100.0g, 1.13 mol). After addition was complete, the reaction mixture was stirred under nitrogen for 1 hr. at room temperature. Formation of the hemiacetal was slightly exothermic, raising the reaction mixture to 30°C. The hemiacetal mixture was added slowly at a controlled rate to a refluxing toluene (150 ml) solution containing 0.24g p-toluenesulfonic acid. Under a nitrogen purge, about 13 ml of water was collected in a Dean-Stark trap. The toluene solution of the product was fractionated through

Table I

Monomers	Yield	bp, °C/mm	References
IA	80	125-127	11
IB	64	66-67/0.06	17
IC	55	76-77/0.3	18
ID	86	59-60/27	19,22
IE	36	53-55/13	18
IF	20	53-55/4.5	11
II	70	60-61/1.0	11,18
III[a]	--	67-69.5/1.0	21

[a] NMR data showed a mixture of II/III with ca. 65% III.

a Vigreux column at reduced pressure to obtain a 92.0g (64% yield) of crude IB, bp 66-76°C/0.06mm. Redistillation over potassium carbonate gave polymerization grade monomer, with gc showing >99% purity: bp 66-67°C/0.06 mm [lit.:105°C/100mm (17), 170-170.6°C/735 mm (11)]. The NMR spectrum ($CDCl_3$) gave peaks at δ1.6-2.2 (-CH<), δ0.8-1.5ppm ($-CH_3$), and no exchangeable protons.

Synthesis of 4,7-Dihydro-2-phenyl-1,3-dioxepin (IC). Benzaldehyde (212g, 2.0 mol) was combined with cis-2-butene-1,4-diol, (88g, 1.0 mol), 0.5g of p-toluenesulfonic acid and 150ml toluene. The mixture was heated (ca. 5 hr.) at reflux under a nitrogen purge until absence of the toluene/water azeotrope collected in the Dean-Stark trap. Rapid distillation at reduced pressure gave 96g (55% yield) of crude IC, bp 75-77°C/0.3mm [lit.: 79°C/0.6mm (14)]. The NMR spectrum gave peaks at δ7.5 (C_6H_5), δ5.8 (-O-CH-O-) δ5.65 (-CH=CH-), δ4.30ppm ($-CH_2O-$), and no exchangeable protons.

Synthesis of 4,7-Dihydro-2,2-dimethyl-1,3-dioxepin (ID). 2,2-Dimethoxypropane (213g, 2.0 mol) was combined with cis-2-butene-1, 4-diol (146.4g, 1.5 mol) and 0.2g of p-toluenesulfonic acid. An endothermic reaction occurred, followed by formation of a single liquid phase. The mixture was distilled at atmospheric pressure through a short Vigreux column until the head temperature reached 60°C, after which 75 ml of hexane was added and distillation continued until the head temperature reached 69°C. Continued distillation at reduced pressure gave 151g (86% yield) of crude ID, bp 46-48°C/15mm. Redistillation over potassium carbonate gave polymerization grade monomer, with gc showing >99.5% purity: bp 59-60°C/27mm [lit.: 41°C/6.0mm (19), 108-109°C/60mm (2), 144.5-147°C/755mm (20)]. $N_d^{22.5}$ 1.4472 [lit.: $N_d^{24.5}$ 1.4465 (20)]. The NMR spectrum ($CDCl_3$) gave peaks at δ5.6 (-CH=CH-), δ4.2 ($-CH_2O-$) and δ1.4ppm (CH_3).

Synthesis of 4,7-Dihydro-2-vinyl-1,3-dioxepin (IE). Following a published procedure (18), crude IE was prepared from acrolein and cis-2-butene-1,4-diol in a 39% yield. Redistillation gave polymerization grade monomer, bp 53-55°C/13mm. The NMR spectrum showed a complex multiplet at δ5.5-5.1, multiplet at δ4.2-4.05ppm ($-CH_2O-$), and no exchangeable protons.

Synthesis of 4,7-Dihydro-2(2-propenyl) 1,3-dioxepins (IF). Using the Brannock and Lappen (11) procedure, crotonaldehyde was condensed with cis-2-butene-1,4-diol to obtain a 20% yield of IF, bp 53-55°C/4.5mm [lit.: 54-55°C/4.5mm (11)].

Synthesis of 7,12-Dioxaspiro (5,6)-dodec-9-ene (II). Cyclohexanone (49g, 0.5 mol) was combined with cis-2-butene-1,-4-diol, (44.0g, 0.5 mol), 1.0g p-toluenesulfonic acid and 300ml toluene. The mixture was heated(5 hr.) at reflux under a nitrogen purge with approximately 8.5 ml water collected in a Dean-Stark trap. After cooling, the acid was neutralized with sodium methoxide, solution filtered, toluene evaporated and the product distilled to obtain 59g (70% yield) of crude II, bp 59-61°C/1mm. Redistillation gave polymerization grade monomer, bp 60-61°C/1mm [lit.: 54°C/0.2mm (18,19), 94°C/10mm (11), 76-76.5°C/8mm (21)], N_d^{21} 1.4878 [lit.:

N_d^{20} 1.4876 (11)]. The NMR spectrum gave peaks at δ5.68 (-CH=CH-), δ4.25 ($-CH_2O-$) and a complex multiplet at δ1.9-1.3ppm (10 methylene protons).

Synthesis of 7,12-Dioxaspiro (5,6)-dodec-8-ene (III). Monomer III (50.4g, 0.3 mol), was combined with 30 ml tert-butyl alcohol and 15g (0.14 mol) of potassium tert-butoxide. The mixture was heated in a sealed glass reactor for 6 hr. at 120°C. After cooling, the solution was diluted with diethyl ether and extracted with water to remove base. Evaporation of the diethyl ether and distillation gave a crude mixture of II and III as a water white liquid, bp 67-69.5°C/1mm [lit.: 76-76.5°C/8mm for III (21)]. The NMR spectrum with peaks at δ6.07, δ5.66, δ4.83, δ4.24, δ3.85, δ2.28 and δ1.9-1.3ppm, indicated that the II/III mixture contained ca. 65% III.

Attempted Homopolymerization of 1,3-Dioxepins. The required amount of monomer and initiator (or monomer, solvent and initiator) were weighed into a glass serum bottle, the system purged with nitrogen and the bottle sealed with a Neoprene stopper. Polymerizations were run in a shaker bath for 6.5 hr. at 75°C. Combinations examined were IA (10.0g, 0.1 mol) with 0.1g AIBN, IB (7.25g, 0.05 mol) with 0.15g AIBN, IC (12.0g, 0.06 mol) with 0.18g AIBN, ID (7.0g, 0.05 mol) with 5.0ml MEK and 0.1g AIBN, ID (12.6g, 0.09 mol) with 9.0 ml toluene and 0.1g AIBN, and IE (8.0g, 0.06 mol) with 8.0ml toluene and 0.16g AIBN. Reactions were terminated by cooling and combining the reaction mixtures with a 50/50 (v/v) hexane/diethyl ether mixture. Polymer failed to form in all systems.

MA-1,3-Dioxepin Copolymerizations. Copolymerizations in Tables II and III were run under nitrogen in glass flasks fitted with a mechanical stirrer, reflux condenser, nitrogen inlet and a thermometer. Where incremental addition of initiator was employed, the flasks were also fitted with an addition funnel. Equimolar amounts of 1,3-dioxepin, MA, solvent and initiator were weighed into the flasks and polymerizations run at the time/temperature conditions shown. In some runs the initiator was dissolved in the minimum amount of solvent and added at a controlled rate during polymerization. After cooling and dilution with acetone, the copolymers were precipitated with diethyl ether. The polymers were purified by dissolving in acetone, reprecipitating into diethyl ether and drying *in vacuo* at 58°C. The new polymers were white powders, which were soluble in acetone, chloroform, water (with reaction), EA, DMF, DMSO and THF. The NMR spectrum (acetone-d_6 or DMSO-d_6) of each copolymer exhibited peaks for polymerized MA residues at δ3-4, cyclic acetal or ketal peaks at δ3-4 (not well resolved due to overlap), and no free MA peaks were shown at δ7.04 or δ7.27ppm. In addition, the NMR spectra of the IA, IB, IC, ID and II copolymers had resonance peaks, respectively, at 4.7 (OCH_2), 0.9 ($2CH_3$), 7.35 (C_6H_5), 1.3 (CH_3) and δ1.5ppm (10 methylene protons). The IR spectrum of each copolymer included absorption bands at 1855, 1775, 1100 and 925 cm^{-1}. In some cases the strong band at 1775 had a weak shoulder at 1720-1725 cm^{-1}, with NMR showing active protons for these situations.

Table II

4,7-Dihydro-1,3-dioxepin (IA)-MA Copolymerizations

IA,g	MA,g	Solvent (g)	Init. (wt%)	Temp. °C	Time, hr.	Yield %	$\bar{M}_w$ ($\bar{M}_w/\bar{M}_n$)
7.51	7.35	EA (10)	AIBN (1.5)[a]	75	7.0	62	---------
7.51	7.35	DCE (10)[i]	AIBN (1.5)[a]	80	7.0	75	1800 (1.61)
7.51	7.35	Toluene (10)	AIBN (1.5)[a]	80	5.0	55	----e-----
7.51	7.35	MEK (10)	AIBN (1.5)[a]	75	8.0	60	----------
7.51	7.35	MEK (10)	BPO (1.5)[c]	80	20.0	54	----------
10.20	10.00	DCE (10)[f,i]	AIBN (2.0)[d]	75	7.0	64	2850 (1.95)
7.01	6.86	EA/DCE (10)[g]	AIBN (3.0)[a]	75	6.0	76	1700 (1.60)[h]
7.51	7.35	MEK (10)	AIBN (1.5)[b]	80	8.0	65	----------
7.51	7.35	MEK (10)	AIBN (1.5)[a]	75	7.0	52	----------
7.51	7.35	CH (10)	DTBP (1.5)[c]	130	7.0	32	----------
10.20	10.00	MEK (13.5)	AIBN (2.0)[c]	80	17.0	71	1264 (1.44)
5.10	5.00	MEK (6.7)	----------	80	17.0	--	----------

a Solution of initiator added incrementally
b Slow (controlled) addition of initiator solution during course of polymerization
c One shot addition of initiator
d AIBN (1.5 wt%) added at start and 0.5 wt% AIBN added after 4 hr.
e η_{inh} 0.23 dl/g (DMF).
f $ZnCl_2$ (1%, wt/wt) added at start of reaction
g EA/DCE (50/50,v/v) used
h η_{inh} 0.34 dl/g (DMF)
i Copolymer precipitated as formed in DCE

Table III

Assorted 1,3-Dioxepin-MA Copolymerizations

Dioxepin (g)	MA, (g)	Solvent (g)	AIBN wt.%	Temp. °C	Time, hr.	Yield %	M_w (M_w/M_n)
IB (7.25)	5.00	---------	2.0[a]	75-80	7.0	47	2000 (1.35)[j]
IB (7.25)	5.00	MEK (7.55)	2.0[a]	75	22.0	53	1600 (1.27)
IB (21.33)	14.71	$CHCl_3$ (10)	1.5[a]	70-75	8.0	41	1650 (1.52)
IB (6.82)	4.71	DCE (10)[k]	3.0[b]	75	7.0	65	2100 (1.61)
IC (13.77)	7.35	DCE (14)[k]	1.5[b]	75-80	6.0	65	3400 (2.00)[d,e]
ID (32.04)	24.52	MEK (33)	2.0[c]	75-80	8.0	60	6400 (2.09)[f]
ID (25.13)	19.14	EA/DCE (96)	2.0[b]	75-80	20.0	46	-----g-----
ID (6.82)	4.71	DCE (12)[k]	3.0[b]	75-80	7.0	65	2100 (1.61)
ID (9.61)	7.35	DCE (5)[k]	1.5[b]	75-80	7.0	75	8100 (1.80)[h]
ID (98.12)	73.55	DCE (78)[k]	1.0[b]	75	6.5	55	-----i-----
IE (9.46)	7.35	DCE (8)[k]	1.5[b]	70-75	5.5	--	Hard Gel
IF (7.01)	4.90	MEK (10)	3.0[a]	75	3.0	--	Hqrd Gel
II (16.08)	9.80	DCE (15)[k]	1.5[a]	80	7.0	80	6200 (1.69)
III (16.08)	9.80	DCE (15)[k]	1.5[a]	80	7.0	69	----------

[a] Initiator added at start of reaction
[b] Incremental addition of initiator solution
[c] 1% AIBN added at start and 1% AIBN booster added after 6 hr.
[d] Acid No. 510 showing ca. 1:1 copolymer
[e] η_{inh} 0.05 dl/g (DMF)
[f] η_{inh} 0.12 dl/g (DMF)
[g] η_{inh} 0.09 dl/g (DMF)
[h] η_{inh} 0.08 dl/g (DMF)
[i] η_{inh} 0.08 dl/g (DMF)
[j] η_{inh} 0.03 dl/g (DMF)
[k] Copolymer precipitated as formed in DCE

For copolymerizations in Table IV, monomers (75%) solids), DCE and AIBN (2.0 wt%) were combined and sealed (Neoprene stopper) under nitrogen in 50ml glass serum bottles. The polymerizations were run in a shaker bath for 5 hr. at 70°C. Polymer recovery and purification was the same as previously described.

Hydrolysis of MA-1,3-Dioxepin Copolymers. IA-MA copolymer (2.00g) was combined with 50ml deionized water and stirred at room temperature for 48 hr. After 12 hr. the dispersion became a homogeneous solution. Evaporation of the water and prolonged drying *in vacuo* over phosphorus pentoxide at room temperature gave a 2.25g yield of white polymer soluble in THF, MEK, DMF and DMSO. GPC showed $\overline{M}_w$ 1853 and dispersivity of 1.4. The IR spectrum had absorptions at 3500, 1720 (weak shoulder at 1760), four peaks in area 1200-1000 and no anhydride absorptions at 1815 and 1775 cm^{-1}. TGA showed onset of decomposition at 75°C, with 20% weight loss from 75-300°C (theory for loss of CH_2O and H_2O to form lactone 22%).

Anal. Calcd. for $(C_9H_{12}O_6)_n$: C, 50.0; H, 5.55. Found: C, 49.50; H, 5.66.

Copolymers IA-MA (3.0g) and IB-MA (3.0g) were suspended in 50 ml of deionized water with 1 drop conc. hydrochloric acid. After 12 hr. the suspensions changed to clear solutions. The solutions were heated for 1 hr. at reflux, insuring complete hydrolysis and the water was evaporated at ambient conditions and *in vacuo* at room temperature. The residues were taken up in MEK and the polymers precipitated into a large volume of diethyl ether. The IR spectra of the dried copolymers were almost identical, with both polymers showing absorptions at 3650-3300, 2980-2900, 1720, 1390, 1135, 965 and 845 $cm.^{-1}$

Anal. Calcd. for hydrolyzed IA-MA $(C_8H_{12}O_6)_n$: C, 47.05; H, 5.88. Found: C, 46.65; H, 5.98.

Anal. Calcd. for hydrolyzed IB-MA $(C_8H_{12}O_6)_n$: C, 47.05; H, 5.88. Found: C,46.72; H, 5.97.

Acid Rearrangement of 1,3-Dioxepin-MA Copolymers. IA-MA, IB-MA and ID-MA copolymers (3.0g) were combined with 0.1N hydrochloric acid (80ml). The polymer suspensions were stirred and heated for 12 hr. at 50°C and 4 hr. at 70°C. Isobutyraldehyde odor was detected coming from the IB-MA system. The polymers were collected, washed with deionized water and dried *in vacuo* over phosphorus pentoxide at room temperature to obtain slightly yellow colored materials soluble in DMF and DMSO. The IR spectrum of each copolymer exhibited absorptions at 3350, 2650, 1760 (with a shoulder at 1720-1700), 1630, 1165 and 1120 cm^{-1}, with no anhydride absorptions at 1855 and 1775 cm^{-1} and no cyclic acetal absorptions in the 1200-1000 cm^{-1} region. The ^{13}C NMR spectra, though not well resolved, clearly showed no methyl protons for the recovered IB-MA and ID-MA copolymers.

Copolymer ID-MA (10.2g) was combined with 10ml deionized water, 40ml isopropyl alcohol and 0.5g conc. hydrochloric acid. The sus-

Table IV

Feed Ratio Influence on ID-MA Copolymerizations

IDg (mol)	MAg (mol)	Mole % MA Feed	Mole % MA Polymer[a]	Yield %	ηinh[b] dl/g	Acid No.
9.62 (0.075)	7.36 (0.075)	50.0	51.0[c]	71	0.07	504
12.82 (0.10)	4.90 (0.05)	33.5	48.0	62	0.08	490
6.43 (0.05)	9,81 (0.10)	66.6	53.0	51	0.06	534
3.85 *0.03)	11.77 (0.12)	79.9	48.5	25	0.06	495
15.38 (0.12)	2.94 (0.03)	20.0	----	37	0.07	---

[a] Estimated from Acid Numbers
[b] Determined in DMF (0.4g/100 ml, 30°C)
[c] Anal. Calcd. for 1:1 copolymer $(C_{11}H_{14}O_5)_n$: C, 58.40; H, 6.19. Found: 57.90; H, 6.09

pension was stirred for 24 hr. at room temperature, polymer recovered, washed with a 50/50 isopropyl alcohol/water mixture and dried in vacuo at room temperature, giving a 7.6g yield of polymer soluble in DMF and DMSO. The IR spectrum exhibited expected strong absorptions at 1760 and 1165 cm^{-1} and no peaks at 1855 and 1775 $cm.^{-1}$

Anal. Calcd. for 50% lactone formation, $(C_8H_{10}O_5)_n$: C, 51.61; H, 5.37. Found: C, 53.82; H, 5.32.

A concentrated hydrochloric acid (20ml) slurry of ID-MA copolymer (5.0g) was stirred and heated for 6 hr. at 50°C. The polymer was collected, air dried, dissolved in DMF, reprecipitated from diethyl ether and dried in vacuo to obtain a 3.66g yield of tan colored material. The IR spectrum exhibited expected cyclic lactone absorption bands at 1760 and 1165 cm^{-1} region.

Anal. Calcd. for ca. 75% lactone formation $(C_{24}H_{28}O_{14})_n$: C, 53.33; H, 5.18. Found: C, 53.40; H, 5.19.

The ID-MA acid treated copolymer was heated under nitrogen for 2 hr. at 200°C. The IR spectrum of the DMF and DMSO soluble polymer exhibited sharper absorption bands at 1760 and 1165 cm^{-1}, with only a hint of a shoulder at 1720 cm^{-1} and only very weak absorption at 3500 cm^{-1} region.

Anal. Calcd. for $(C_8H_8O_4)_n$: C, 57.14; H, 4.76. Found: C, 56.82; H, 4.91.

Copolymers IA-MA (4.0g) and ID-MA (10.0g) were dissolved in MEK (20 and 40 ml, respectively) and the stirred solutions heated under nitrogen at 70-75°C. Concentrated hydrochloric acid (1.2g) was added and heating continued, with cloudiness and polymer precipitation noted after ca. 2 min. After 3 hr. at 75°C, the precipitated polymer was collected, washed with MEK and diethyl ether and dried in vacuo at room temperature, to obtain a 2.61g and 8.06g yield, respectively, of light tan-colored materials. The IR spectrum of each polymer included strong absorptions at 1760 and 1165 $cm.^{-1}$ The NMR spectrum (DMSO-d_6), with peaks at δ2.7-4.4 (α-methylene protons) and δ2.03-3.0ppm for methine protons, was not well resolved. The ^{13}C NMR spectrum which was also not well resolved, showed no methyl group protons. The η_{inh} of recovered polymer was 0.12 dl/g (DMF, 30°C).

Base Hydrolysis of 1,3-Dioxepin-MA Copolymers. ID-MA copolymer (3.0g) was combined with 5ml water and 10ml acetone. The solution, which became homogeneous after addition of 0.2g triethylamine, was allowed to sit for 24 hr. at room temperature. After evaporation of the water and acetone, the recovered polymer was purified by precipitation from a 50/50 (v/v) DMF/acetone mixture as solvent and diethyl ether as a precipitant and dried in vacuo. The IR spectrum exhibited an absorption band at 1725 cm^{-1}, attributed to acid.

Anal. Calcd. for $(C_{11}H_{16}O_6)_n$: C, 54.09; H, 6.56. Found: C, 53.62; H, 6.60.

Thermal Rearrangement of 1,3-Dioxepin-MA Copolymers. A cyclohexanone (40ml) soluton of IA-MA copolymer (4.0g) was heater under nitrogen for 48 hr. at reflux. The IR spectrum of the recovered polymer retained absorptions at 1855, 1775 and 1200-1000 cm^{-1}, with no new absorption at 1760 $cm.^{-1}$ The reaction was repeated with IB-MA copolymer, with the IR spectrum of the recovered material exhibiting absorptions at 1760 and 1165 $cm.^{-1}$

Copolymer IB-MA (1.0g) was heated under nitrogen for 6 hr. at 235-240°C, with off gas collection in a mineral oil trap. Isobutyraldehyde was confirmed in the mineral oil by IR, with 1740 cm^{-1} absorption and odor. The recovered polymer (0.73g), which was no longer soluble in water, EA, MEK and THF, was soluble in DMF, DMSO and γ-butyrolactone. Heating 1.5g of IA-MA copolymer under the same conditions produced similar results, with 1.39g polymer recovered.

Copolymer IB-MA (4.89g) was heated for 4 hr. at 240°C with a slow nitrogen stream carrying off gas into a 2,4-dinitrophenylhydrazine (DNPH) solution (23) trap. After 10 min. the 2,4-DNPH derivative of isobutyraldehyde started precipitating from solution, m.p. 181-182°C [lit.: mp 182°C (23)]. The complex IR spectrum of the recovered polymer included strong absorption bands at 1760 and 1165 $cm.^{-1}$

Anal. Calcd. for $(C_{24}H_{28}O_{14})_n$, estimated 75% conv. to lactone: C, 53.33; H, 5.18. Found: C, 53.30; H, 5.19.

Samples of IA-MA and ID-MA copolymers were thermolyzed under the same conditions. The 2,4-DNPH derivatives of formaldehyde and acetone were also formed, with IR spectra at the recovered copolymers showing absorptions at 1760 and 1165 $cm.^{-1}$

4,7-Dihydro-1,3-dioxepin-N-Phenylmaleimide Copolymerization. A 50ml glass serum bottle was charged with 7.51 (0.075 mol) IA, 12.99g (0.075 mol) N-phenylmaleimide, 6.8g MEK, 6.8g THF and 0.44g (1.8 mol%) AIBN. The bottle and contents were purged with nitrogen, sealed and placed in a shaker bath at 60°C. After 21 hr. the viscous solution was diluted with 20ml THF and the polymer precipitated from a diethyl ether/pentane (4/1, v/v) mixed solvent. The polymer was purified by dissolving in acetone and precipitation from diethyl ether. After drying *in vacuo* overnight at 40°C, the slightly yellow colored copolymer was obtained in a 15.2g (74%) yield, having GPC estimated $\overline{M}_w$ 3,900 ($\overline{M}_w/\overline{M}_n$ 2.4) and η_{inh} 0.07dl/g (DMF at 30°C). DTA and TGA, with a single-stage decomposition profile, showed the copolymer stable in nitrogen to >300°C.

Synthesis of Head-to-Tail Poly (γ-crotonolactone). A 50ml glass serum bottle was purged with nitrogen, charged with 5.0g (0.05 mol) γ-crotonolactone (24) and sealed under nitrogen with a Neoprene seal. Triethylamine (0.15g), a reported polymerization initiator (25, 26), was added with a syringe and the bottle placed in a shaker bath at 25°C. After 24 hr., the viscous solution was diluted with DMSO and the polymer precipitated from diethyl ether. After drying *in vacuo*, the white polymer was collected in a 1.5g (30%) yield. The polymer was purified by dissolving in DMF and precipitation from diethyl ether and dried for 24 hr. *in vacuo* at 40°C, η_{inh} 0.02dl/g (γ-butyrolactone at 25°C). The IR spectrum

spectrum exhibited 1765 and 1165 cm^{-1} absorptions and no absorption at 1675 $cm.^{-1}$ The NMR spectrum (acetone-d_6), with γ-methylene protons in the δ2.7-4.4 region and methine protons at about δ2.0-3.0 ppm region, not well resolved, supported structure (27). DTA and TGA (N_2) showed softening at ca. 160°C and decomposition onset at ca. 250°C [lit.: 150°C (26)].

Amide Derivatives of MA-1,3-Dioxepin Copolymers. ID-MA copolymer (4.0g) was combined with 30ml aniline and stirred under nitrogen for 48 hr. at room temperature. After 24 hr. all the dispersed copolymer had dissolved in the aniline solution. The amide copolymer was precipitated with diethyl ether. The copolymer was purified by dissolving in methanol/acetone (50/50, v/v), reprecipitation from diethyl ether and drying *in vacuo* for 24 hr. at 58°C. The IR spectrum included absorption bands at 1720-1725, 1700, 1560 and 1100-1200 cm^{-1}, with the aniline modified material having 3.86% nitrogen (theoretical for a $C_{17}H_{21}NO_5$ copolymer is 4.39% nitrogen).

Amide Derivatives of Head-to-Head Poly (γ-crotonolactone). Head-to-head poly (γ-crotonolactone), 4.0g, prepared in MEK with conc. hydrochloric acid, was combined with 40 ml n-butylamine and the stirred mixture heated under nitrogen for 72 hr. at reflux (78°C). After cooling and dilution with 20 ml acetone, the copolymer was precipitated with diethyl ether, collected and dried *in vacuo* for 24 hr. at 60°C, obtaining a 5.24g yield of slightly yellow colored amide derivative with 6.97% N (theory, 8.91%). The complex IR spectrum included absorptions at both 1765 and 1690 $cm.^{-1}$

The same poly (γ-crotonolactone), 10g, was combined with 45ml aniline and heated with stirring under nitrogen for 24 hr. at 125-130°C. Recovery of copolymer as above gave an aniline modified material having 4.86% N (theoretical for 50% conversion of lactone moieties, 5.05% N). As above, absorptions in the IR spectrum included bands at both 1765 and 1700 $cm.^{-1}$

Results and Discussion. It is claimed in the patent literature that 1,3-dioxepins will copolymerize with a variety of vinyl monomers (4,29) and function well as chain transfer agents (29) in the polymerization of vinyl monomers. In view of these reports and prior to any copolymerization studies, we wished to know if any of the 4,7-dihydro-1,3-dioxepins prepared (Table I) would undergo free-radical initiated homopolymerization. Free-radical polymerization of IA-IE were not observed. Using IA, Yokoyama and Hall (10) confirmed these results.

As background information, an attempt was made to copolymerize cis-2-butene-1,4-diol diacetate with MA. Heating equimolar mixtures of the two monomers with AIBN (2 wt%) for 15 hr. at 75°C produced only a very low yield (<10%) of polymeric material. Using the same conditions, equimolar mixtures of IA-di-t-butyl fumarate and IA-dimethyl maleate gave only a trace of copolymer. Yokayama and Hall (10) also repored that diethyl maleate and diethyl fumarate undergo free-radical copolymerization with IA to give very low yields of non-equimolar copolymer.

Since 4,7-dihydro-1,3-dioxepin (IA) is both the simplest and one of the more easily prepared 1,3-dioxepin monomers (Table I), copolymerization of equimolar mixtures of IA-MA were first examined under a variety of conditions (Table II). As shown, yields of 1:1 alternating IA-MA copolymer are highly dependent upon polymerization conditions, with highest conversions obtained in 1,2-dichloroethane solvent. Also, incremental or controlled addition of initiator improves yields of copolymer. Using DCE solvent, the copolymers precipitated during polymerization. Copolymerization without added initiator was not observed.

Several other 4,7-dihydro- (IB-IF and II) and 6,7-dihydro-1,3-dioxepin (III) monomers were also found to undergo free-radical initiated copolymerization with MA, giving low to moderate molecular weight copolymers (Table III). As expected, conversions for the IE-MA and IF-MA copolymerizations were highest in DCE solvent with incremental or controlled addition of AIBN initiator. Conversions also varied with 1,3-dioxepin structures.

To determine the effects of monomer feed ratio on conversion, molecular weight and structure of copolymer, various feed ratios (1:4, 1:2, 1:1, 2:1 and 4:1) of MA-ID were examined (Table IV). Conversions were 37, 62, 71, 51 and 25%, respectively. Assuming only 1:1 copolymer produced, the corresponding conversions are 98, 96, 71, 72 and 60%. The composition of the copolymers tends to be independent of the monomer feed ratios. These facts are coincident with the alternative structures of the resultant copolymers. Molecular weights also tend to be independent of feed ratios, with comparable low molecular weights obtained for all feed ratios.

The IR spectrum of each 1,3-dioxepin-MA copolymer, with anhydride bands at 1775 and 1855 cm^{-1}, acetal or ketal three band pattern for C-O-C stretching located around 1100 cm^{-1} and the acetal bending band at 925 cm^{-1} was supportive of materials with MA and 1,3-dioxepin residues. In some cases acid functionality was indicated by a weak carbonyl band at 1725 cm^{-1}, appearing as a distinct shoulder of the 1775 band. The free acid hydroxyl was less evident in the 2400-3600 cm^{-1} region. The IR spectrum band shapes in the CH stretching regions could also be correlated with isopropyl, methyl, phenyl and methylene substituents at the 2-position of the various 1,3-dioxepin residues.

The NMR spectrum of each 1,3-dioxepin-MA copolymer was supportive of materials containing both the MA and the respective 1,3-dioxepin moieties. However, copolymer compositions were generally difficult to authenticate by NMR, due to overlapping resonances of the two monomer moieties. Further, the proton spectrum of the IB-MA, IC-MA, ID-MA and II-MA copolymers suggested that these materials were not strictly 1:1 alternating, i.e., each copolymer contained a slight excess of MA. For example, NMR suggested the MA content of the IC-MA copolymer was 57%, based on the integrated area of the aromatic and aliphatic protons. Within experimental error, nonaqueous titration of the copolymer in acetone gave an acid number supporting a IC-MA ratio in the copolymer of 1:1. Elemental analyses of several of the copolymers also indicated the materials possibly contained a very slight excess of MA.

The proposed structure of the IA-ID 1,3-dioxepin-MA copolymers (IV) is consistent with the above facts.

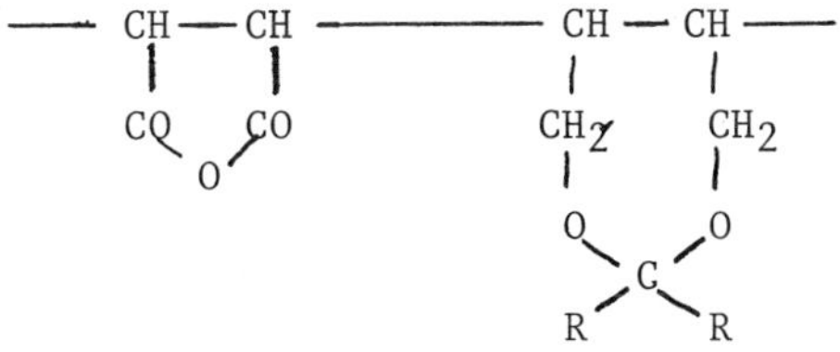

```
—— CH——CH ——————— CH — CH ——
    |    |          |     |
    CO   CO        CH2   CH2
      \ /           |     |
       O            O     O
                     \   /
                       C
                     /   \
                    R     R
```

IV

This brief study shows that IA-ID monomers tend to copolymerize in a 1:1 fashion with MA, regardless of conditions and feed ratios. This suggests the possibility that a charge interaction facilitate these alternating copolymerizations. Attempts were made to determine the CTC equilibrium constants for the IA-MA and IB-MA comonomer pairs.

Chloroform solutions of IA-IF, II, III and MA are colorless, but mixtures of the 1,3-dioxepin and MA solutions, such as IA-MA, IB-MA, etc., often become slightly yellow colored. It was thought that the color development could be attributed to the formation of a 1,3-dioxepin electron donor-MA electron acceptor CTC. However, H'NMR studies, using the Hanna-Ashbaught (30) procedure with $CDCl_3$ solvent, failed to show any significant interaction between IA-MA or IB-MA. Yokoyama and Hall (10) confirm this finding for the IA-MA system. Since an equilibrium constant (K) could not be determined for the IA-MA and IB-MA system, K must have a negligible small ($K \ll 0.1$) value. Kokubo, Iwatsuke and Yamashita (31) claim that spontaneous alternating copolymerization occurs between an electron donor and an electron acceptor monomer having a CTC value about 0.1 l/mol. Since 1,3-dioxepin-MA mixtures fail to undergo spontaneous copolymerization, support exists for the thought that any CTC formation between 1,3-dioxepin-MA is too weak to produce a measurable K value.

It was shown that 1,3-dioxepins will also copolymerize with other strong electron acceptor monomers, such as maleimides. Copolymerization of IA-N-phenylmaleimide occur readily, in the presence of typical free-radical initiators, to give good yields of moderate molecular weight, nonequimolar copolymer. NMR confirmed the presence of the two monomer residues, with the copolymer containing about 80 mol % N-phenylmaleimide. In contrast to the 1,3-dioxepin-MA copolymers, which start to undergo thermal rearrangement at about 200°C, the IA-N-phenylmaleimide copolymer was stable in nitrogen to >300°C.

Copolymerization of 1,3-dioxepins with MA and maleimides are shown to produce only low to moderate molecular weight copolymers. This was expected, since MA is difficult to homopolymerize and generally copolymerizes to give low molecular weight copolymers. 1,3-Dioxepins were reported by Pawloski and Sterling (29) to be good chain transfer agents for vinyl polymerizations. Considering the Pawloski and Sterling (29) explanation for transfer reaction of 1,3-dioxepins, a growing polymer having an electron poor MA radical moiety at its end should interact fairly readily with an electron rich IA-ID, etc., monomer to end chain growth and generate a 1,3-dioxepin based radical.

Hydrolysis of the IA-MA, IB-MA and ID-MA copolymers were studied in neutral, acidic and basic solutions. The copolymers slowly solubilize on prolonged standing in deionized water, due to partial or complete hydrolysis of the anhydride moieties. On hydrolysis the copolymers become acetone insoluble. The IR and NMR spectra of these copolymers, dried for prolonged periods *in vacuo* at room temperature over phosphorus pentoxide, was supportive of materials with maleic acid and cyclic acetal or ketal residues. Elemental analysis for the partially hydrolyzed IA-MA copolymer was approximately correct for structure V. When heated (>75°C) for prolonged periods the materials slowly convert to new

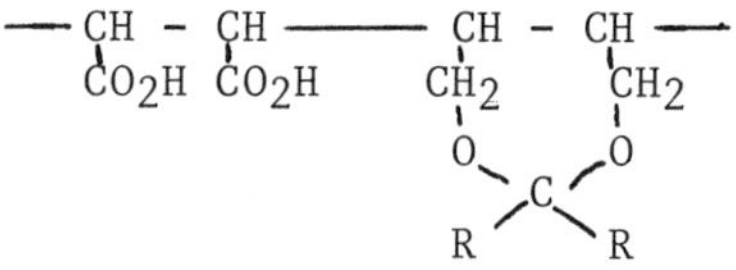

$$\text{—CH(CO}_2\text{H)—CH(CO}_2\text{H)—CH(CH}_2\text{O)—CH(CH}_2\text{O)—} \quad \text{(both O linked to } CR_2\text{)}$$

V

polymers with a strong lactone carbonyl absorptions in the IR at 1760 cm.$^{-1}$ This is not surprising, since Pawloski (1) reported that carboxylic acids react with 1,3-dioxepins to produce 4-hydroxy-2-butenyl esters.

Hydrolysis of IA-MA and IB-MA copolymers in deionized water with only a trace amount of hydrochloric acid gave complete hydrolysis of both the anhydride and acetal moieties. Elemental, IR and NMR analyses were supportive of structure VI, head-to-head poly (4-hydroxycrotonic acid), production. These results were confirmed by Yokoyama and Hall (10), using the IA-MA alternating copolymer.

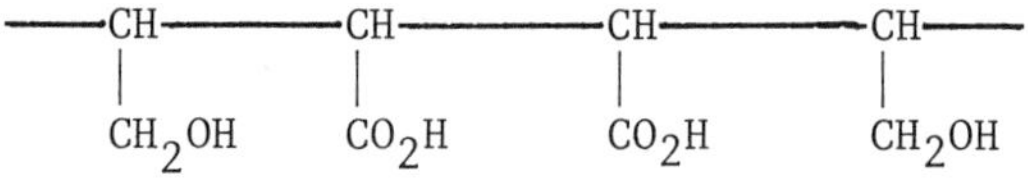

$$\text{—CH(CH}_2\text{OH)—CH(CO}_2\text{H)—CH(CO}_2\text{H)—CH(CH}_2\text{OH)—}$$

VI

Treatment of IA-MA, IB-MA and ID-MA copolymers for prolonged periods with hot hydrochloric acid produced DMF, DMSO and γ-butyrolactone soluble materials with a strong lactone carbonyl absorption band in the IR at 1760 cm^{-1} and weak bands for hydroxyl and carboxylic acid residues. A heterogeneous mixture of the ID-MA copolymer in a water/isopropyl alcohol mixture, containing hydrochloric acid, also produced after 12 hr. reaction a DMF soluble copolymer. Similar to the previous results, the new polymer exhibited a strong lactone carbonyl absorption in the IR at 1760 $cm.^{-1}$ The ^{13}C NMR spectra of the copolymer confirmed absence of methyl protons for the recovered IB-MA and ID-MA polymers. Elemental analyses data for the various copolymers was supportive of materials being produced with considerable lactone rings.

The acid treated ID-MA copolymer, with lactone, acid and hydroxyl absorptions in the IR, was heated for 2 hr. at 200°C. The IR spectrum of the recovered polymer showed considerable sharpening of the 1760 cm^{-1} band and elemental analysis supportive of head-to-head poly (γ-crotonolactone), VII, production. However, absorptions in the same spectrum suggested the polymer retained a small amount of carboxylic acid and hydroxyl functionality.

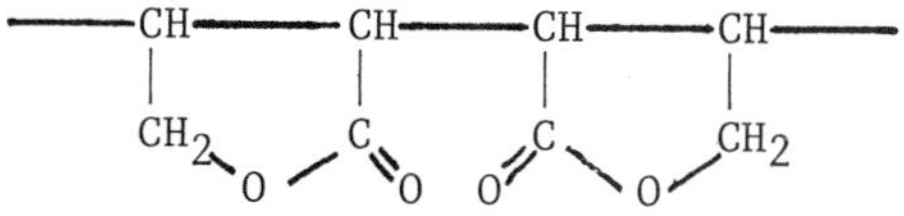

VII

The thermal properties of the ID-MA, IB-MA and ID-MA copolymers were examined, after discovery that these copolymers rearrange substantially to lactone ring containing materials on exposure to strong acid and finding from IR studies that these materials often contained absorptions indicative of carboxylic acid residues.

Copolymer IB-MA was heated under nitrogen at 235-240°C with the off gases collected in a cold trap. The cold trap contained isobutyraldehyde, identified both by odor and IR. In a second experiment, the off gases were also absorbed into 2,4-dinitrophenylhydrazine (DNPH) solution. After about ten minutes the DNPH derivative started forming in the collection trap. A 26.4% weight loss was observed during heating (6 hr.), compared with a theoretical weight loss for isobutyraldehyde coming from a 1:1 copolymer of 23.7%. The IR spectrum of the light tan colored material, which was soluble only in polar solvents, exhibited a strong lactone carbonyl absorption at 1760 $cm.^{-1}$ Under similar conditions, copolymers IA-MA and ID-MA gave off, respectively, formaldehyde and acetone, identified as their 2,4-dinitrophenylhydrozones. The IR spectra of these two thermolyzed copolymers also exhibited strong lactone absorptions at 1760 and 1165 $cm.^{-1}$

Cyclohexanone solutions of IA-MA and IB-MA copolymer were heated at reflux for extended periods. The IR spectrum of the recovered IA-MA copolymer was substantially the same as the IR of the starting material. In contrast, absorptions in the IR of the recovered IB-MA copolymer included new bands in the 1760 and 1165 cm^{-1} region, indicative of lactone moieties.

A small amount of conc. hydrochloric acid was added to non-aqueous solutions of 1,3-dioxepin-alt-MA copolymers and the mixtures heated to obtain materials with lactone rings. For example, a MEK solution of copolymer ID-MA (25% solids) was combined with 1% by wt. conc. HCl and the mixture heated at 70-75°C. The IR spectrum, whith absorptions at 1760 and 1165 cm^{-1}, showed the recovered copolymer was substantially converted to a lactone material. NMR spectral changes were also supportive of lactone formation. The ^{13}C NMR spectrum showed no methyl protons.

Low molecular weight head-to-tail poly (γ-crotonolactone), VIII, was prepared for comparison with the lactone polymers prepared in this study. The IR spectrum of the anionically prepared material (VIII), with strong absorptions in the region 1760 and

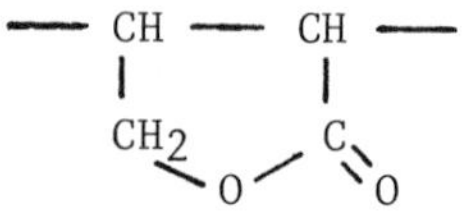

— CH — CH —
CH_2 C
O O

VIII

1165 cm^{-1}, was very similar to the IR spectra of head-to-head poly(γ-crotonolactone), VII, prepared from 1,3-dioxepin-MA copolymers. The DTA and TGA profiles of VII and VIII were also very similar, with both showing onset of decomposition >300°C and solubility in DMF, DMSO and γ-butyrolactone. The NMR spectra of VII and VIII were also similar, with each showing absorptions for methylene and methine protons. Concluding, all evidence collected to date suggests a path has been found to previously unreported head-to-head poly (4-hydroxy-crotonic acid), VI, and head-to-head poly (γ-crotonolactone), VII, via hydrolysis and rearrangement reactions of 1:1 alternating 1,3-dioxepin-MA copolymers.

Amide derivatives of the 1,3-dioxepin-MA and head-to-head poly (γ-crotonolactone) polymers were readily prepared. However, the cursory studies conducted fail to supply an answer as to why amide formation is below theoretical.

It was pointed out earlier that acid and anhydride containing polymers have many possible applications. With the exception of curing epoxide resins and examination of possible biological activity, little has been done to explore uses for the copolymers developed in this study.

A standard epoxide resin (Epon 828, 25.2g) was combined with 15 ml MEK, 15.0g ID-alt-MA copolymer and 0.2g dimethylbenzylamine, giving a homogeneous mixture with 73% solids. Films (3 mil) were drawn on glass and stainless steel plates. The films were air dried and cured at 300°F for 60 min. The cured films exhibited high gloss, high adhesion and excellent solvent resistance on both surfaces.

A sample of poly (2,2-dimethyl-4,7-dihydro-1,3-dioxepin-alt-maleic anhydride) polymers was fractionated and briefly examined for antiviral properties (31). The copolymer was fractionated through Amicon filters into a 1,000-10,000 and 10,000-30,000 molecular weight cuts. Both molecular weight fractions were examined for effect against mice inoculated (ip) with 10^6 Ehrlich Ascites tumor cells. Five days after inoculation the animals were sacrificed and total peritoneal exudate cells were counted with a hemocytometer. Under these conditions, Ottenbrite (31) showed that the 1,000-10,000 molecular weight fraction of the ID-MA copolymer was as effective as Pyran (control in experiment) for control of Ehrlich Ascites tumor cells. Pyran, the copolymer of divinyl ether-MA is a well known antitumor agent (32) and interferon inducer (33).

Acknowledgement Appreciation is expressed to Ashland Chemical Co. management for permission to publish. For analytical (NMR, DTA, TGA, GPC, IR, etc.) support, appreciation is also expressed to A. J. DiGioia, D. R. DeGarmo, C. S. Wu and W. MacCaughey of the Ashland Chemical Analytical staff.

Literature Cited

1. Pawloski, C.E. Chem. Heterocycl. Compounds 1972, 26, 319-411
2. Thuy, V.M.; Maitte, P. Bull. Soc. Chem. Fr. 1975, 11-12 (2), 2558-2560.
3. Soulier, J.; Farines, M.; Laguerre, A; Bonafos-Bastouill, A. Bull. Soc. Chem. Fr. 1976, 1-2 (2), 307-11.
4. Sterling, G.B. U.S. Patent 3,219,629, 1965; Chem. Abstr. 1966, 64, 5262.
5. Murahashi, S.; Nozakura, S.; Yosufuku, K. Bull, Chem. Soc. Japan 1966, 39 (6), 1338.
6. Kawai, W.; J. Polym. Sci. 1968A, 6 (7), 1945.
7. Kimbrough, R.D., Jr.; Dickson, W.P.; Wilkerson, J.M. III J. Polym. Sci. 1964, B2 (1), 85.
8. Tanaki, H.; Otsu, T., J. Macromol. Chem. 1977, All (9), 1663.
9. Culbertson, B.M.; Aulabaugh, A.E. Polym. Prepr., Am. Chem. Soc, Div. Polym. Chem. 1981, 22 (1), 28-29.
10. Yokoyama, Y.; Hall, H.K., Jr. Macromolecules 1981, 14 (3), 471-5.
11. Brannock, J.C.; Lappin, G.R. J. Org. Chem. 1956, 21, 1366-68.
12. Wibert, K.B. "Laboratory Techniques in Organic Chemistry"; McGraw-Hill, New York, 1960.
13. Cava, M.P.; Deana, A.A.; Muth, K.; Mitchell, M.J. Org. Syntheses 1973, Coll. Vol. 5, 944-946.
14. Sterling, G.B.; Watson, E.J.; Pawoloski, C.E.; U.S. Patent 3,116,298, 1963.
15. Pattison, D.B.; J. Org. Chem. 1957, 22, 662-664.
16. Friebolin, H.; Kabus, S. Nucl. Magnetic Resonance Chem. Proc., Symp. Cagliari, Italy, 1964, 125-32.

17. Sturzenegger, A.; Zelauskas, J.J. U.S. Patent 3,410,871, 1968; Chem. Abstr. 1970, 73, 87950p.
18. Tinsley, S.W., Jr.; MacPeak, D. L. U. S. Patent 3,337,587, 1967.
19. Sterling, G.B.; Watson, E.J.; Pawloski, C.E. U. S. Patent 3,116,299, 1963.
20. Kimel, W.; Leimgruber, W. Fr. Patent 1,384,099, 1965; Chem. Abstr. 1965, 63, 4263d.
21. Thuy, V.M.; C.R. Acad. Sci., Ser. C 1971, 273 (23), 1655-57; Chem. Abstr. 1972, 76, 85801d.
22. Lorette, N.B.; Howard, W.L. J. Org. Chem. 1960, 25, 521.
23. Shriner, R.L.; Fuson, R.C.; Curtin, D.Y. "The Systematic Identification of Organic Compounds"; John Wiley & Sons, Inc., New York, 1959, pp. 111, 112, 283, 316.
24. Price C.C.; Judge, J.M. Org. Syntheses 1973, Coll. Vol. 5, 255-257.
25. Cherdron, J.H.O.; Ohse, H.; Palm, R.A.; Korte, F. German Patent 1,228,065, 1966.
26. Palm. R.; Ohse, H.; Cherdron, H. Angew. Chem., Int. Ed. Engl. 1966, 5 (12), 994-1000.
27. Komiyama, M.; Hirai, H.J. Polym. Sci, 1976, 14, 1993-2007.
28. Raphael, R.A. J. Chem. Soc. 1962, 401-5.
29. Pawloski, C.E.; Sterling, G.B. U.S. Patent 3,280,148, 1966.
30. Hanna, M.W.; Ashbaugh, A.L. J. Phys. Chem. 1964, 68, 811.
31. Ottenbrite, R.M. Virginia Commonwealth University, private communication.
32. Butler, G.B. J. Macromol. Sci., Chem. 1971, A5, 219.
33. Merigan, T.C. Nature (London) 1967, 214, 416.

RECEIVED February 25, 1982.

30

Synthesis of Polyspiroketals Containing Five-, Six-, Seven-, and Eight-Membered Rings

WILLIAM J. BAILEY, CHARLES F. BEAM, JR., EDMUND D. CAPPUCCILLI, IBRAHIM HADDAD, and ANGELO A. VOLPE

University of Maryland, Department of Chemistry, College Park, MD 20742

Since the study of model compounds had shown that ketals containing five-, six-, seven-, and eight-membered rings could be formed in very high yields by heating a ketone or diketone with a diol in the presence of an acid catalyst with removal of the water that was formed by azeotropic distillation, these reactions were extended to produce a series of new analogous polyspiroketals. Furthermore since the formation of the intermediate hemiketal is not a stable product and it is necessary to proceed all the way to the ketal before a stable unit is formed, this reaction appeared ideal for the synthesis of soluble linear polyspiroketals containing only cyclicized structures. This reaction also represents a condensation polymerization analog of the free radical inter-intramolecular polymerization to produce linear polymers. Thus it was shown that linear soluble polymers could be prepared from the following polyfunctional materials: a five-membered ring-containing polyketal from 1,2,4,5-tetrahydroxy-cyclohexane and 1,4-cyclohexanedione; six-membered ring-containing polyketal from 1,1,4,4-tetrakis(hydroxymethyl)cyclohexane, and 1,1,6,6-tetrakis(hydroxymethyl) cyclohexanone; six-membered ring-containing polyketal from 1,3-dihydroxyacetone, as well as 1,10-cyclooctadecanedione and pentaerythritol; a seven-membered ring-containing polyketal from 1,4-cyclohexanedione and 1,2,4,5-tetrakis(hydroxymethyl)-

0097-6156/82/0195-0391$06.00/0
© 1982 American Chemical Society

cyclohexane; and an eight-membered ring-containing polyketal from 1,4,5,8-tetrakis-(hydroxymethyl)-tetralin and 1,4-cyclohexanedione.

Although the condensation polymerization of a trifunctional or tetrafunctional monomer with a difunctional monomer, such as the condensation of glycerol with phthalic anydride, will usually produce a highly cross linked, insoluble product, there are many exceptions. Just as in the case of inter-intramolecular free radical addition polymerization, if stable five- or six-membered rings can form, a soluble polymer usually results. Ketal formation between a cyclic diketone and a tetraol is a good illustration of such a reaction (1,2,3). The resulting spiro ketal polymers which are double-stranded polymers and are therefore related in properties to ladder polymers are of additional interest because they have a high degree of chemical and thermal stability since they resist changes in molecular weight and because they have extremely high melting points or glass transitions since they have high restricted rotation (4). A number of other spiro polymers have been reported recently including polyspirobenzothiazolines (5), the condensation products of dianhydrides with tetraamines (6), spiro polyallenes (7), and spiro tetrasulfides (8).

The ketal formation is an ideal reaction for this study since it produces either a stable spiro ring or no new bond at all. The condensation proceeds stepwise, but since the hemiketal intermediate is unstable, no condensation would be expected to take place until the stable ketal is formed.

$$\text{C=O} + \begin{matrix}\text{HO}\\\text{HO}\end{matrix} \rightleftarrows \text{C}\begin{matrix}\text{O}\\\text{OH}\end{matrix}\ \text{OH} \xrightarrow[-H_2O]{H^+} \text{C}\begin{matrix}\text{O}\\\text{O}\end{matrix}$$

Pentaerythritol was chosen since it forms spiro compounds very readily and gives high enough yields to be used in a polymerization reaction with dialdehydes (9,10). Thus it was reasoned that a similar reaction with a cyclic diketone should produce a polyspiroketal. Thus, when 1,4-cyclohexanedione was heated in benzene with pentaerythritol plus a trace of *p*-toluenesulfonic acid (PTSA) and the resulting water (93%) was removed by distillation, a white polymer precipitated. The spiro polymer I was shown by X-ray analysis to be highly crystalline and decomposed at 350°C without melting. It was almost completely insoluble in most solvents but was completely soluble in hot hexafluoroisopropanol.

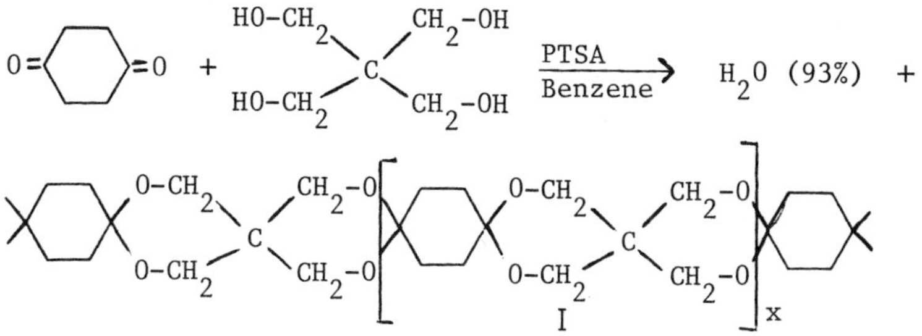

```
                HO-CH2     CH2-OH
O=<   >=O   +         \  /          PTSA
                        C         ------->   H2O (93%)   +
                      /  \         Benzene
                HO-CH2     CH2-OH

        O-CH2       CH2-O [      O-CH2      CH2-O ]
  X   X      \     /     [ X   X      \    /     ] X   X
              C          [             C         ]
        O-CH2 /     \     [      O-CH2 /    \     ]
                  CH2-O [             CH2-O  ]x
                                   I
```

End group analysis indicated that the molecular weight was about 20,000. Addition of a small amount of cyclohexanone as a chain stopper with the 1,4-cyclohexanedione produced lower molecular weight polyspiroketals that were more soluble.

It can be noted that this reaction is closely analogous to the free radical inter-intra-molecular polymerization discovered by Butler and Angelo (11) in which a difunctional monomer undergoes such efficient ring closure that a soluble linear polymer is obtained instead of a crosslinked or three-dimensional product. In the case of the polyketals the condensation of a tetraol with a diketone might at first glance be expected to form a cross-linked polymer, but the ring closure to the six-membered ring is so efficient that a linear soluble product results. Thus the ketal formation can be viewed as a special case of inter-intra-molecular condensation polymerization.

Further investigation showed that a variety of polyspiroketals can be prepared by the condensation of various cyclic diketones and various tetraols involving the formation of a six-membered ketal. These include the condensation of 1,10-cyclooctadecanedione with pentaerythritol to produce a more soluble polymer II.

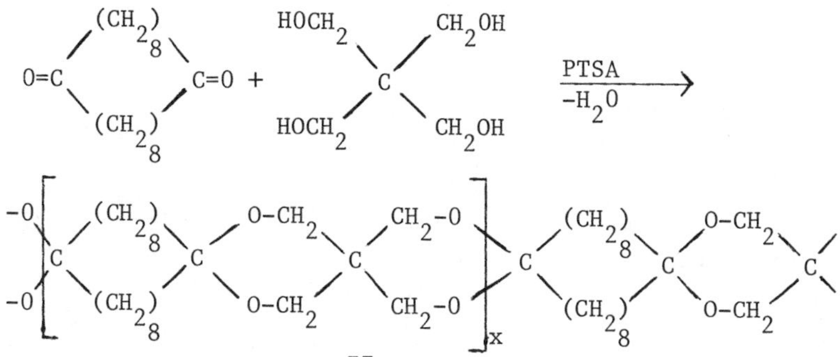

```
      (CH2)8             HOCH2       CH2OH
     /      \                 \     /          PTSA
O=C          C=O  +             C           -------->
     \      /                 /     \          -H2O
      (CH2)8             HOCH2       CH2OH

-O [   (CH2)8        O-CH2       CH2-O ]     (CH2)8       O-CH2
    C          C            C           C           C            C
-O [   (CH2)8        O-CH2       CH2-O ]x    (CH2)8       O-CH2
                              II
```

It was of interest to extend the reactions of tetrols with cyclic diketones to produce a variety of spiro polymers with less oxygen in the base chain and with other functional groups present in the molecule. Therefore the synthesis of a number of tetraols which would be useful in preparing polyspiroketals was undertaken. For this reason, 1,1,4,4-tetrakis-(hydroxymethyl)-cyclohexane was prepared by the following sequence of reactions (12):

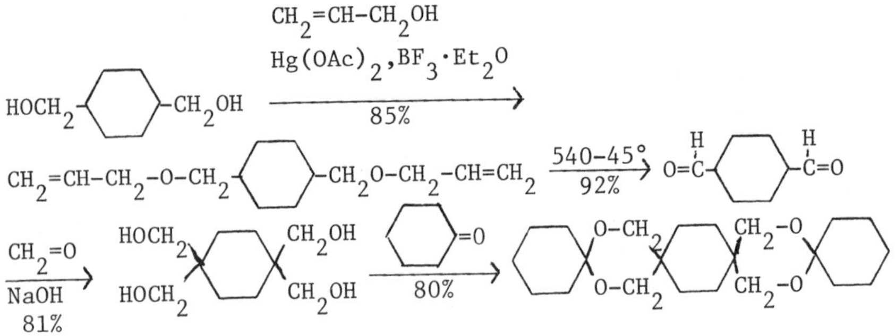

CH_2=CH-CH_2OH
Hg(OAc)$_2$,BF_3·Et_2O
$HOCH_2$- -CH_2OH
85%
CH_2=CH-CH_2-O-CH_2- -CH_2O-CH_2-CH=CH_2
540-45°
92%
O=C(H)- -C(H)=O
CH_2=O
NaOH
81%
$HOCH_2$ CH_2OH
$HOCH_2$ CH_2OH
=O
80%
O-CH_2 CH_2-O
O-CH_2 CH_2-O

When cyclohexanone was condensed with the tetrakis-(hydroxymethyl)-cyclohexane, an 80% yield of the model spiro ketal was formed. Under very similar conditions, then, the tetrol plus 1,4-cyclohexanedione was condensed to produce a 70% yield of the corresponding polyspiroketal III. This material III

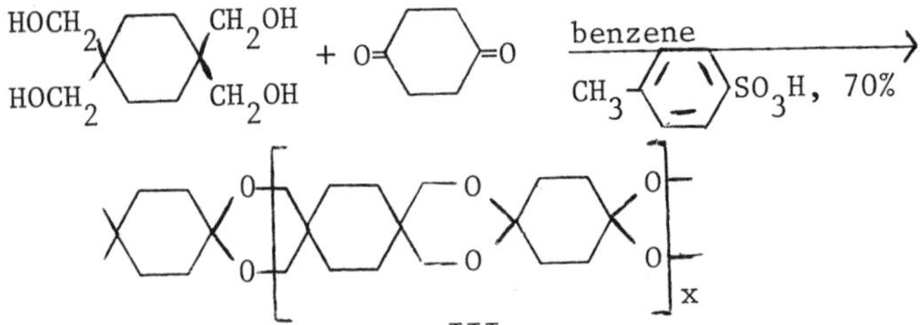

$HOCH_2$ CH_2OH
$HOCH_2$ CH_2OH
+ O= =O
benzene
CH_3- -SO_3H, 70%
O O O
O O O
x
III

was insoluble in most common solvents, but was soluble to a limited extent in hexafluoroisopropanol (0.2 g/40 ml) after being stirred for a 12-hour period. This polymer did not melt and had an [η] of 0.064 dl/g (25°, hexafluoroisopropanol).

The condensation of cyclohexanone (13) and cyclopentanone (14) with formaldehyde produced the corresponding tetrakis-(hydroxymethyl)-ketones in 37 and 90% yields, respectively. When the tetramethylolcyclohexanone was treated with cyclohexanone, a 77% yield of the model keto spiro ketal was formed. In this compound the carbonyl group located between the two quaternary carbon atoms is so hindered that no ketal formation apparently takes place.

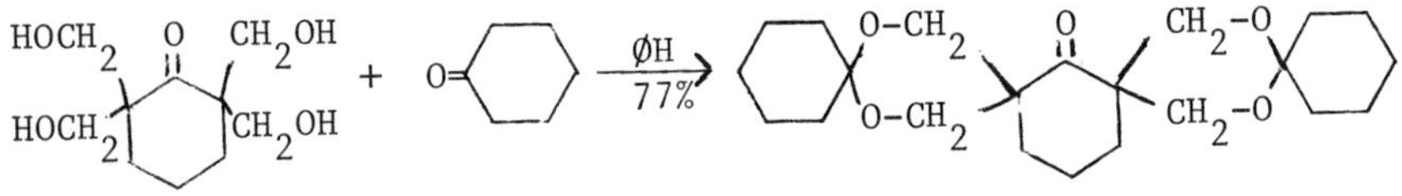

$HOCH_2$ O CH_2OH
$HOCH_2$ CH_2OH
+ O=
ØH
77%
O-CH_2 O CH_2-O
O-CH_2 CH_2-O

In a similar manner, when the tetramethylolcyclohexanone was condensed with 1,4-cyclohexanedione, a 68% yield of the corresponding polymer was obtained. This polymer was swollen by hot benzene and was quite soluble in hexafluoroisopropanol to give an [η] of 0.052 dl/g (25°, hexafluoroisopropanol). The polymer appeared to soften at 255-290°, evidence that the assymetry of the tetrol portion tends to reduce the melting point.

In a similar fashion the tetramethylolcyclopentanone plus cyclohexane gave a 72% yield of the model spiroketal.

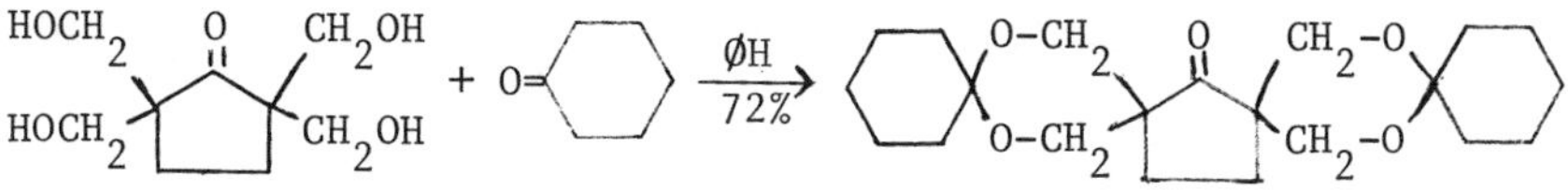

ϕH, 72%

When the polymerization was carried out with the cyclopentanone derivative and 1,4-cyclohexanedione, a 74% yield of the polyspiroketal was formed.

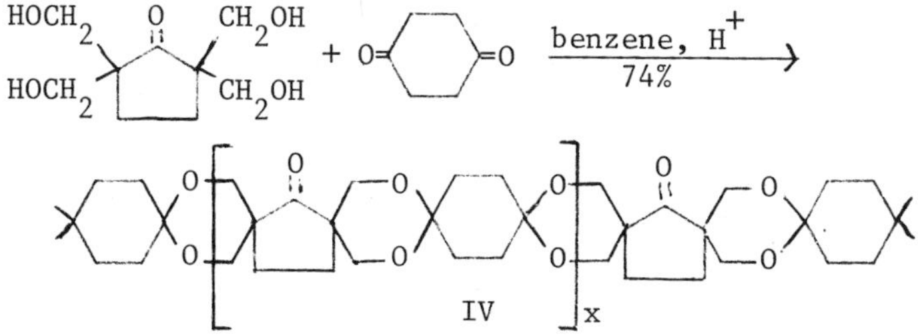

benzene, H^+, 74%

IV

Again, this polymer was soluble in hexafluoroisopropanol with an [η] of 0.045 dl/g and was swollen by a number of solvents. A strong carbonyl absorbtion at 1698 cm^{-1} indicated that the carbonyl group of the tetrol did not participate in the polymerization reaction. Again, this polymer appeared to melt at 370-395° indicative of the fact that the asymmetry increased the solubility and decreased the melting point compared to the symmetrical polyspiroketals.

In an attempt to find the simplest compound that would form a polyspiroketal, our attention was directed to dihydroxyacetone.

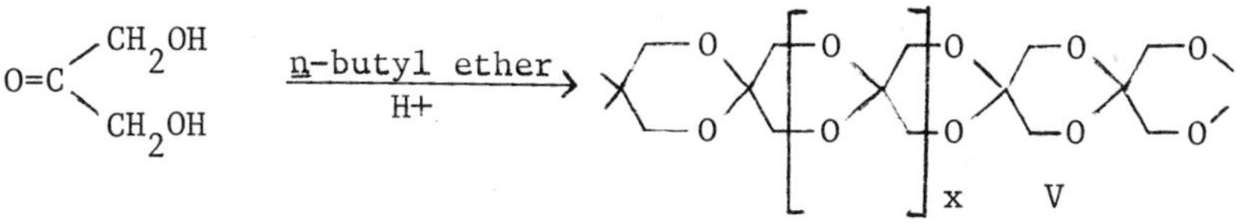

n-butyl ether, H+

V

When dihydroxyacetone was heated in the presence of an acid catalyst in benzene, the theoretical amount of water was collected and a white polymer was obtained. This polymer V did not melt and was insoluble in all solvents tried including hexafluoroisopropanol. The x-ray powder pattern, however, indicated that the material was crystalline, and therefore probably was not cross-linked, but just so highly polar that it did not dissolve.

Since all the previous polyspiroketals that have been discussed involve the formation of six-membered rings, it was of interest to study the extension of this reaction to other size rings. In view of the volume of literature indicating that the formation of five-membered cyclic ketals would be rather facile, the formation of model compounds from both cis- and trans-1,2-cyclohexanediol with 1,4-cyclohexanedione was investigated. When these materials were heated in benzene with a trace of p-toluenesulfonic acid as a catalyst, an 86% yield of the desired cyclic ketal was formed in both cases.

OH OH + O= =O benzene, H+, 86%

cis or trans → cis, cis or trans, trans

In the extension of the reaction to polymers 1,2,4,5-cyclohexanetetrol was prepared in a 12% over-all yield from 1,4-cyclohexadiene by the oxidation with osmium tetroxide. The meso (12/45) diastereomer of (cis/cis) 1,2,4,5-tetrahydroxy-cyclohexane (15) was condensed with 1,4-cyclohexanedione to give a 95% yield of a crystalline spiro polymer VI. This material did not melt, but was soluble in hexafluoroisopropanol with an [η] of 0.056 dl/g (25°, hexafluoroisopropanol).

HO HO OH OH + O= =O n-butyl ether, 95%

cis/cis → VI, x

We next became interested in the possibility of extending this to a seven-membered cyclic ketal. It was immediately recognized that if the yield of the cyclic compound were not nearly quantitative, the resulting polymer would be cross-linked. In our previous work (16) the cis- and trans-1,2-bis (hydroxymethyl)-cyclohexane were prepared. When either the cis- or trans-diol was condensed with 1,4-cyclohexanedione, a 92% yield of the corresponding model compound was obtained.

CH_2OH CH_2OH + O= =O benzene, H+, 92%

cis or trans → cis, cis or trans, trans

Also in our previous work, the tetrol, 1,2,4,5-tetrakis-(hydroxymethyl) cyclohexane (17,18) of unknown stereochemistry was obtained. When this tetrol was condensed with cyclohexanone, a 92% yield of the bicyclic ketal containing the seven-membered ring was obtained.

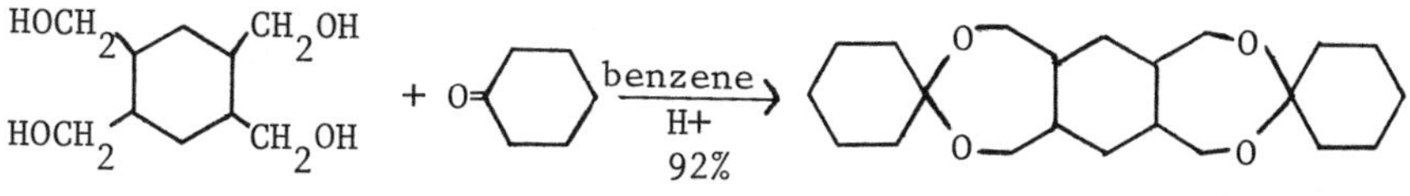

When this same tetrol was condensed with 1,4-cyclohexanedione in n-butyl ether in the presence of a trace of p-toluenesulfonic acid, a nearly quantitative yield of a white polymer VII was obtained. This material did not melt, but was soluble in hexafluoroisopropanol and had an [η] of 0.04 dl/g (25°, hexafluoroisopropanol). This polymer VII was found to be highly crystalline by X-ray powder pattern and had a $T^{\circ}_{1/2}$ in the DTA apparatus of 400°.

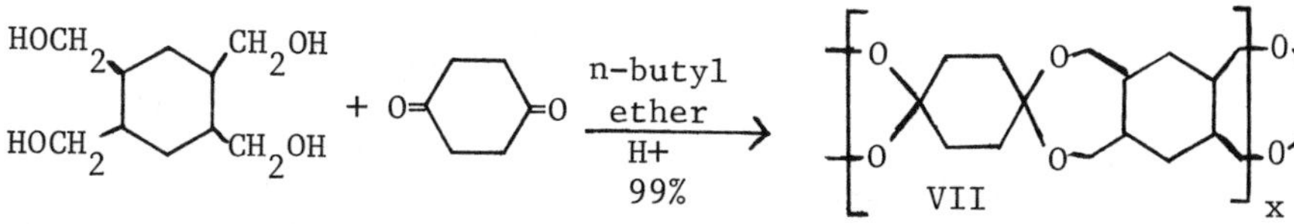

A very similar polymer was prepared from this same tetrakis-(hydroxymethyl) cyclohexane and 1,10-cyclooctadecadione. A white solid was obtained in a 93% yield which was soluble in hexafluoroisopropanol and slightly soluble in hot dioxane. It was shown to be highly crystalline by an X-ray powder diffraction and did not melt. Its [η] in hexafluoroisopropanol was 0.062 dl/g (25°). The fact that these two polymers containing seven-membered spiro ketals were highly crystalline and soluble, indicates that the ketal formation proceeded in essentially a quantitative manner.

In order to reduce the crystallinity of the polyketal an unsymmetrical cyclic diketone was prepared from 2,7-dihydroxynaphthalene. Reduction of the naphthalene nucleus with Raney nickel and hydrogen gave a 90% yield of the 2,7-decalindiol, which on oxidation with chromic acid gave a 50% yield of the 2,7-decalindione. When this diketone was condensed with the tetrakis-(hydroxymethyl) cyclohexane, an 89% yield of a white polymer VIII was obtained which did not melt and was soluble only in hexafluoroisopropanol.

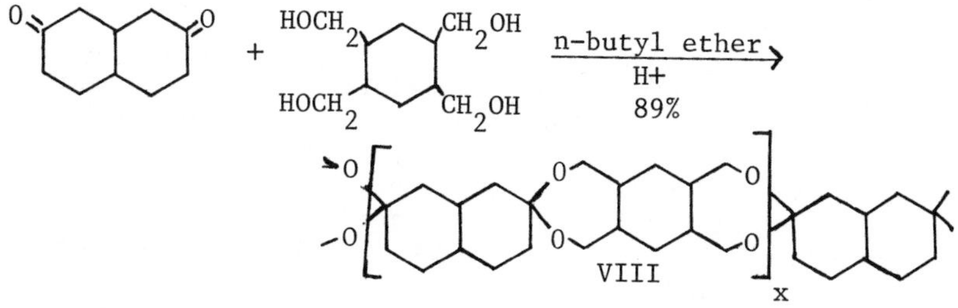

Since the seven-membered rings formed so readily to produce the ketals, an investigation was undertaken to see if eight-membered cyclic ketals would form as readily. A model compound was prepared from 1,8-bis(hydroxymethyl) naphthalene and cyclohexanone in a nearly quantitative yield of the cyclic ketal containing the eight-membered ring.

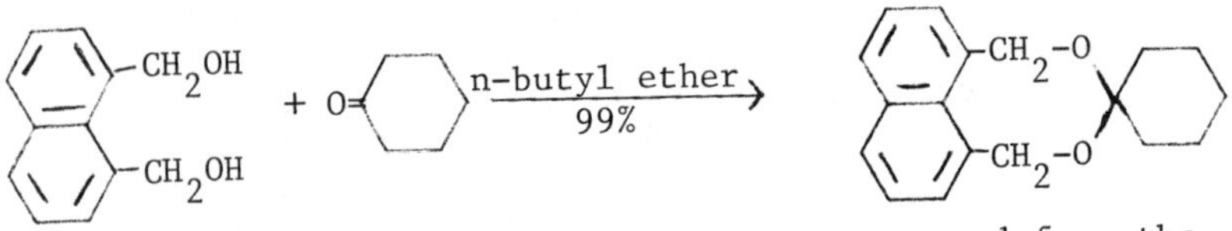

$-CH_2OH$ $-CH_2OH$ + O= $\xrightarrow[99\%]{\text{n-butyl ether}}$ $-CH_2-O$ $-CH_2-O$

When the corresponding polymer was prepared from the 1,4,5,8-tetrakis (hydroxymethyl) naphthalene and 1,4-cyclohexanedione, a low molecular weight polymer resulted. This was probably due to the near insolubility of the tetrol in most of the solvents. If more vigorous conditions were used to increase the solubility, the corresponding cyclic diether was formed. When the naphthalene ring was saturated so that both of the adjacent alcohol groups were not benzylic in nature, no cyclic ether was produced and a somewhat higher molecular weight polymer IX resulted.

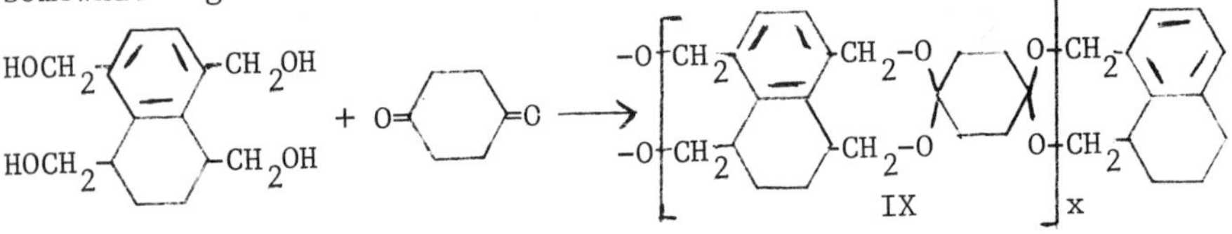

$HOCH_2$ CH_2OH $HOCH_2$ CH_2OH + O= =O $\longrightarrow$ $-O-CH_2$ CH_2-O $O-CH_2$ $-O-CH_2$ CH_2-O $O-CH_2$ IX x

It thus appears that a wide variety of tetrols and cyclic diketones can be used to prepare a large variety of soluble polyspiroketals. It appears that the reaction can be made to go readily when the ketal contains either five-, six-, seven-, or eight-membered rings. Some of the polymers containing the five-, seven- and eight membered rings can, of course, be considered to be a hybrid between ladder polymers and spiro polymers. An effort will be made to see what other cyclic ketals are useful for the formation of spiro polymers and what other types of reactions besides the ketal formation will produce similar polymers.

Experimental

7,14,21,26-Tetraoxatetraspiro [5.2.2.2.5.2.2.2] hexacosane. Into a 500-ml, round-bottomed flask equipped with a magnetic stirrer, a Dean-Stark trap and a reflux condenser was added 2.04 g (0.01 mol) of 1,1,4,4-tetrakis(hydroxymethyl) cyclohexane, mp 223-224° [reported (12) mp 219-222°C], 2.16 g (0.022 mol) of cyclohexanone, 0.1 g of p-toluenesulfonic acid, and 300 ml of benzene. After the mixture had been heated under reflux until no more water collected in the Dean-Stark trap, the benzene was removed by distillation under reduced pressure. The resulting solid was dissolved in abs ethanol and treated with several grams of de-

colorizing charcoal. The volume of the solution was reduced to 10-15 ml by heating on a steam bath. When the mixture was cooled for 1 hr, crystallization occurred. One recrystallization from a dilute soln of ethanol gave 2.91 g (80%) of 7,14,21,26-tetraoxatetraspiro [5.2.2.2.5.2.2.2] hexacosane, mp 188°C.
Anal. Calcd for $C_{22}H_{36}O_4$: C, 72.48; H, 9.95; m/e, 364.2614.
Found: C, 72.58; H, 9.87; m/e, 364.2616.

An IR spectrum of this compound had neither a hydroxyl absorption at 3400 cm^{-1} nor a carbonyl absorption at 1725 cm^{-1}.

Polyspiroketal III by the Reaction of 1,4-Cyclohexanedione with 1,1,4,4-Tetrakis(hydroxymethyl)cyclohexane. A mixture of 10.2 g (0.05 mol) of 1,1,4,4-tetrakis(hydroxymethyl) cyclohexane, 5.6 g (0.05 mol) of 1,4-cyclohexanedione and 0.5 g of p-toluenesulfonic acid was heated under reflux in 200 ml of benezene (reagent grade). After the mixture had been heated with rapid stirring for 5 hr, another 150 ml of benzene was added and the mixture was heated until no further water was collected in the Dean-Stark trap. The hot mixture was filtered through a Buchner funnel, and the solid was washed sequentially with 400-ml portions of boiling benzene, methanol, carbon tetrachloride, chloroform, and N,N-dimethylformamide. Finally the polymer was washed with 200 ml of benzene, air dried, and then vacuum desiccated for 6 hours to give 10.0 g (70%) of polymer III which showed no loss of water at 100°C and no change of color until 400°C. Some shrinkage was observed to start at 370°C and melting with decomposition resulted between 410-450°C.
Anal. Calcd for $(C_{16}H_{24}O_4)_n$: C, 68.54; H, 8.63. Found: C, 68.38; H, 8.37.

After a 0.2-g sample was added to 40 ml of hexafluoroisopropanol, the mixture was stirred overnight and filtered. A residue of approximately 2 mg remained undissolved. An ir spectrum of the vacuum desiccated sample showed trace absorptions at 3420 and 1699 cm^{-1} which was attributed to entrapped starting material or end-groups on the polymer chain. An ir spectrum of the same sample although nondesiccated, showed a pronounced increase in the hydroxyl absorption at 3420 cm^{-1} indicative of the hydroscopic nature of the polymer. A detailed ir (Nujol) spectrum contained the following absorptions: 1360 (m), 1345 (m), 1270 (m), 1240 (w), 1225 (w), 1169 (m), 1144 (s), 1132 (m), 1110 (s), 1101 (s), 1079 (m), 1059 (m), 1050 (m), 1038 (m), 1025 (m), 988 (w), 977 (m), 948 (m), 929 (w), 908 (s), and 891 (m) cm^{-1}.

7,14,21,25-Tetraoxatetraspiro[5.2.2.2.5.2.1.2] pentacosan-23-one. Into a 500 -ml flask equipped with a magnetic stirrer, a Dean-Stark trap and a reflux condenser was added 2.04 g (0.01 mol) of 2,2,5,5-tetrakis(hydroxymethyl)cyclopentanone, mp 142-143°C [reported (14) mp 143°C], 2.10 g (0.021 mol) of cyclohexanone, 150 ml of benzene and 0.1 g p-toluenesulfonic acid. The mixture was heated under reflux until no more water was collected in a Dean-Stark trap. The mixture had a blue residue which was removed by filtration, and the filtrate was washed with 10 ml of a 5% sodium bicarbonate soln followed by 10 ml of water. After the soln was dried ($MgSO_4$), filtered, and concentrated, the residue was recrystallized twice from abs ethanol to give 2.62 g (72%) of 7,14,21,25-tetraoxatetraspiro(5.2.2.2.5.2.1.2)pentacosan-23-one, mp 198-199°C.

Anal. Calcd for $C_{21}H_{32}O_5$: C, 69.20; H, 8.85. Found: C, 69.06; H, 8.81.

An IR spectrum of this compound showed no hydroxyl absorption, and a carbonyl absorption was noted at 1698 cm^{-1}.

Polyspiroketal from 2,2,5,5-Tetrakis(hydroxymethyl)cyclopentanone and 1,4-Cyclohexanedione. A mixture consisting of 1.12 g (0.01 mol) of 1,4-cyclohexanedione, 2.04 g (0.01 mol) of 2,2,5,5-tetrakis-(hydroxymethyl)cyclopentanone, 0.1 g of p-toluenesulfonic acid, and 300 ml of benzene were heated under reflux with stirring until no more water was collected in a Dean-Stark trap. The time required was approximately 12 hr. The hot mixture was filtered through a Buchner funnel and the polymeric residue was washed with 300-ml portions of boiling benzene, methanol and carbon tetrachloride. The residue was then vacuum desiccated for several hours to yield 2.07 g (74%) of polymer, mp 370-395°C.

Anal. Calcd for $(C_{15}H_{20}O_5)_n$: C, 64.27; H, 7.19. Found: C, 62.40; H, 7.16.

An IR spectrum of this polymer showed mere trace absorption for hydroxyl group at 3400 cm^{-1}. The carbonyl absorption peak at 1700 cm^{-1} had no shoulders. A detailed IR (Nujol) spectrum contained the following absorptions: 1700 (s), 1321 (w), 1263 (w), 1239 (m), 1195-1201 (m), 1161-1170 (w), 1138 (s), 1109 (s), 1060 (m), 1020 (w), 985 (w), 961 (m), 952 (m), 931 (m), 905 (m), 885 (w), 760 (m), 750 (m), and 713 (m) cm^{-1}.

cis, cis-2,3,10,11-Bistetramethylene-1,4,9,12-tetraoxadispiro-[4.2.4.2] tetradecane. Into a 50 -ml flask were added 1.5 g of cis-1,2-cyclohexanediol, 0.56 g of 1,4-cyclohexanedione, 0.1 g of p-toluenesulfonic acid and 30 ml of benzene. The contents of the flask were heated under reflux overnight during which time the theoretical amount of water separated. After the mixture was cooled to room temperature, the benzene solution was washed with water. After the solution was dried over sodium sulfate, most of the solvent was removed by evaporation. The solid that separated was collected by filtration and washed with ether. A repetition of this process gave a second crop or a total of 1.5 g (86%). The solid which was recrystallized from chilled ether melted at 173-173.5°C.; IR: 2935 (s), 2860 (m), 1450 (m), 1380 (m), 1300 (w), 1208 (m), 1187 (m), 1123 (s), 1100 (s), 1045 (m) 980 (m), 970 (m), 920 (s), 890 (w), 857 (w), 825 (w), 675 (w), 468 (w) cm^{-1}.

Anal. Calcd for $C_{18}H_{28}O_4$: C, 70.10; H, 9.10. Found: C, 70.20; H, 9.27.

Polyspiroketal from meso (12/45) Diastereomer of (cis/cis)-1,2,4,5-Cyclohexanetetrol and 1,4-Cyclohexanedione. Into a 50 -ml flask equipped with a magnetic stirrer and capped with a Dean-Stark trap, a reflux condenser and a drying tube were added 0.148 g of 1,2,4,5-cyclohexanetetrol, 0.112 g of 1,4-cyclohexanedione and 50 ml of n-butyl ether. After the mixture was heated under reflux for 2 hours, 0.01 g of p-toluenesulfonic acid was added. Heating was continued overnight at the end of which time the flask was cooled to room temperature and the resulting solid was collected by filtration. After it was washed well with water, acetone and ether, the solid was dried over P_2O_5 under 0.2 mm pressure at about 70°C for 24 hours. The dry polymer V did not melt, was soluble only in hexafluoroisopropanol, and had an [η] of 0.056 dl/g.

Anal. Calcd for $(C_{12}H_{16}O_4)_n$: C, 64.20; H, 7.14. Found: C, 63.70; H, 7.14.

Polymer VII from 1,4-Cyclohexanedione and 1,2,4,5-Tetrakis-(hydroxymethyl) cyclohexane. Into a 100-ml flask equipped with a mag. stirrer, a Dean-Stark trap, a reflux condenser and a drying tube were added 50 ml of dry n-butyl ether, 2.04 g of 1,2,4,5-tetrakis(hydroxymethyl)cyclohexane and 1.12 g of 1,4-cyclohexanedione. The solution-suspension was heated under reflux for 12 hours followed by the addition of 0.1 g of p-toluenesulfonic acid. After the mixture was heated overnight, it was cooled to room temperature. The mixture was filtered with suction and the resulting solid was washed well with n-butyl ether. After most of the ether was removed, the solid was washed with water, acetone and finally ether. The solid, which was dried over P_2O_5 and under 0.2 mm pressure, weighted 3.1 g (99%). The solid was soluble only in hexafluoroisopropanol and had an [η] of 0.04 dl/g at 25°C. The polymer was found to be crystalline by X-ray and did not melt.

Anal. Calcd for $(C_{16}H_{24}O_4)_n$: C, 68.54; H, 8.63. Found: C, 68.30; H, 8.49.

A portion of the polymer VII, which was refluxed in hexafluoroisopropanol to test for hydrolysis, was recovered unchanged as shown by IR, TGA and analysis. The infrared spectrum showed bands at 2930 (m), 2865 (m), 1450 (w), 1370 (w), 1250 (w), 1120 (s), 1057 (w), 1035 (m), 946 (w), and 928 (w) cm^{-1}.

The authors are grateful to the Army Office of Research-Durham, the Goodyear Tire and Rubber Company, and the Naval Surface Weapon Center for partial support of this work.

This paper is dedicated to Professor George B. Butler, who initiated all of the interest and research in the field of inter-intramolecular polymerization, on the occasion of his 65th birthday.

Literature Cited

1. Bailey, W. J.; Volpe, A.A. J. Polym. Sci., Part A-1 1970, 8, 2109-17.
2. Bailey, W. J.; Hinrichs, R.I. Am. Chem. Soc., Div. Polym. Chem., Preprints 1970, 11(2), 500.
3. Bailey, W. J.; Beam, C. F., Jr.; Haddad, I. Am. Chem. Soc., Div. Polymer Chem., Preprints 1971, 12, 169-74.
4. Bailey, W. J. in Bikalis, N. "Encyclopedia of Polymer Science and Technology" Vol. 8; John Wiley and Sons, Inc.: New York, NY, 1968; pp. 97-120.
5. Augl, J. M.; Wrasidlo, W.J. J. Polym. Sci., Part A-1 1970, 8, 63-66.
6. Hodgkin, J. H.; Heller, J. J. Macromol. Sci.-Chem. 1969, 3, 1067-71.
7. Rinehard, R. E. (Uniroyal, Inc.) Ger. Offen. 1,943,624 (Mar. 5, 1970); Chem. Abstr. 1970, 72, 101, 242x.
8. Tokarzewski, L.; Plonka, S. Polimery 1969, 14(12), 596; Chem. Abstr. 1970, 73, 25,925a.
9. Cohen, S.M.; Lavin, E. J. Appl. Polym. Sci. 1962, 6, 503-7.
10. Cohen, S. M.; Hunt, C. F.; Kass, R. E.; Markhart, A. H., J. Appl. Polym. Sci. 1962, 6, 508-11.
11. Butler, G. B.; Angelo, R. J. J. Am. Chem. Soc. 1957, 79, 3128-31.
12. Beam, C. F., Jr.; Bailey, W. J. J. Chem. Soc., C 1971, 2730-2.
13. Mannich, C.; Brose, W. Ber. 1923, 566, 833.
14. Ray, G. C. (Phillips Petroleum Co.) U. S. Patent 2,500,570 (Mar. 14, 1950); Chem. Abstr. 1950, 34, 5385.
15. McCasland, G. E.; Furtra, S.; Johnson, L. F.; Shoolery, G. N. J. Org. Chem. 1963, 28, 894.
16. Bailey, W. J.; Golden, H. R. J. Am. Chem. Soc. 1953, 75, 4780.
17. Bailey, W. J.; Fetter, E. J.; Economy, J. E. J. Org. Chem. 1962, 27, 3469.
18. Longone, D. T. J. Org. Chem. 1963, 28, 1770.

RECEIVED January 22, 1982.

Hydrogen Bonding Ring Containing Polymers: Poly(enamine–Ketones)

SAMUEL J. HUANG, BRIAN BENICEWICZ, and JOSEPH A. PAVLISKO

University of Connecticut, Department of Chemistry and Institute of Materials Science, Storrs, CT 06268

Monomeric and polymeric enamine-ketones were synthesized by the condensation reaction of 1,3-diketones with amines. These compounds exhibit liquid crystalline properties when heated to appropriate temperatures. The transition temperatures of these compounds are much lower than those of the aromatic hydrocarbon analogs. High solubilities in organic solvents of these compounds together with the low transition temperatures suggest possible processability for the poly(enamine-ketones) as high strength materials.

Hydrogen bonds are often responsible for holding natural proteins in specific conformations. Segments of the protein chain and hydrogen bonds form ring systems that are the main part of the protein helix. The hydrogen bonding rings formed between the base pairs are responsible for the unique double helical structure of the DNA molecules. However, intramolecular hydrogen bonds in synthetic polymers remain as a relatively unexplored area of research. We have recently synthesized several polymers containing hydrogen bonding rings in the polymer backbone with very interesting properties. We report here our recent findings on poly(enamine-ketones) (1-3).

Regularity and rigidity in a polymer chain enhance the ability for the polymer to crystallize. High crystallinity and orientation order of the crystalline regions result in high strength/modulus and thermal stability of polymeric materials. Recent studies on rigid chain polymers have resulted in a few useful high strength/modulus materials (4-9). Traditionally aromatic hydrocarbon rings are incorporated into a polymer chain backbone to provide rigidity for the polymer chain. As the number of aromatic hydrocarbon rings increases in the polymer chain the melting point of the polymeric material increases and the solubility of the material decreases. Many

0097-6156/82/0195-0403$06.00/0
© 1982 American Chemical Society

high temperature stable rigid polymers containing aromatic hydrocarbon rings are infusible and insoluble in common solvents. These properties limited the usefulness of the polymers because of the lack of effective ways to process them into useful end products. We have been interested in the possibility of replacing aromatic hydrocarbon rings with other rigid structural units to obtain processable materials with high strength/modulus and thermal stability. Six-membered hydrogen bonding rings such as carbonyl-enamines and carbonyl-hydrazones are of special interests to us since the conjugated double bonds and the hydrogen bonding system make up the near planar six-membered rings. We expect these systems to be reasonably rigid and thermally stable.

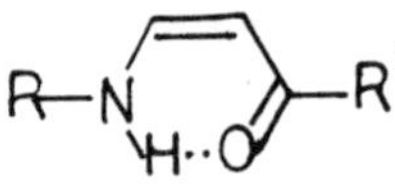

R–N H··O –R'

Carbonyl-enamines

R–N N= H··O –R'

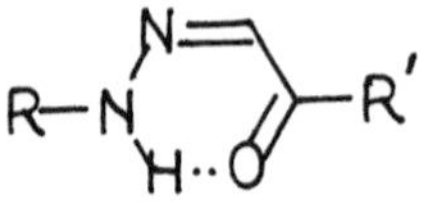

Carbonyl-hydrazones

A polymer containing one or more of these ring systems in the backbone can be expected to be rigid and yet having lower thermal transition temperatures than that of the analogous polymers containing only aromatic hydrocarbon rings. The presence of hetero atoms and the low symmetry of the hydrogen bonding rings should also provide the polymer with higher solubility than those of the analogous polymers containing only hydrocarbon rings. Low thermal transition temperatures and high solubility should provide processability for the hydrogen-bonding ring containing polymers. We approach the research by firstly studying the synthesis and properties of monomeric model components.

EXPERIMENTAL

General. Thermal analysis was carried out with a DuPont 990 Thermal Analyzer with a DTA/DSC cell and a DuPont 950 DTA unit. The thermal optical analysis (70A) unit used was constructed in our laboratories and consists of a Leitz Dialux Pol polarizing microscope fitted with a hot stage. Infrared spectra were recorded with a Perkin Elmer 283 grafting spectrophotometer. NMR were recorded with a Brucker WH90FT NMR spectrophotometer and a Varian EM360 NMR spectrophotometer with a Varian EM3630 lock/ decoupler. Inherent viscosity measurements were made with a Ubbelohde viscometer in a constant temperature bath. Microanalysis were performed by Baron Consulting Co. of Orange, Conn. All solvents and reagents were purified before used.

4,4'-Diacetylbiphenyl. Anhydrous aluminum chloride 26.7 g (200 mmoles) was covered with a mixture of 100 mL CS_2 and 15.4 g (100 mmoles) biphenyl. Acetyl chloride 15.7 g (200 mmoles) was slowly added and evolution of HCl allowed to become complete by subsequent heating for 12 hours on a water bath. The CS_2 was evaporated, the mixture decomposed on ice-HCl and the resulting white solid collected and dried. Fractional crystallization from alcohol gave phenyl acetophenone and 4,4'-diacetylbiphenyl. The latter was purified by recrystallization from CCl_4 to give 9.95 g (41.5%) of colorless leaflets, mp 193-194°C. The infrared spectra (KBr) showed absorptions at 1685 (νnC=O), 1600 (νnC=C) and 815 (νSC-H). The proton NMR showed peaks at 2.65δ singlet (3) and 7.8δ multiplet (4). Anal: Calcd C, 80.61, H, 6.00. Found: C, 80.67; H, 5.88.

Acetylacetophenone (Benzoylacetone). To an ice cooled solution of 352 g (4 moles) ethylacetate and 100 g sodium ethoxide (1.47 mole) was added dropwise 240 mL of acetophenone (2 moles) extended over a period of one hour. After addition was completed the reaction mixture was allowed to warm to room temperature and let stand 20 hours. The reaction mixture was poured into 1.5 of water and stirred until solution. The aqueous solution was washed with two 250 mL portions of ethyl ether, and acidified with 100 mL of glacial acetic acid to precipitate a yellow solid which was collected and dried overnight. The yellow solid was crystallized from alcohol to afford 175.6 g (72.2%) of pale yellow needles, mp 54-55°C. The infrared spectra showed absorptions at 3450 (νO-H), 1615 (νC=O), and 3005 (νC-H). The proton NMR showed peaks at 2.5δ singlet (3), 6.2δ singlet (1), and 7.6δ multiplet (5). Anal: Calcd C, 74.07; H, 6.17. Found: C, 73.86; H, 6.40.

(1,1'-p-Phenylene di)1,3-Butanedione(Terephthaloyldiacetone). Sodium ethoxide 6.8 g (100 mmoles) was covered with a mixture of 30.8 g (350 mmoles) of ethyl acetate and 8.1 g (50 mmoles) of p-diacetylbenzene. The reaction mixture was allowed to stand for 18.5 hours and then poured into 750 mL of water and extracted with two 100 mL portions of ethyl ether. The aqueous dark brown solution was then acidified with 20 mL of glacial acetic acid to precipitate a yellow solid which was collected and dried. The yellow material was recrystallized several times from chloroform to give 6.5 g (52.9%) of white prisms, mp 183.5-184.5°C. The infrared spectrum (KBr) showed absorptions at 3425 (νO-H), and 1620 (νC=O). The proton NMR showed peaks at 2.5δ singlet (3), 6.18δ singlet (1), and 7.9δ singlet (2). Anal: Calcd C, 68.29; H, 5.69. Found: C, 68.24; H, 5.81.

(1,1'-p-Biphenylene di)1,3-Butanedione(p,p'-Dibenzoyldiacetone). Sodium ethoxide 8 g (220 mmoles) was covered with a mixture of 20.7 g (235 mmoles) of ethyl acetate and 14 g (59 mmoles) of diacetylbiphenyl. The reaction mixture was allowed to stand for 23 hours and then poured into 2.5 L of water. Solution did not occur and upon subsequent heating appeared as a finely suspended orange solid. This suspension was acidified with 75 mL of glacial acetic acid and stirred for a one-hour period with mild heating. The previously orange solid now appeared as a yellow material which was filtered and dried. The yellow solid was crystallized from choloroform followed by recrystallization from a chloroform-CCl_4 mixture to give 8.5 g (44.9%) of white platelets, mp 221.5-223°C. The infrared spectrum (KBr) showed absorptions at 3425 (νN-H), 1620 (νC=0). The proton NMR showed peaks at 2.3δ singlets (3), 6.18δ singlet (1), 7.9δ, 7.4δ-8.1δ multiplet (4). Anal: Calcd C, 74.53; H, 5.59. Found: C, 74.37; H, 5.48.

1,4-Bis(2-benzoyl-1-methyl-vinylamino)benzene. A typical synthesis of a model compound follows: To a solution of 5 mL N-methylpyrolidene (NMP) and 1 mL trifluoroacetic acid (TFA) (10^{-1} *M* in NMP) was added 0.5 g (3.1 mmoles) of acetylacetophenone and 0.154 g (1.5 mmoles) of *p*-phenylenediamine. The solution was heated to 125°C for a one-hour period and then poured into 50 mL of water precipitating a yellow solid. The yellow material was collected, dried, and crystallized from a chloroform-alcohol mixture to give 0.53 g (86%) of yellow platelets, mp 224-225°C. The infrared spectra (KBr) showed absorptions at 3400 (νN-H), 3005 (νC-H), 1600 (νC=0), 1580 (νC=C). The proton NMR showed peaks at 2.2δ singlet (3), 5.8δ singlet (1), 7.3δ, 7.8δ multiplet (7), and 13.5δ singlet (1). Anal: Calcd C, 78.79; H, 6.99; N, 7.07. Found: C, 77.56; H, 6.72; N, 6.82.

Poly[1,4-Bis(3-m-xylylamino-2-butenoyl)benzene]. A typical polymer synthesis is given below.

To a solution of 5 mL NMP and 1 mL TFA (10^{-1}*M* in NMP) was added 0.25 g (1 mmole) of terephthaloyldiacetone and 0.12 g (1 mmole) of xylyenediamine. The reaction mixture was stirred, kept under nitrogen, and heated to 135°C for a period of 8 hours. The reaction mixture was then poured into 75 mL of water to precipitate a yellow solid which was collected and dried to give 0.32 g (92%) of a yellow powder. The infrared spectrum (KBr) showed absorptions at 3400 (νN-H), 1600 (νC=0), 1580 (νC=C), and 3002 (νC-H). Anal. Calcd. for repeating unit $C_{22}H_{22}N_2O_2$: C, 76.27; H, 6.40; N, 8.09. Found: C, 75.63; H, 6.10; N, 8.47.

RESULTS AND DISCUSSION

Monomeric Models

Although ordinary enamines are easily subjected to hydrolysis and oxidation carbonyl-enamines such as methyl 3-aminocrotonate and 3-aminocrotonamide are stable materials. In solutions of inert solvents the cis-carbonyl-enamine form is generally the most important detectable form presence by proton and carbon-13 nmr spectroscopy. The 4-electron π system and bonding electron of the N-H---0 system together make up a six-membered six-electron "aromatic-like" system. This apparently provides sufficient stabilization to make the cis-carbonyl-enamine the most stable form among the possible tautomeric forms (others include carbonyl-imine, trans-carbonyl-enamine, and emine-enol).

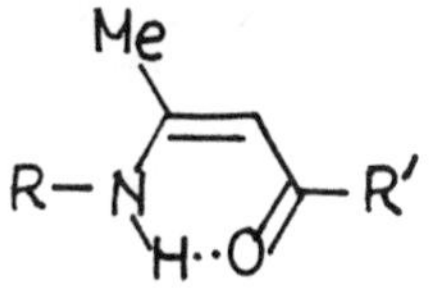

cis-carbonyl-enamine

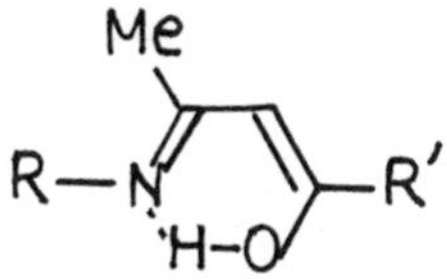

emine-enol

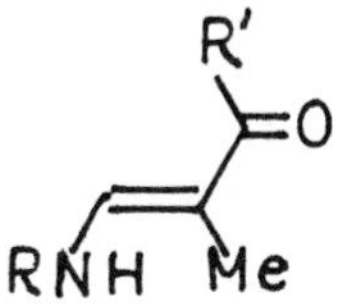

trans-carbonyl-enamine

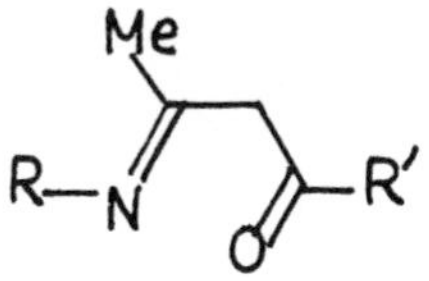

carbonyl-imine

The proton nmr of a monomeric enamine-ketone prepared from benzoylacetone and m-xylylenediamine, 1, shows a internal hydrogen bonding signal at 11.5-13.5. Its C^{13} nmr signals are listed in Table I. The nmr signals are in agreement with an enamine-ketone structure.

Table I
Carbon-13 NMR Assignment of Enamine-Ketone, 1

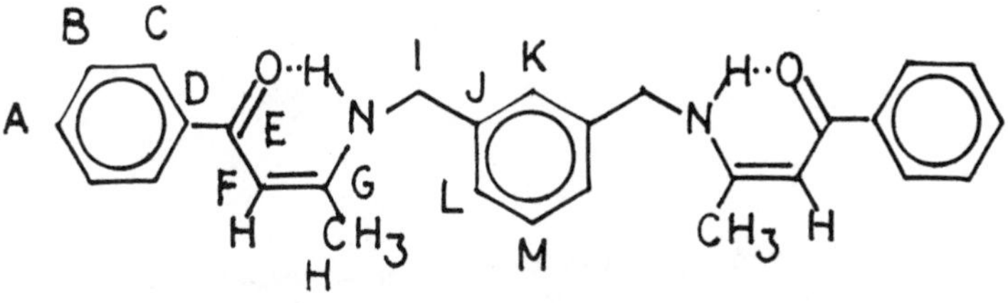

A.	128.34 ppm[a]	H.	19.38
B.	127.14	I.	46.93
C.	130.71	J.	138.81
D.	140.55	K.	125.61
E.	188.36	L.	126.27
F.	92.84	M.	129.69
G.	165.91		

We first investigated the feasibility of using carbonyl-enamine rings as rigid mesogenic units by studying monomeric compounds containing enamine-ketone units. Recently, the melting behavior of some oligomeric polyphenyls was investigated and the nature of their mesophases was verified (10-11). The first compound in the series to display a nematic mesophase was p-quinque-phenyl. In general, these compounds exhibit very high transition temperatures, in the case of p-sexiphenyl, it decomposed at 500°C in the nematic mesophase before forming an isotropic liquid. They are also practically insoluble. We reasoned that the replacement of one or more of the phenyl rings with an enamine-ketone hydrogen bonding ring should lower the transitions of the compounds and increase their solubility in common solvents. A series of enamine-ketones were prepared by the condensation of β-diketones with arylamines and arylene diamines (2, 12). The

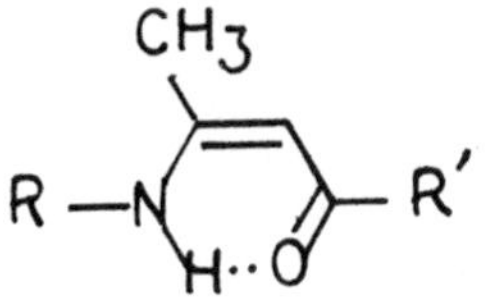

enamine-ketones

Table II
Melting Behavior of Some Oliglomeric Polyphenyls and Enamine-Ketones

Aromatic Hydrocarbons	Enamine-Ketone
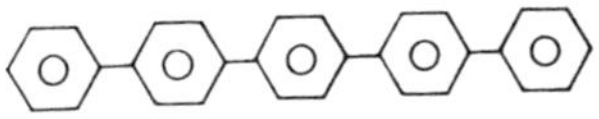	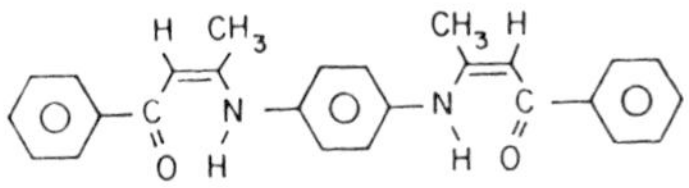
K 386 n 415 i	K 229 i
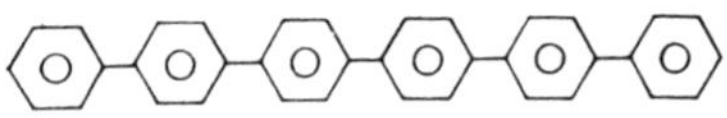	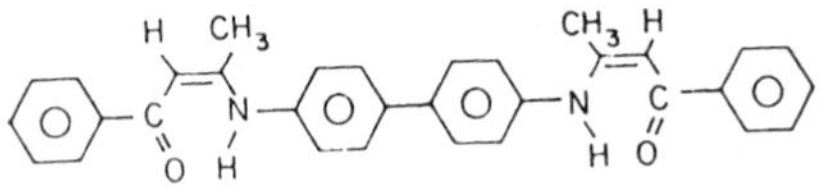
K 434 s_A 464 n 500 d	K 239 n 244 i

melting behavior of the enamine-ketones was compared with that of the polyphenyls, Table II. Compared with the polyphenyls, the enamine-ketones have much lower transition temperatures, with the six-membered system being the first one in the series that displayed a nematic mesophase. Also, as expected, all the enamine-ketones were found to be soluble in common organic solvents. In order to obtain more proof that the enamine-ketone ring system is sufficiently rigid to be used as mesogenic structural unit we prepared three series of enamine-ketones with alkyl, alkoxy, and carbalkoxy flexible end groups. All three series were found to exhibit nematic mesomorphic phases. Some of the higher alkoxy members studied also display sematic phases. The thermal data of the alkyl series are shown in Table III and Fig. 1. The thermal behavior of the enamine-ketones are similar to that of the monomeric liquid crystals containing only hydrocarbon rings. With the knowledge that enamine-ketone ring is sufficiently rigid to act as mesogenic unit we proceeded to prepare polymeric enamine-ketones.

Poly(enamine-ketones)

Several polymers with various combinations of enamine-ketone and phenyl rings as the rigid case connected with alkylene and m-xylylene groups were prepared by condensation reaction of diamines with bis-(β-diketones). High yields of poly(enamine-ketones)I-VIII with moderate molecular weight were obtained, Table IV. Polymers with C_{10}, C_{12} alkylene and m-xylylene connecting

$CH_3COCH_2COArCOCH_2COCH_3$ + $H_2N\text{-}R\text{-}NH_2$ ⟶

CH3 CH3
[Ar — (ring) N — R — N (ring)]n
O··H H··O

Ar = (p-phenylene), (4,4'-biphenylene) R = $(CH_2)_n$, (m-xylylene)

groups between the rings were found to exhibit thermotropic nematic mesophases whereas those with short alkylene connecting groups only exhibit isotropic melt on heating, Table V.

Although poly(carbonyl-enamines) have been previously reported (14-19), to our knowledge we were the first to detect the thermotropic liquid crystalline properties of poly(carbonyl-enamines) (3). Since 1975 several thermotropic liquid crystalline polymers containing aromatic rings as the main mesogenic unit have been reported. These include polyalkenoates of 4,4'-dihydroxy-α,α'-dimethylbenzalazine (20), polyalkanoates of biphenols (21), polycarbonates of 4,4'-dihydroxy-α,α'-dimethylbenzalazine (22), copolymers containing 4-hydroxybenzoate and ethylene terephthalate segments (23, 24, 27), poly-

Table III
Thermodynamic Data for the
bis[3-(p-n-alkylanilino)-2-butenoyl] benzenes

Compound	Transition	T,°C	$\Delta H x 10^4$,cal/mole	ΔS,cal/mole/°K
C_1	K→n			
	n→i			
C_2	$K_I \rightarrow K_{II}$	173	.211	4.73
	$K_{II} \rightarrow n$	184	.628	13.7
	n→i	222	.039	.788
C_3	K→n	173	1.12	25.2
	n→i	224	.050	1.01
C_4	K→n	150	.767	18.1
	n→i	225	.035	.703
C_5	K→n	150	.826	19.5
	n→i	220	.040	.811
C_6	$K_I \rightarrow K_{II}$	134	.274	6.72
	$K_{II} \rightarrow n$	138	.510	12.4
	n→i	198	.038	.807
C_7	K→n	145	1.22	29.2
	n→i	188	.044	.954
C_8	K→n	157	1.38	32.1
	n→i	175	.042	.937

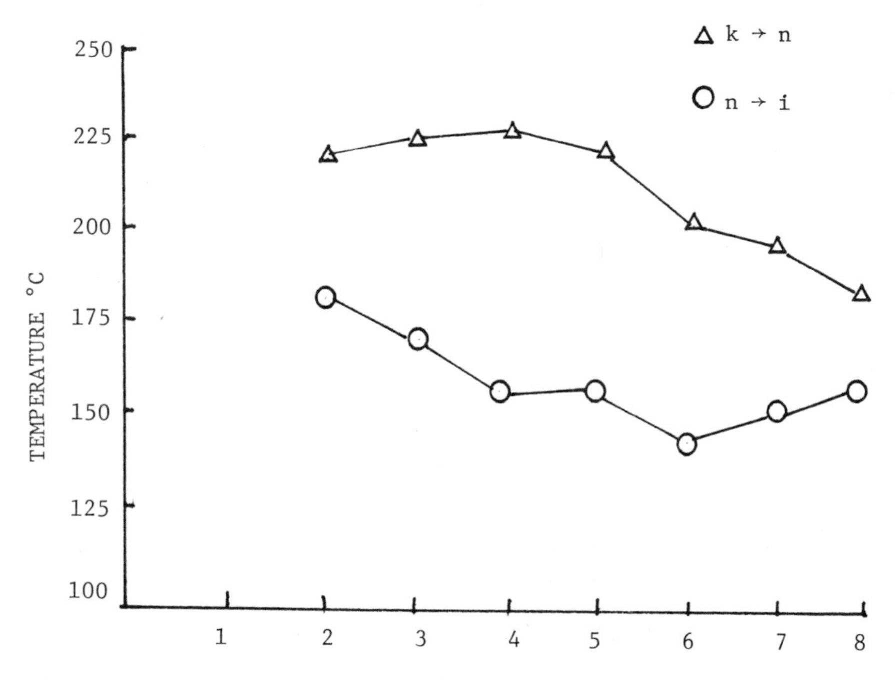

Figure 1. Transition temperatures versus the number of carbon atoms in the alkyl chain of the C_n series.

Table IV
Poly(1,4-bis(3-hydrocarbylamino-2-butanoyl)benzene)

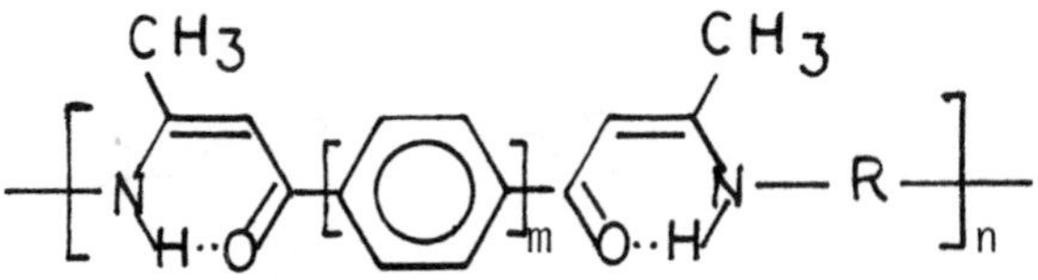

Polymer	m	R	Conditions solvent/catalyst/time	Yield(%)	η^{a}_{inh} dL/g
I_A	1	MXD[b]	HMPA/TFA[c]/4 hrs.	97	0.18
I_B	1	MXD	DMAc/TFA/4 hrs.	88	0.16
I_C	1	MXD	NMP/TFA/4 hrs.	97	0.16
I_D	1	MXD	NMP/TFA/8 hrs.	97	0.21
I_E	1	MXD	NMP/TFA/16 hrs.	89	0.18
I_F	1	MXD	NMP/TFA/24 hrs.	82	0.18
II	1	$(CH_2)_{10}$	NMP/TFA/8 hrs.	76	0.18
III	1	$(CH_2)_6$	NMP/TFA/8 hrs.	75	0.39
IV	1	$(CH_2)_4$	NMP/TFA/8 hrs.	87	0.45
V	2	MXD[b]	NMP/TFA/8 hrs.	83	0.22
VI	2	$(CH_2)_{12}$	NMP/TFA/4 hrs.[d]	75	0.26
VII	2	$(CH_2)_{10}$	NMP/TFA/6 hrs.[d]	89	0.76
VIII	2	$(CH_2)_6$	NMP/TFA/8 hrs.	83	0.17

a. Concentration of 0.5g/dl measured in m-cresol at 35°C.

b. MXD = m-xylylene.

c. TFA concentration of $1.6 \cdot 10^{-2}$M.

d. Precipitated from reaction solution.

Table V
Thermal Transition of Poly(enamine-ketones)[a]

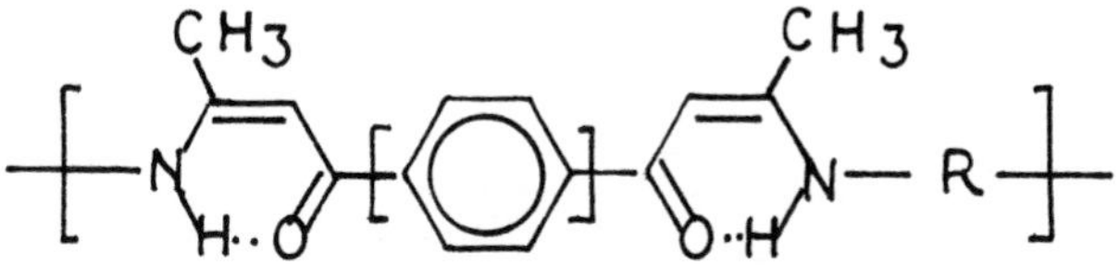

Polymer	m	R	Transition Temperature (°C)
I_D	1	MXD[b]	$K_2 \rightarrow K_1$(165), $K_1 \rightarrow n$ (175), $n \rightarrow i$(235), d250
II	1	$(CH_2)_{10}$	$K_2 \rightarrow K_1$(95), $K_1 \rightarrow n$(120), $n \rightarrow i$(185), d235
III	1	$(CH_2)_6$	$K \rightarrow i$(170), d245
IV	1	$(CH_2)_4$	$K \rightarrow i$(185), d235
V	2	MXD	$K_2 \rightarrow K_1$(220), $K_1 \rightarrow n$(240), $n \rightarrow i$(290), d295
VI	2	$(CH_2)_{12}$	$K_2 \rightarrow K_1$(225), $K_1 \rightarrow n$(260), $n \rightarrow i$(290), d295
VII	2	$(CH_2)_{10}$	$K_2 \rightarrow K_1$(230), $K_1 \rightarrow n$(240), $n \rightarrow i$(295), d300
VIII	2	$(CH_2)_6$	$K \rightarrow i$(285), d 330

a. DSC and Thermal Optical Microscopy, heating rate $5^{\circ}C$/min.

b. MXD = m-xylylene.

Table VI
TGA of Poly(enamine-ketones)[a]

Polymer	m	R'	10% Wt. Loss Temp °C	Wt. Residue at 500°C(%)
I_D	1	MXD[b]	380	59
II	1	$(CH_2)_{10}$	250	30
III	1	$(CH_2)_6$	360	28
IV	1	$(CH_2)_4$	340	39
V	2	MXD	360	47
VI	2	$(CH_2)_{12}$	390	33
VII	2	$(CH_2)_{10}$	370	32
VIII	2	$(CH_2)_6$	385	30

[a]Heating rate at 15°C/min in N_2 atmosphere.

b. MXD = m-xylylene.

azomethines (25), copolyester with poly(ethylene terephthalate) and terphenylene segments (26), poly (Schiff bases) (27, 28), polyester analogs of 4-alkoxyphenyl-4'-alkoxybenzoates (29), and polyesters derived from terephthalic acid and 4,4'-dihydroxy-α,w-diphenoxyalkanes (30). These polymers contain flexible connecting groups between rigid aromatic mesogenic units, substituted aromatic mesogenic units, or mesogenic units of different sizes and shapes for the purpose of decreasing the symmetry of the polymer backbone so that the packing order of the chain will be somewhat reduced. Otherwise the transition temperatures of the polymers would have been too high for the polymers to be thermally processable. Compared with those reported polymers our poly(enamine-ketones) have relatively low transition temperatures and much higher solubilities in commonly used solvents (m-cresol, NMP, DMA, etc.). These polymers were also found to have reasonable thermal stability as indicated by the TGA analysis, Table VI.

We are now studying both the lyotropic and thermotropic liquid crystalline properties of poly(enamine-ketones). Also in progress is the synthesis of poly(enamine-ketones) derived from diaminoarenes, polymers containing all ring systems. We hope to obtain rod-like polymers with both lyotropic and thermotropic liquid crystalline properties that permit their processing into useful end products.

Acknowledgement

Partial financial support from the National Science Foundation (DMR8013689) is gratefully acknowledged.

Literature Cited

1. Huang, S. J.; Benicewicz, B. C.; Pavlisko, J. A.; Quinga, E. Polym. Preprints 1981, 22, No. 1, 56.
2. Benicewicz, B. C.; Huang, S. J.; Johnson, J. F. Polymer Preprints 1980, 21, No. 2, 268.
3. Huang, S. J.; Pavlisko, J. A.; Hong, E. Polym. Preprints 1978, 19, No. 2, 57.
4. Benicewicz, B. C. Ph.D. Dissertation, University of Connecticut, 1981, pp. 20-31.
5. Blumstein, A. Polymer News 1979, 5, 254.
6. Samulski, E. T.; Dupre, D. B. in "Advance in Liquid Crystals" Vol. 4, Brown, G. H., Ed.; Academic Press, New York, N. Y., 1979; p. 121-145.
7. Blumstein, A., Ed.; "Liquid Crystalline Order in Polymers"; Academic Press, New York, N. Y., 1978.
8. Cifferi, A.; Wards, I. M., Eds.; "Ultrahigh Modulus Polymers"; Applied Science, Barking, Essex, U. K., 1979.

9. Black, W. B.; Preston, J.; "High Modulus Aromatic Polymers"; Dekker, New York, N. Y., 1970.
10. Smith, G. W. Mol. Cryst. Liq. Cryst. 1979, 49, 207.
11. Lewis, I. C.; Kovac, C. A. Mol. Cryst. Liq. Cryst. 1979, 51, 173.
12. Benicewicz, B. C.; Huang, S. J.; Johnson, J. F. Mol. Cryst. Liq. Cryst., in press.
13. Gray, G. W. "Molecular Structure and Properties of Liquid Crystals"; Academic Press, New York, N. Y., 1962.
14. Moore, J. A.; Kochanowski, J. E. Macromolecules 1975, 8, 121.
15. Moore, J. A.; Mitchell, T. D. Polymer Preprints 1978, 19, No. 2, 13.
16. Ueda, M.; Sakai, N.; Imai, Y. Makromol. Chem. 1979, 180, 2813.
17. Ueda, M.; Otaira, K.; Imai, Y. J. Polym. Sci.: Polym. Chem. Ed. 1978, 16, 2809.
18. Ueda, M.; Kino, K.; Hirono, T.; Imai, Y. J. Polym. Sci.: Polym. Chem. Ed. 1976, 14, 931.
19. Imai, Y.; Ueda, M.; Otaira, K. J. Polym. Sci.: Polym. Chem. Ed. 1977, 15, 145-7.
20. Roviello, A.; Sirigu, A. J. Polym. Sci.: Polym. Lett. Ed. 1975, 13, 455.
21. Blumstein, A.; Sivaramakrisknan, K. N.; Clough, S. B.; Blumstein, R. B. Mol. Cryst. Liq. Cryst. Lett. 1978, 49, 119.
22. Roviello, A.; Sirigu, A. Europ. Polym. J. 1979, 15, 423.
23. MacFarlane, F. E.; Nicely, V. A.; Davis, T. G. Contemp. Top. Polym. Sci. 1977, 2, 109.
24. Jackson, W. J., Jr.; Kuhfuss, H. F. J. Appl. Polym. Sci. 1980, 25, 1685.
25. Morgan, P. German Patent 2, 620, 351 (Nov. 18, 1976).
26. Fayolle, B.; Noel, C.; Billard, J. J. Phys. (Paris), Colloq., 1979. C3, 485.
27. Guillim, D.; Skoulious, A. Mol. Cryst. Liq. Cryst. Lett. 1978, 49, 119.
28. Samulski, E. T.; Wiercinski, R. W. Bull. Am. Phys. Soc. 1978, 23, 296.
29. Griffin, A. C.; Havens, S. J. Mol. Cryst. Liq. Cryst. Lett., 1979, 49, 239.
30. Antoun, J.-I. S.; Ober, C.; Lenz, R. Brit. Polym. J. 1980, 132.

RECEIVED December 7, 1981.

Cyclopolymerization of Carbon Suboxide: Mechanism and Polymer Properties

NAN-LOH YANG and H. HAUBENSTOCK

City University of New York, College of Staten Island, Department of Chemistry, Staten Island, NY 10301

A. SNOW

Naval Research Laboratory, Washington, DC 20375

The thermal polymerization of carbon suboxide (C_3O_2) was shown not to propagate through a free radical path. A mechanism involving a Zwitterionic intermediate with competing propagation or termination steps is proposed. Solution polymerization of C_3O_2 in DMF exhibited autocatalytic behavior as was also observed in heterogeneous systems. Substantial interactions between the polymer of strong ionic character and the highly polarized monomer are considered to play an important role. The polymer shows interesting paramagnetic properties including high spin density, strong molecular field and photosensitivity. The polymer paramagnetism can be reversibly enhanced by visible light with high quantum efficiency. Studies of ^{13}C labelled polymer confirmed the photo-generated ESR signal to be identical to that originally present. A square root dependence of the steady-state ESR signal strength in incident light intensity at different temperatures indicated a bimolecular electron transfer process with an activation energy for recombination equal to 0.44 Kcal per mole of paramagnetic center.

Carbon suboxide, O=C=C=C=O, an unusually reactive bisketene, functions as an efficient double electrophile, in the synthesis of heterocyclic compounds (1). One would expect that the molecule, having four cumulative π bonds, should be of considerable interest in the field of polymer chemistry and should have been investigated vigorously. On the contrary, not only has the structure of its polymer not been unequivocally established, but also the physical properties of the polymer and the kinetics and mechanism of the polymerization process have not been studied to any significant extent, even though its polymerization product

0097-6156/82/0195-0417$06.50/0
© 1982 American Chemical Society

was discovered at the beginning of this century (2). There was some work on the carbon suboxide polymer in the nineteen sixties (3-7), focusing on its being a by-product formed in the radiolysis of carbon monoxide generated in graphite-moderated carbon dioxide-cooled nuclear reactors.

Results of spectroscopic and x-ray studies (3,5,8) suggest that the polymer is basically a polycyclic chain of six-membered lactone rings with conjugated double bonds through a ladder structure (Figure 1-b).

Polymerization of carbon suboxide in heterogeneous systems, i.e. polymer precipitating from the reaction solution, by various initiators was examined qualitatively by Hegar (9). It was found that the best catalysts in cyclohexane solution (at 0°C., then room temperature) were triethylamine and pyridine. Boron trifluoride was somewhat less effective, while sulfuric acid was inefficient and slow. With aluminum chloride the catalyst particles were rapidly coated with polymer, and no significant amount of polymer could be obtained.

The kinetic aspect of the polymerization process has been studied only for the gaseous phase polymerization (6,7). Smith (6), by following polymer optical density at 370 nm, and Blake (7), by monitoring the fall in monomer pressure, found the polymerization to be first order with respect to both monomer and polymer. They observed an induction period when no polymer was initially present, and that the polymer, once formed, served as the substrate for the growth of more polymer. Smith proposed that C_2O, formed by the thermal decomposition of C_3O_2, is responsible for formation of the initial polymer structure. Blake attributed the induction period to a surface poison, which he identified as water, and the initiation was considered as a possible surface free radical mechanism. Neither considered the structures involved in the processes of chain initiation, propagation and termination.

Calorimetric studies of radiation-induced bulk polymerization also confirmed the autocatalytic nature of the process (10, 11).

This paper reports our current studies and summarizes recent findings to bring the subject of carbon suboxide polymer up to date.

Experimental

Methods for the preparation of the monomer are well-documented (12,13). Details of the procedures used in ESR studies have been published previously (14,15).

Solution Polymerization. Dimethyl formamide, DMF, is a good solvent for poly(carbon suboxide) with DP lower than six and hence can serve as an effective medium for the kinetic studies of solution polymerization. The use of a polymer solvent such as

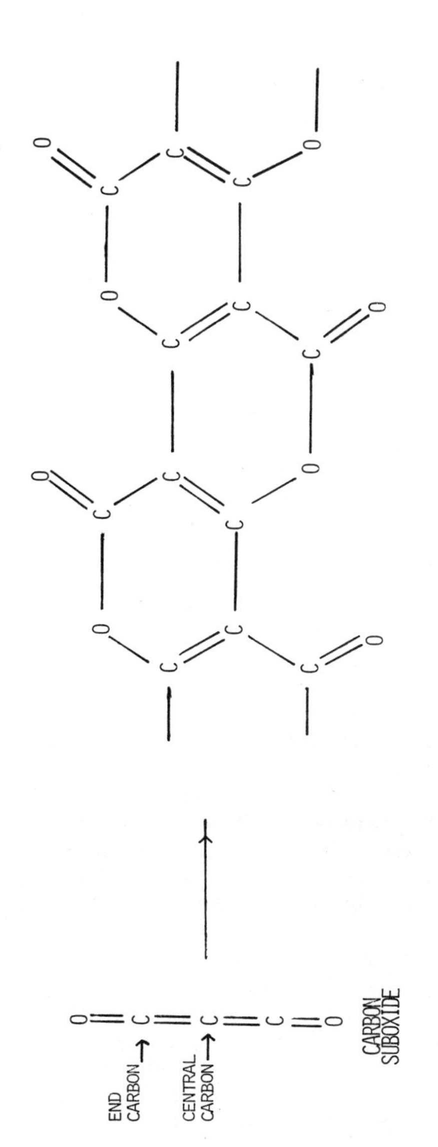

Figure 1. Poly(carbon suboxide): a polycyclic chain of six-membered lactone rings with polyconjugated double bonds.

DMF precludes using ESR to follow the polymerization because of high dielectric loss due to the solvent. However, the polymerization can be followed by optical density since the system is homogeneous and the monomer absorption does not interfere in the visible region.

The optical spectrum of a DMF solution polymerization was found to be different from that of the gas deposited film in having two pairs of maxima centered at 420 and 320 nm instead of the one at 370 nm. The spectrum was independent of conversion, and a simple log A, vs. time plot (A = Absorbance) was not linear indicating the rate law was no longer first order with respect to polymer.

DMF was dried and purified by stirring over P_2O_5, then over KOH pellets and distilled at reduced pressure. The polymerization was run in a one cm optical cell connected to a wide bore stopcock having a closed volume of 9.51 ml. The cell was filled with C_3O_2 at pressures of 590, 430 and 340 mm Hg (C_3O_2 vapor pressures of 0, -7 and -12.5°C). After condensing the C_3O_2 with dry ice, a calibrated volume of 3.98 ml of DMF or DMF-toluene mixture was added. The cell was agitated by tipping but not inverting for about 20 seconds to insure mixing and placed in the Cary 118C visible-ultraviolet spectrometer. The absorbance at 420, 500 and 600 nm against a DMF reference was recorded consecutively as a function of time at room temperature.

Photoparamagnetism. Poly(carbon suboxide) thin films with controlled thickness were formed by deposition from a gaseous monomer phase on to heated ESR cells with flat surfaces specially designed for photoparamagnetism studies (14). Light sources used for photochemical reactions were a 750 watt tungsten lamp for the visible region, 200 watt xenon-mercury lamp and 1000 watt xenon lamp for the ultraviolet. Wavelength selectors employed were combinations of filters and a Bausch and Lomb grating monochrometer (#33-86-4). Light intensity measurements were made with a YSI-Kettering Model 65 Radiometer.

ESR measurements were performed using a JEOL model JES-Me-3X, X-band, ESR spectrometer with provisions for the irradiation of sample *in situ* of the microwave cavity. A JEOL model JES-VCT-2AX variable temperature adapter was used for temperature control for ESR measurement.

In the ESR absorptions, the lack of a change in spectrum shape or line width upon irradiation indicates the photogenerated spin centers are identical to those initially present or residual spin centers (*i.e.* same g-value and relaxation time). Since there is no change in line width or shape, the peak-to-peak signal intensity, D, is proportional to the signal intensity, and the residual spin density, n^R, can be used as a yardstick to measure the number of photogenerated electrons, n^P. This is particularly useful in studying the temperature dependence of the photosensitivity. The wavelength dependence of the photo-excitation was examined by measuring the intensity of the ESR photosignal as the difference between the total and residual peak-to-peak intensities, D^T and D^R, as a function of wavelength. The incident irradiation was monochromatized to a band pass of 20 nm.

The steady-state quantum yield for spin generation is obtained by measurement of the light intensity absorbed at a known wavelength, the effective exposure area and the number of photogenerated spins. Light from the 1000 watt xenon lamp was monochromatized to a 20 nm band pass; and its intensity was measured in $erg/cm^2 sec.$ by placing the probe of a YSI-Kettering Model 65 Radiometer immediately in front of the ESR cavity window. The effective exposure area was taken from the 4 x 25 mm dimensions of the cell as 1.0 cm^2. The number of photogenerated spins was obtained by comparison of the photosignal with the signal intensity corresponding to residual spin density. The quantity of poly(carbon suboxide) composing the film was found to be 0.33 mg. The steady-state quantum yield, Φ_{ss}, is then given by

$$\Phi_{ss} = \frac{\text{photogenerated spins}}{\text{photons absorbed}} = \frac{N^e}{N^p_{abs.}}$$

The dependence of the steady-state photosignal on the light intensity was investigated with neutral density filters. The transmission of the neutral density filters and combinations of them were calibrated with a Cary Model 118-C ultraviolet-visible spectrophotometer. The neutral density filters were placed immediately in front of the cavity window, and the photosignal intensity was measured as the difference between the peak-to-peak intensities of the light and dark signals. Measurements were made monochromatically with a band pass of 20 nm using the 547 and 577 mercury lines of the 200 watt mercury-xenon lamp.

Results and Discussion

Kinetics from ESR Studies. The polymerization process for carbon suboxide is accompanied by an autoaccelerated growth of a paramagnetic center which is an intrinsic property of the solid polymer (14,15,16). Following the ESR absorption of the polymerization system has led to important information on its kinetics and mechanism. The polymerization rate was followed by monitoring the peak to peak intensity of the ESR signal. The line width and the line shape remained constant throughout the conversion. The reaction order with respect to polymer was first order. It was found that signal intensity is proportional to quantity of polymer (*vide infra*). An Arrhenius plot was linear over the 0 to 50°C temperature range yielding an activation energy of 9.9 Kcal/mole and a frequency factor of 2.4×10^3 sec^{-1}. The activation energy found here is higher than the previously obtained value of 6.23 Kcal/mole in the gaseous phase polymerization (6,7). The very low preexponential factor for the polymerization is typical for solid surface catalyzed reactions (17) and should be expected for the present cyclopolymerization involving stringent topological requirements for the monomer to add on to the cyclic chain. The initial monomer concentration was varied by the addition of toluene as an inert diluent. The polymer is insoluble in toluene. The overall rate law was established to be:

$$\frac{d[P]}{dt} = k[M][P]$$

where [M] refers to the C_3O_2 monomer concentration and [P], the quantity of C_3O_2 thermopolymer.

This is the same rate law obtained by Smith (6) and Blake (7) for the gas phase polymerization. This supports the conclusion that the paramagnetism is a property of the carbon suboxide polymer in the solid state and is not due to formation of a side product. It was further confirmed by parallel measurements of the ESR signal intensity and optical absorbances of the polymer at 370 nm and 4.6 μm at various spin concentrations. The spin density was related linearly to the optical densities. The possibility that the paramagnetism is due to a buried radical intermediate is precluded by evidence that propagation does not involve a radical intermediate (*vide infra*).

Retardation and Inhibition. In order to obtain additional information on the polymerization process, various potential inhibitors or retarding agents were added to carbon suboxide, and the rate of polymer deposition from the gas phase was quantitatively studied at 100°C. To the monomer at a pressure of 330 mm Hg was added either an equimolar quantity of inhibitor based on pressure measurements, or room temperature vapor pressure of additive if less than 330 mm Hg. The additives used were oxygen, nitric oxide, 3-methyl-1-butene, 1,3-butadiene, acetone and acetaldehyde. Polymerization rates were followed by ESR measurements.

The presence of oxygen and nitric oxide, when compared with the control, exerted no appreciable effect on the polymerization rate. Nitric oxide had a dramatic effect on the ESR signal's line shape while oxygen did not change the ESR signal features. In the former case, a second signal in the ESR spectrum, having a central line at a lower magnetic field and wing structure between 20 and 30 gauss on either side, overlapped the normally observed signal. The spectrum did not change as polymerization proceeded. Evacuation of the cell for one hour did not change the spectrum indicating it is not a weak association of polymer with the nitric oxide molecule. When preformed polymer was exposed to nitric oxide, a shoulder maximum on the ESR signal slowly developed at a point corresponding to the new line's central maximum. The nature of the interaction between polymer and nitric oxide is not known and no useful information in connection with the polymer or polymerization can be derived as yet. However, the observation is noteworthy because the ESR absorption of this very stable inorganic radical has never been unambiguously observed in any condensed phase (18). The lack of an effect on the rate in the presence of oxygen and nitric oxide had also been observed by Blake (7) who concluded that the reaction does not take place by a gas phase free radical mechanism. Oxygen and nitric oxide are both well established efficient radical scavengers (19,20). Further attempts to trap reactive radicals using phenyl *tert*-butylnitrone led to negative results.

Carbon suboxide's susceptibility to radical polymerization was also examined by an attempted copolymerization of equimolar quantities of carbon suboxide and styrene initiated by azobisisobutyronitrile at 60°C in toluene solution. The infrared spectrum of the polymeric product was identical with that of a styrene homopolymer control, and no carbonyl absorption was detected.

On the basis of these observations it is concluded that the polymerization does not involve a propagating reactive radical species. It is unlikely that carbon suboxide monomer would be attacked by the paramagnetic polymer species since C_3O_2 is inert to radicals as reactive as the styryl radical while the paramagnetic polymer is not reactive enough to attack the nitrone spin trap.

Significant retardation was observed with the alkene additives and particularly with acetone or acetaldehyde. It has been experimentally observed that carbon suboxide prepared by the pyrolysis of diacetyltartaric anhydride is more stable toward polymerization than that prepared by the dehydration of malonic acid. We have observed infrared impurity bands at 3.65, 5.75, 7.2, and 9.0μ in carbon suboxide prepared by the pyrolysis of diacetyltartaric anhydride. These impurity bands are attributed to the presence of acetaldehyde, formed as one of the products of the pyrolysis reaction. The relative stability of carbon suboxide prepared by the pyrolysis method is probably due in part to the presence of acetaldehyde, which retards polymerization of the monomer.

Solution Polymerization (21). Solution polymerization of carbon suboxide in DMF was followed by monitoring the absorbance of the system at 420, 500 and 600 nm. At all three wavelengths a plot of the square root of the absorbance against time was linear indicating a reaction order of 1/2 with respect to polymer (Figure 2). The reaction order with respect to monomer

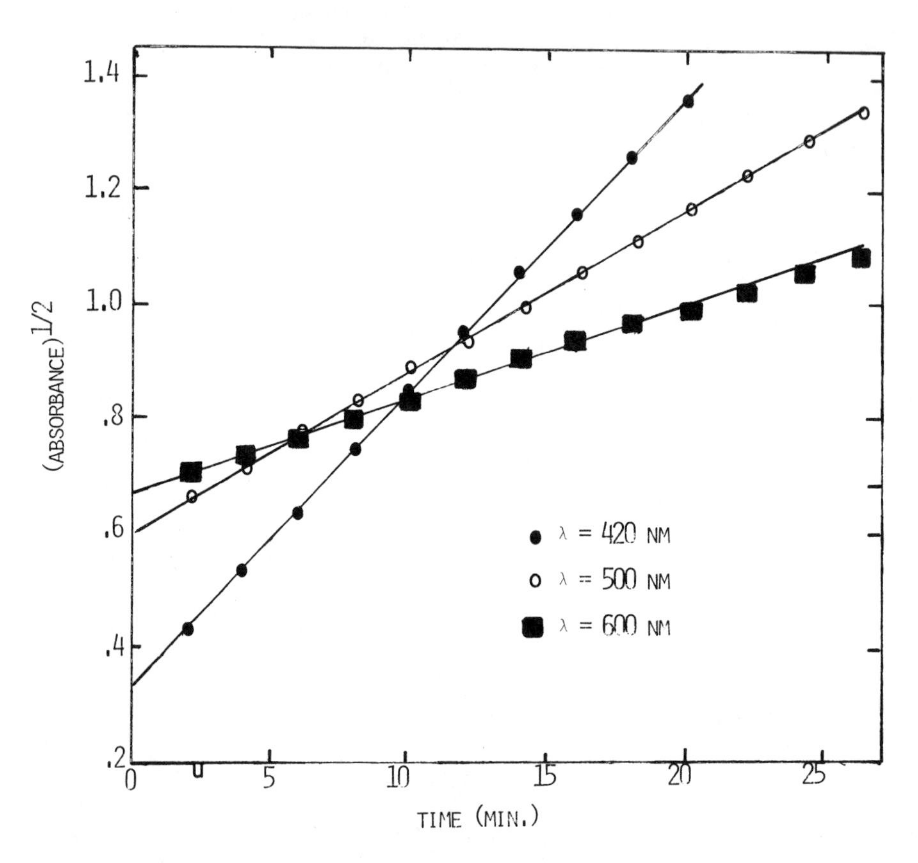

Figure 2. Solution polymerization of carbon suboxide in DMF.

was obtained by the initial rate method where the natural log of the slope of the $A^{1/2}$ vs. time plot was plotted against the log of the initial monomer pressure (Figure 3). Slopes of this plot corresponding to the 420, 500 and 600 nm measurements were 1.14, 1.03 and 1.14 respectively indicating a first order reaction with respect to monomer.

The change in optical spectrum and reaction order with respect to polymer may be due to the formation of a dimeric polymer complex in DMF. The following kinetic scheme is consistent with the change in rate law:

$$P + M \xrightarrow{k} P$$

$$P + P \underset{}{\overset{K}{\rightleftarrows}} C \qquad K = \frac{[C]}{[P]^2}$$

where M is a C_3O_2 monomer, P is a polymer molecule and C is the complex. If the dimeric complex is the major species absorbing at the longer wavelengths where the absorbance was measured, the polymerization rate is given by the following expression assuming the complex equilibrium to be rapid compared to the polymerization rate.

$$\frac{d[P]}{dt} = k[P][M] = \frac{k}{K^{1/2}}[M][C]^{1/2}$$

When the polarity of the medium was decreased by diluting the DMF with toluene, a reduction in the polymerization rate was observed (Figure 4). This decrease varied regularly with decreasing mole fraction of DMF. Since the monomer is a linear molecule and has to assume a bent configuration to fit into the growing polymer chain, the transition state for propagation must involve the monomer assuming a form with an increase in dipole moment. Solvent medium of high polarity is expected to be more favorable for polymerization.

Paramagnetism of the Polymer. The native polymer was found to display interesting paramagnetic properties (14,15). The spin concentration of the polymer increases with higher reaction temperature, reaching 2 x 10^{19} spin/g at a polymerization temperature of 105°C. The paramagnetism of poly(carbon suboxide) follows the Curie-Weiss law, indicating the spin centers are due to a doublet ground state. The polymer obtained from the 105°C polymerization exhibited unusual properties. Its ESR absorption measured at room temperature has a maximum slope line width, ΔH_{msl}, of 0.4 G compared to that of 2.3 G shown by polymer samples from lower polymerization temperatures.

The narrow and broad signals further differ in their temperature dependence. The line width ΔH_{msl} of the broad signal

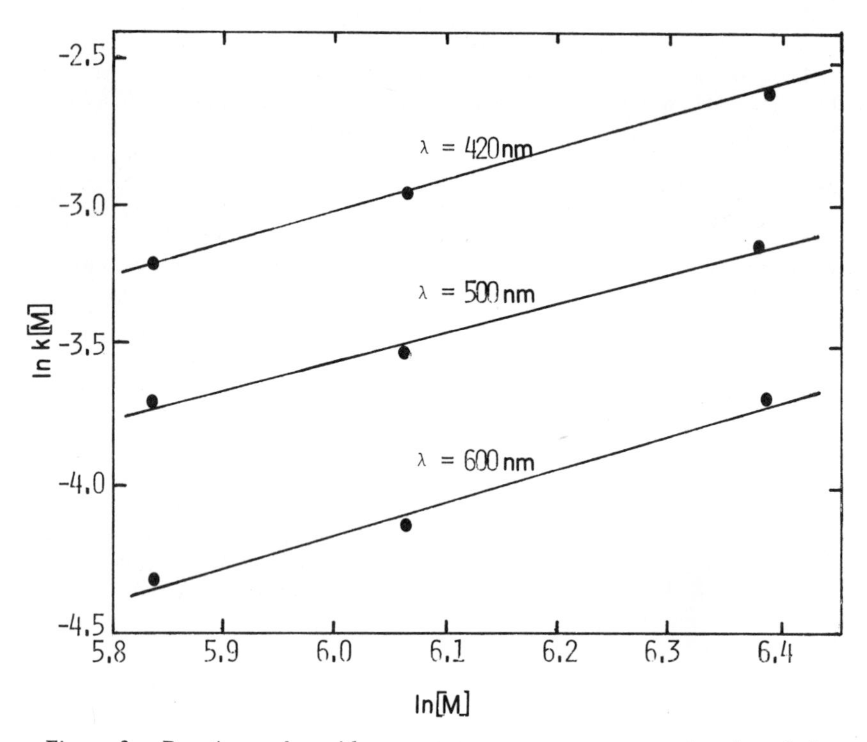

Figure 3. Reaction order with respect to monomer concentration in solution polymerization.

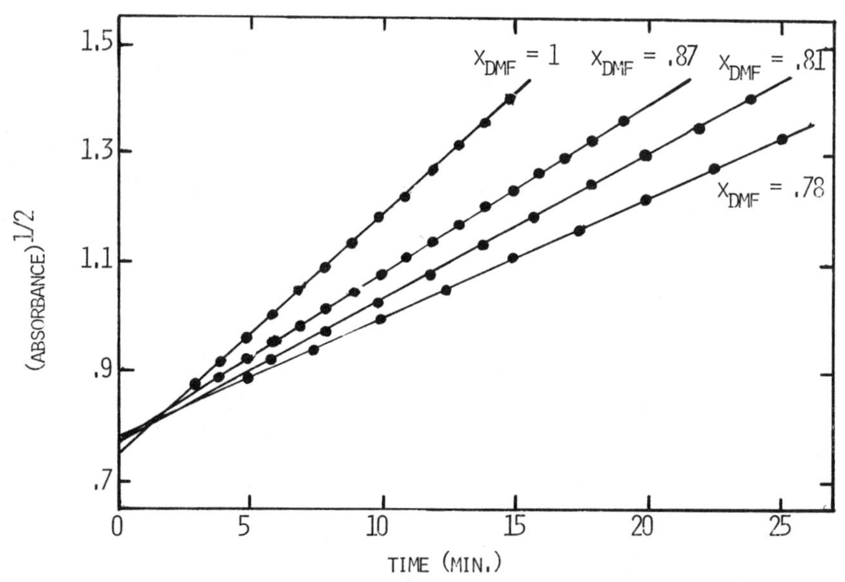

Figure 4. Effect of reaction mediums on solution polymerization.

remains constant at 2.2 to 2.3 G over the temperature range from room temperature down to 5°K. The narrow signal remains at a constant width of 0.40 ± 0.05 G over the temperature range of 298 to 110°K. At 60°K the line width becomes 0.39 G; 19°K, 0.30 G. At 5°K, the signal becomes extremely narrow, ca. 10 mG. Both signals give a straight line on a reciprocal of spin susceptibility vs. temperature plot, i.e. the signals follow the Curie-Weiss law: X = const. $(T + \theta)$, where X is proportional to the ESR signal strength, and θ, the Weiss temperature. The broad signals follow the Curie-Weiss law below 100°C with a θ equal to zero within experimental error. Above 100°C, additional paramagnetic centers are created. As the temperature is further increased, the intensity of ESR absorption is enhanced, accompanied by a slight broadening of ΔH_{ms1} to 2.7 G. The ESR signal intensity of the newly created centers plus those originally present, also follows Curie's law. The narrow signal from polymer obtained at 105° shows, in the Curie-Weiss plot, a Weiss constant equal to 17.5 ± 2.6°K, which is ferromagnetic in sign by convention.

It is very unusual that a polymer obtained under such mild conditions in the absence of initiator should exhibit a spin density as high as 10^{18} to 10^{19} spin/g. The spin concentration for polyphenylacetylene is about 1.5 x 10^{18} spin/g in the high-temperature limit (22); while the maximum radical concentration of carbonaceous materials obtained under the optimum conditions with an optimum heat treatment temperature of 550°C is 6 x 10^{19} spins/g (23). The polymer samples were prepared under contamination free conditions, i.e. pure monomer introduced under high vacuum in the absence of initiator to a pyrex reaction vessel. The direct correlation between the results of ESR and uv absorption studies and the ESR kinetic parameters with literature values obtained by independent methods strongly supports the conclusion that this high concentration of paramagnetic centers is intrinsic to the polymeric system and cannot be associated with chemical impurities. Based on the central ^{13}C splitting constant of 9 G for the 30°C poly(carbon suboxide), the free electron cannot be localized on any single carbon atom and is most likely a π-electron delocalized over the ladder system (16). A localized electron on a carbon-13 atom would show a much higher coupling constant, e.g. the methyl radical has a ^{13}C coupling constant of 38.5 G (24). The ^{13}C coupling constant is not simply proportional to the π-spin density on the same carbon but is also related to spin densities on neighboring atoms (25). A detailed spin assignment is not available for the polymer at the present. However, compared with ^{13}C splittings in anthracene cation and anion (26), the 9-G coupling constant obtained here does clearly point out the delocalized π-electron nature of the paramagnetic center.

The sharp ESR absorption of the 105°C sample, the extremely narrow line width at 5°K and a Weiss constant equal to 17.5°K are interesting observations for the polymer. According to the Weiss molecular field equation, a Weiss temperature of 17.5°K, assuming a simple cubic lattice, leads to an estimated exchange integral of 8.1×10^{-16} erg, i.e. an exchange frequency of 7.6×10^{11} sec^{-1} or an effective exchange field, H_e, of 6.2×10^4G (27). At the spin concentration of 2×10^{19} spin/g, the root mean square dipolar field perturbation H_p due to neighboring spins is estimated to be 0.7 G (27) which is much smaller than the effective exchange field. Accordingly, the ESR absorption should be subject to exchange narrowing and assumes Lorentzian shape. The observation of slowly decreasing line width with decreasing temperature also supports exchange narrowing, i.e. the effective exchange frequency increases slowly as temperature decreases (28). The extremely narrow line width observed below the Weiss temperature at 5°K, indicates that the molecular field overcomes the thermal effects, and that the relaxation times for the spin system become very long. The relaxation time of 10^{-7} sec obtained at room temperature for this signal would dictate a line width of 0.6 G according to the uncertainty principle. A long relaxation time requires the lattice vibrations of the polymer matrix to be suppressed. If the relaxation times are long enough not to be the line width controlling factor, the line width at half-power for the 105°C sample with a spin density of 2×10^{19} spins/g is estimated (27) to be $(10/3)\ H_p^2/H_e = 2.6 \times 10^{-3}$ G. This is in line with the experimental observation. A matrix with low-amplitude lattice vibrations is, of course, conducive to electronic conduction. The exchange frequency can be related to the "trial time in hopping" for a hopping conduction mechanism (29).

Photoparamagnetism (30). Since paramagnetic behavior of the polymer follows the Curie-Weiss law, the paramagnetic center can not be populated by thermal excitation for a polymer formed at temperatures below 90°C. However, we have found that visible light irradiation, in the wavelength region 400 to 750 nm can generate additional paramagnetic centers with an ESR signal identical to that originally present. The photo process is reversible, i.e. the signal returns to its original size upon cessation of irradiation. For a fixed number of incident photons, the steady-state concentration of the photosignal (i.e. ESR absorption after Curie law correction) increases with decrease in temperature. At low temperatures, polymer samples with a residual signal after a finite decay time have to be warmed to room temperature to bring the photosignal to zero concentration.

To ascertain that the photo-generated paramagnetic center in poly(carbon suboxide) is indeed identical to that originally present in the polymer, the effect of light irradiation on the ESR spectrum of ^{13}C-labelled polymer was examined. The ESR spectrum of poly(carbon suboxide) with natural abundance of ^{13}C is a strong singlet absorption with very weak ^{13}C satellite lines

(14,16). To impart distinct fine structures to the ESR absorption of the polymer, monomer with end-atoms enriched to 99%, $O{=}^{13}C{=}C{=}^{13}C{=}O$, was synthesized and polymerized for investigation. The enriched polymer shows an ESR spectra with a clear fine structure (curves I and I' in Figure 5). On irradiation with visible light, the signal is enhanced (curves II and II' in Figure 5) by the same proportion over the entire range of absorption. A comparison of curves I and II and their scale-up displays (I' and II') of a portion of fine structures, shows clearly that the effect of irradiation results in an increase in signal strength without generating any structurally different paramagnetic species. After the identity of the photo-signal has been thus established, the photoparamagnetism study was carried out with polymer film of natural ^{13}C content.

The effect of irradiating the poly(carbon suboxide) film with the full spectrum of a 200 watt mercury-xenon lamp on the ESR spectrum is illustrated in Figure 6. There is a significant and rapid increase in the signal to a steady-state level intensity but no change in spectrum shape or line width. A scan of 500 to 5500 gauss showed no half field transition or absorptions at any other magnetic field.

A square root dependence of the steady-state photosignal strength on incident light intensity was observed and is illustrated in Figure 7. The square root dependence is indicative of a two-electron combination process for the signal decay. Quantum yields of charge accumulation at the steady-state, Φ_{ss}, were obtained at 25°C over the visible wavelength region and are listed in Table I.

TABLE I

STEADY-STATE QUANTUM YIELDS FOR CHARGE ACCUMULATION AT 25°C.

λ(nm.)	N^e(spin)	$N^p_{Abs.}$ (photon/sec.)	Φ_{ss}
450	1.2×10^{14}	2.7×10^{15}	.05
500	1.3×10^{14}	3.9×10^{15}	.03
547	1.1×10^{14}	1.2×10^{15}	.10
577	1.4×10^{14}	1.1×10^{14}	.13
650	1.3×10^{14}	4.1×10^{14}	.32
700	$.98 \times 10^{14}$	2.7×10^{14}	.36

These values of quantum yield are indicative of a very efficient charge generation process, since they do not include corrections for the decay by recombination. Decay curves for the photo-signal can be followed by setting the magnetic field of the ESR spectrometer at the derivative peak maxima of the signal and

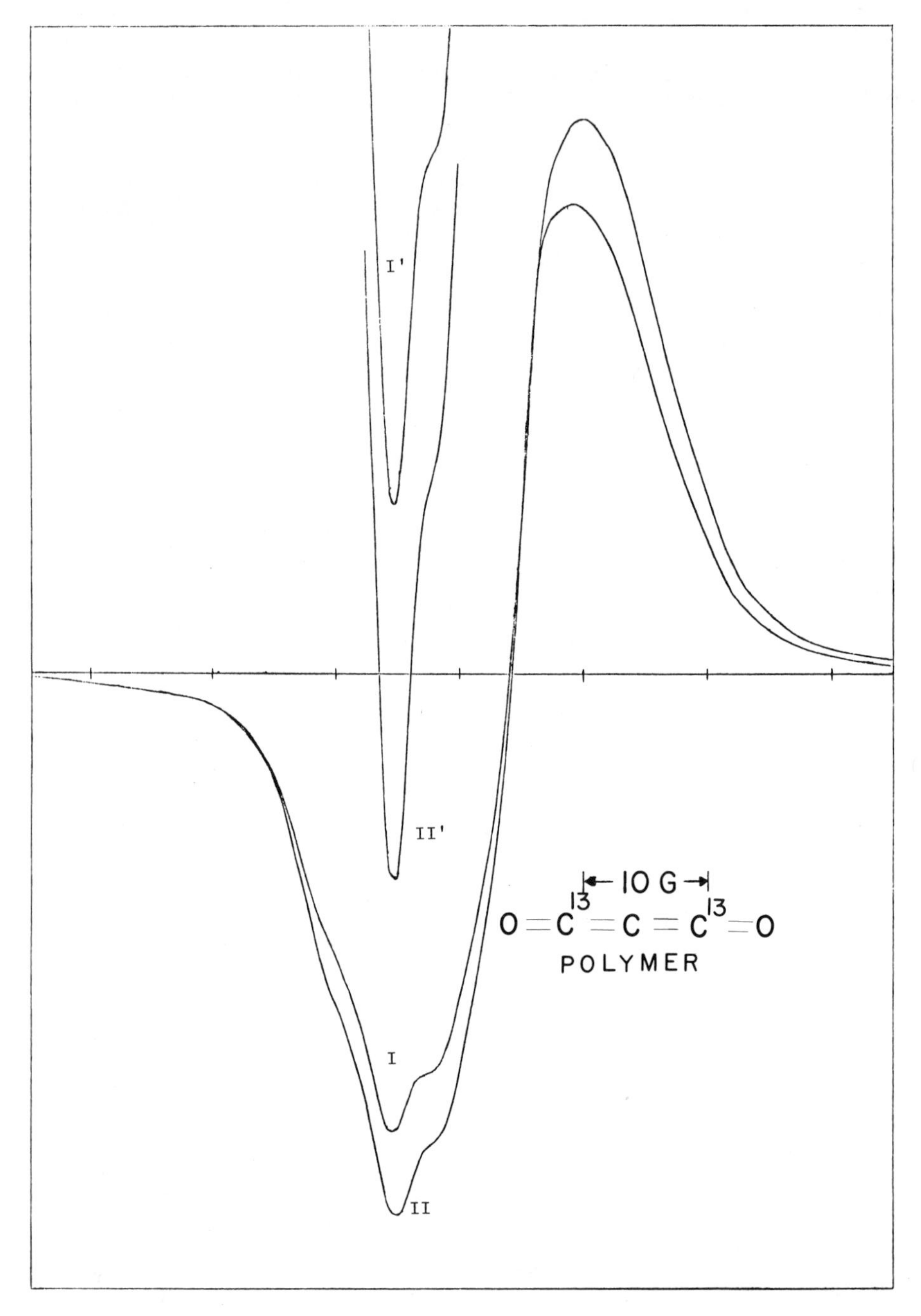

Figure 5. ESR spectra of ^{13}C-labelled poly(carbon suboxide). Key: I and I', signal in absence of irradiation; II and II', signal on irradiation; and I' and II' with gain five times that of I and II.

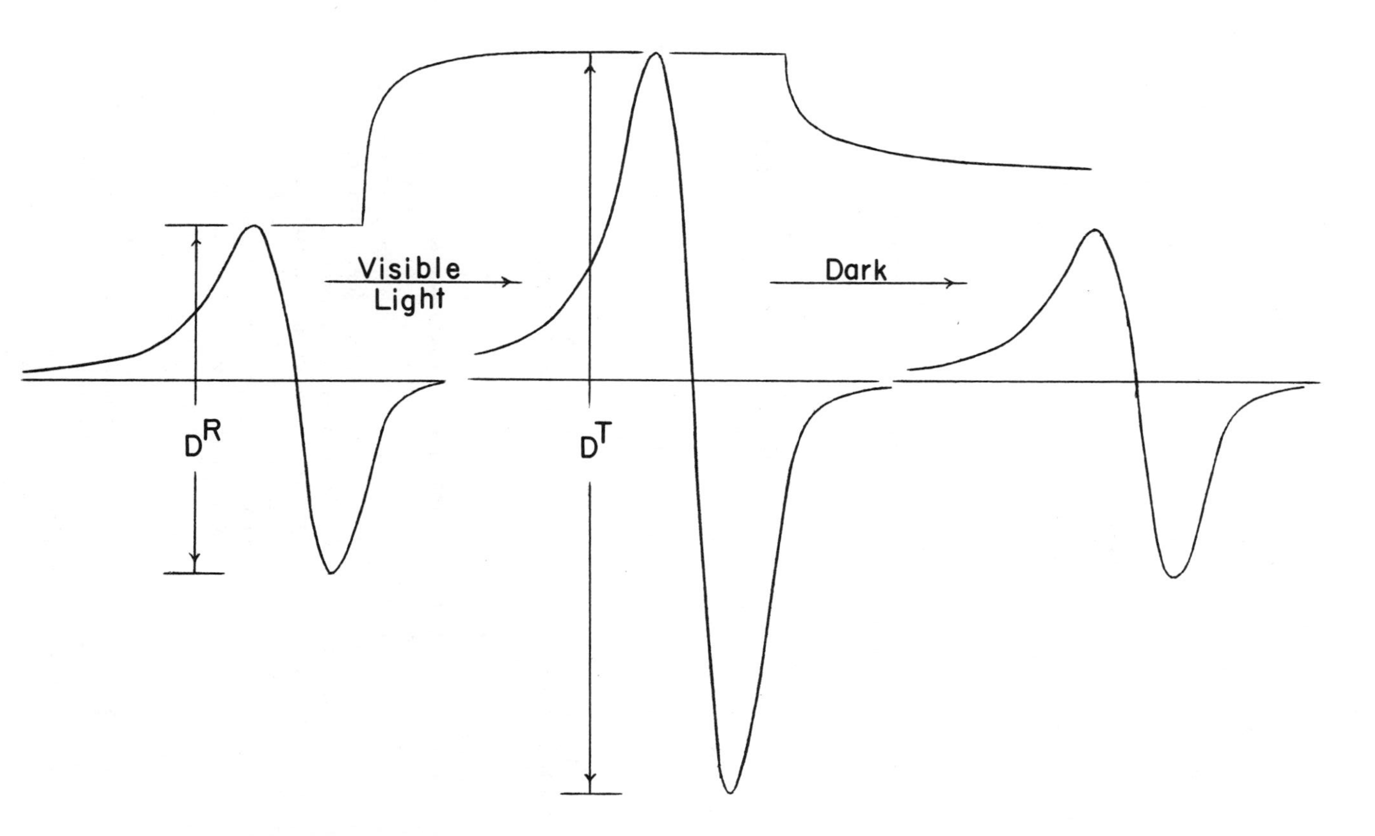

Figure 6. Growth and decay of photosignal in poly(carbon suboxide) film ($D^P = D^T - D^R$). Key: D^P, photosignal; D^T, total signal; and D^R, signal of the native polymer.

recording the signal strength as a function of time. To follow the entire growth kinetic curve accurately requires recording system with faster response than a strip chart recorder. The initial growth is very fast, i.e. reaching 60% of its steady-state concentration in fractions of a second. At a later stage of growth, the decay rate is competitive. The decay curves do not correspond to a simple second order process, because the polymer film, with finite thickness, contains varying concentrations of spin center along the light path due to the requirement of the Beer-Lambert law imposed on the incident light beam. Assuming two electron combination decay, an analysis of the steady-state spin concentration as a function of temperature indicates the activation energy for the decay process is ca. 0.44 Kcal per mole of spin.

Polymer Structure. The direct proportional relationship between ESR signal strength and the 4.6 micron I.R. absorption indicates that the ketenyl group plays an important role in sustaining the paramagnetic center. Since the ketenyl functional groups can readily be converted into acid groups, it should be possible to titrate them. By comparing the molecular weights determined by VPO with the equivalent weight, the number of ketenyl groups per polymer molecule was determined to be two (15).

While no direct molecular weight measurement has been reported for this polymer previously, a calculated x-ray diffraction pattern, based on six C_3O_2 units in the fused ring poly(α-pyrone) structure was found to be in very good agreement with the observed pattern (3). The x-ray diffraction data also suggested an increase in molecular weight with increasing polymerization temperature. Direct molecular weight measurements by VPO in DMF solution at 75°C, for the 0°, 30° and soluble fraction of the 60° polymers led to $\overline{DP}_n$ equal to 5.2, 5.4 and 6.0 respectively. The solubility of the polymer in DMF decreases with an increase in polymerization temperature. At 0°C and 30°C, the polymers are completely soluble. Only 66% of the 60°C polymer is soluble in DMF. At a polymerization temperature of 104°C, the polymer is completely insoluble. The density of the polymer was found to be equal to 1.6 by flotation method.

Results of kinetics of reaction between ketenyl groups of low temperature polymers and nucleophiles with different values of dipole moment indicated that the polymer is of considerable ionic character (15). The contribution of resonance structures with ionic character (5) may be significant as corroborated by data from measurements of heat of formation and polymerization (31).

The paramagnetic properties of the polymer clearly establish a poly-conjugated system and rule out the polyspirocyclobutanedione structure suggested (32) for the polymer.

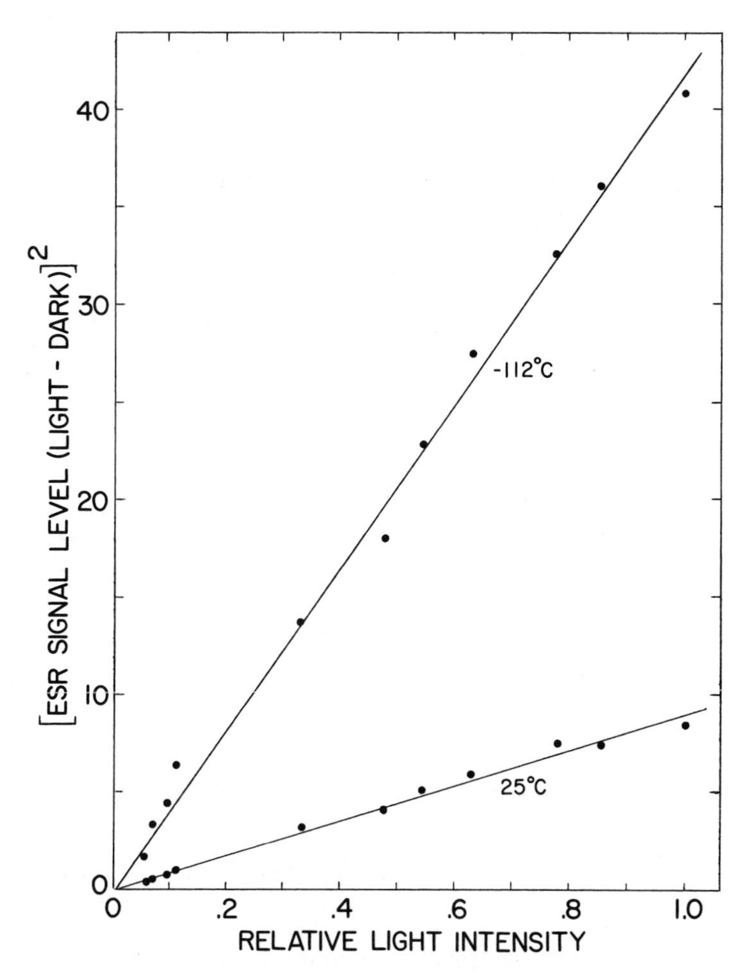

Figure 7. Dependence of steady-state ESR photosignal on the incident light intensity. Temperatures, 25° and —112°C as indicated; λ, 547 ± 10 nm.

Polymerization

Monomer Properties and Reactivities. Carbon suboxide shows unusual electronic properties. The ^{13}C NMR spectrum of carbon suboxide in $CDCl_3$ at -40°C exhibits noted features (33): the estremely high-field position of the central carbon absorption at -14.6 p.p.m. and the carbonyl carbon absorption at a relative low field of 129.7 p.p.m. This central carbon is one of the most shielded carbon atoms known to date. The electronic structure of the molecule was considered to reflect a considerable contribution due to the resonance structure of Figure 8-b (33). This is consistent with an observed low frequency of 63 cm^{-1} for bending at the central carbon (34) and CNDO/2 prediction of low C-Cπ bond order (35). Charge distributions of the molecule derived from experimentally (36) determined ESCA values (Figure 8-d) and theoretical calculations (35,36,37) all show high negative charge on the central carbon atom and the oxygen atoms and a high positive charge on the end carbons. Thus carbon suboxide is a highly polarized molecule. Consequently the molecule is expected to complex strongly with charged surfaces which will also favor the bent structure with the unshared pair electron in an sp^2 orbital (Figure 8-c).

When a carbon suboxide molecule reacts with two sites of comparable nucleophilicity, ring closure can take place if a favorable mutual spatial orientation of the four reaction centers prevails. For example, carbon suboxide can react with N-aryl-substituted benzamidines to give six-membered ring products quantitatively (38):

$$R-C(=\ddot{N}Ar)-\ddot{N}H_2 \; + \; O=C^{\delta+}=C=C^{\delta+}=O \longrightarrow R-C\langle N{=}C(OH){=}C{-}C(=O){-}N \rangle$$

In the polymerization of carbon suboxide, the monomer molecules are incorporated into the polymer ladder as a part of a six-membered ring (Figure 1-b). Here both the electrophilic and the nucleophilic sites of the carbon suboxide molecule must participate in the reaction. In the initiation step, the first monomer is also required to assume a bent geometry, since two linear molecules colliding to form a six-membered ring is very unlikely due to the unfavorable spatial orientation.

Mechanism of Polymerization. A mechanism for polymerization of carbon suboxide was suggested by Ziegler (8) involving initiation by water or acetic acid, with the formation of an

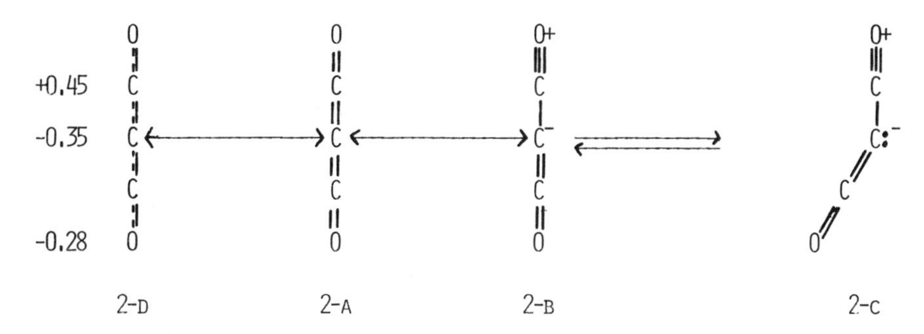

Figure 8. Electronic structures for carbon suboxide.

acetone dicarboxylic anhydride intermediate. This mechanism would predict the formation of ladder polymers of inordinately high molecular weight, or would require appreciable quantities of water or acetic acid to be present. It was recognized by Ziegler that "high-temperature" polymerizations may be exceptions to this mechanism (1).

It was suggested by Ulrich (39) that the polymer could arise from an initial noncatalyzed formation of a dimer which would further react with carbon suboxide to give polymer. A conjugated carbon backbone may be formed by [4 + 2] cycloaddition of a growing polymeric α-acylketene with monomer.

A difficulty with the cycloaddition mechanism of Ulrich is that a reactive acylketene is generated at each step, there is no obvious termination step, and it is hard to see why high molecular weight polymer is not produced.

An alternative mechanism which can be considered involves a zwitterionic intermediate (Figure 9). In the initial step in this mechanism a complex is formed between the monomer in a bent configuration, and a site provided by polymer, or perhaps less effectively, the container surface.

Poorer complexing ability of a surface site compared with polymer can account for the observed induction periods. There is evidence for an attraction between monomer and polymer. The infrared spectrum of a polymer film was obtained by polymerizing C_3O_2 onto the NaCl windows of a 100 mm gas cell at room temperature followed by pumping the monomer from the cell once sufficient polymer had formed. The most intense band of the monomer at 4.46μ which is well resolved from the polymer's 4.6μ ketenyl band, required 30 minutes of pumping with a mechanical pump to remove. This mechanism can accommodate the observed first order rate dependencies on monomer and polymer if it is assumed that either the complexation step of polymer with monomer is rate limiting, or alternatively, in the presence of excess monomer where essentially all of the polymer molecules are

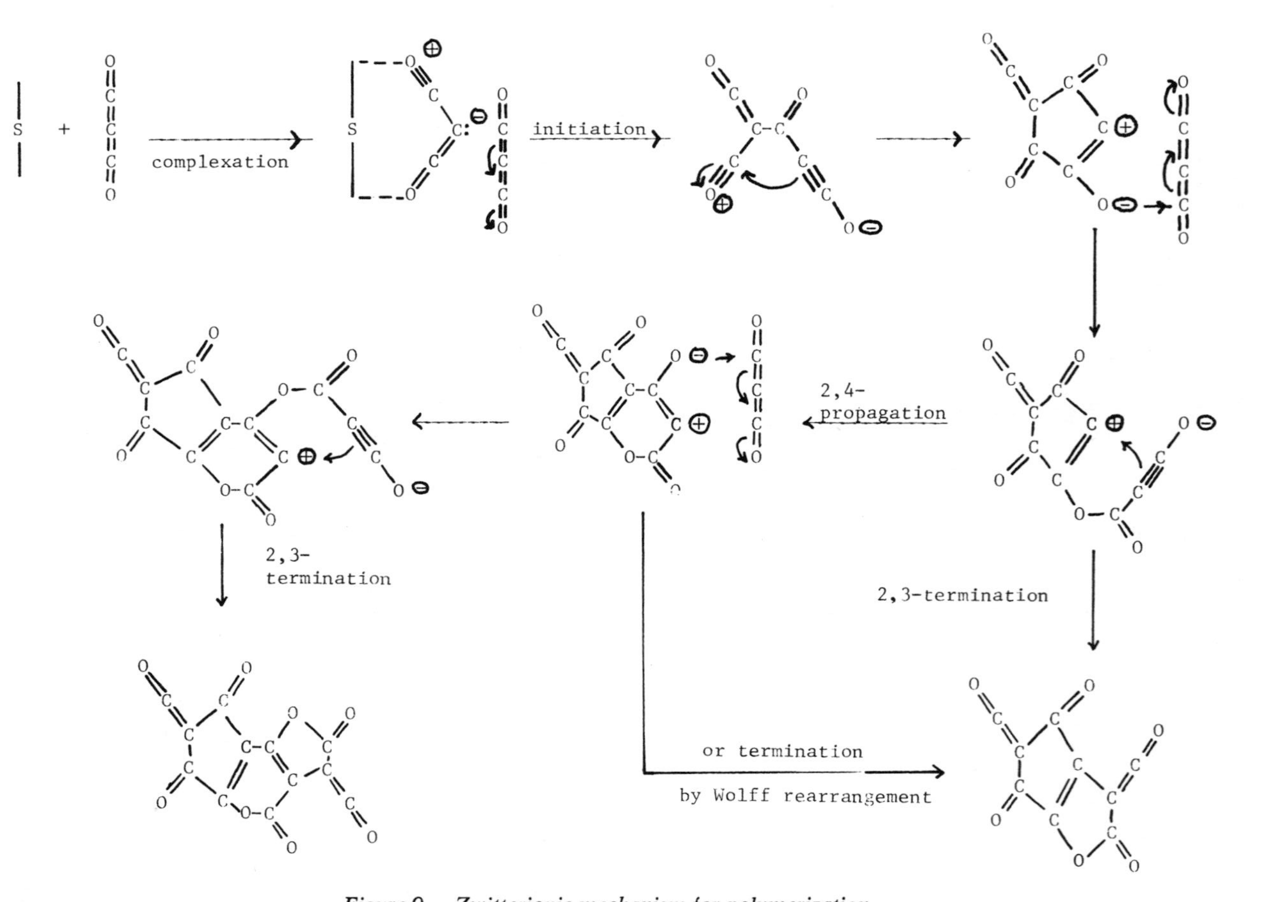

Figure 9. Zwitterionic mechanism for polymerization.

complexed, the slow step may be reaction of the complex with monomer (the second step in the reaction sequence). In the zwitterionic mechanism, termination can arise by a competitive Wolff rearrangement.

While ketocarbenes are known to readily undergo Wolff rearrangement, they have also been postulated to react *via* 1,3-dipolar addition as zwitterions. For example the uncatalyzed reaction of ketene with diazoacetophenone (40) has been interpreted as a 1,3-dipolar addition of a zwitterionic ketocarbene (41).

The inability of oxygen to retard the polymerization of C_3O_2 does not rule out ketocarbenes as intermediates. While carbenes are known to react with oxygen (42), in a study of the fate of ketocarbenes in the presence of oxygen, it was found that the substituents present on the ketocarbene determined whether reaction with oxygen occurred, or whether the ketocarbene followed other pathways, including zwitterionic cycloaddition (43). The ketocarbene intermediate may be expected to react with additives containing multiple C,C or C,O bonds, including those compounds which have been observed to retard the polymerization of carbon suboxide.

It is clear that further work will be required to elucidate the details of the mechanistic pathway for the polymerization of carbon suboxide. Both the mechanism proposed by Ulrich (39) and the above zwitterionic mechanism lead to a functionality of two ketenyl groups per molecule.

The ionic properties of the polymer are not only conducive to initiation, i.e. formation of species depicted in Figure 8-c, but also to the propagation steps, i.e. the assumptions of charged, aigzag state by the monomer. This is supported by the observed autoacceleration effect of the polymerization processes both in bulk and in solution.

Conclusions

Significant progress has been made in the understanding of the mechanism of polymerization of carbon suboxide based on the finding of the intrinsic paramagnetic properties of the polymer. The itinerant nature of the unpaired electrons associated with the polymer supports a polyconjugated structure for the polymer. Further direct structural evidence should be readily available from mass spectrometry and ^{13}C NMR studies. Preliminary mass spectrometry investigation indicated the preponderance of C_3O_2 fragments. For the unlabelled polymer, no ^{13}C NMR absorption was observed, perhaps due to its paramagnetism or high solution viscosity. Polymers with ^{13}C labelled at central and end carbons are being investigated.

Poly(carbon suboxide) is the only conjugated polymer known to form thin films of readily controlled thickness. The findings of its high spin density together with its photoparamagnetism open avenues for investigation in spin waves and photoelectric properties of organic thin films.

In view of these recent findings, the applications of carbon suboxide in polymer science should be examined more closely, e.g. block and graft copolymers, etc. (44,45).

Acknowledgement

The following sources of support are gratefully acknowledged: PSC-BHE Award Program, the City University of New York, Grant Numbers 1620, 11084, 11420, and 12292; a Petroleum Research Fund administered by the American Chemical Society; and a National Science Foundation Grant (NSFGU3957).

Literature Cited

1. Kappe, T; Ziegler, E. Angew. Chem. (International edition) 13, 491 (1974).
2. Diels, O.; Wolf, B. Ber. 39, 689 (1906).
3. Blake, A.R.; Eeles, W.T.; Jennings, P.P. Trans. Faraday Soc. 60, 691 (1964).
4. Blake, A.R.; Hyde, A.F. Trans. Faraday Soc. 60, 1775 (1964).
5. Smith, R.N.; Young, D.A.; Smith, E.N.; Carter, C.C. Inorg. Chem. 2, 829 (1963).
6. Smith, R.N. Trans. Faraday Soc. 62, 1881 (1966).
7. Blake, A.R. J. Chem. Soc. 3866 (1965).
8. Sterk, H.; Tritthart, P.; Ziegler, E. Monatsh. Chem. 101, 1851 (1970).
9. Hegar, G. "Beitrag zur Chemie des Kohlensuboxides," Ph.D. thesis, Zurich (1961).
10. Barkalov, I.M.; Kim, I.P.; Mikhailov, A.I.; Kiryukhin, D.P. J. Polymer Sci., Polymer Chem. Ed., 18, 1551 (1980).
11. Kiryukhin, D.P.; Barkalov, I.M.; Goldanskii, V.I. Vysokomol. soeyed. A18, 759 (1976).
12. Brauer, G. "Handbook of Preparative Inorganic Chemistry," Second edition, p. 648, Academic Press, N.Y. (1963).
13. Gemelins Handbuch des Anorganischen Chemie. Kohlenstoff. C. Section 1, Syst. No. 14, pp. 75-99, Verlag Chemie, Weinheim, 1970.
14. Yang, N.-L.; Snow, A.; Haubenstock, H.; Bramwell, F. J. Polymer Sci., Polymer Chem. Ed., 16, 1909 (1978).
15. Snow, A.W.; Haubenstock, H.; Yang, N.-L. Macromolecules, 11, 77 (1978).
16. Snow, A.; Yang, N.-L.; Haubenstock, H. Carbon, 14, 177 (1976).
17. Somorjai, G.A. Account Chem. Res., 9, 248 (1976).
18. Beringer, R.; Castle, J.G. Phys. Rev., 78, 581 (1950).
19. Kolthoff, I.M.; Bovey, F.A. Chem. Revs., 42, 491 (1948).
20. Vollmert, B. "Polymer Chemistry," Springer-Verlag, New York, 1973, p. 61.

21. Yang, N.-L.; Snow, A.W.; Haubenstock, H. Polym. Prep. Am. Chem. Soc., Div. Polym. Chem., 22, 52 (1981).
22. Ehrlich, P. J. Macromol. Sci.-Phys., 2 152 (1968); Holob, G.M.; Ehrlich, P.; Allendoerfer, R.D. Macromolecules, 5, 569 (1972).
23. Mrozowski, S. "Proceedings of the Fourth Carbon Conference," Pergamon, London, 1960, p. 271.
24. Fessenden, R.W.; Schuler, R.H. J. Chem. Phys., 43, 2704 (1965).
25. Fraenkel, G. Pure Appl. Chem., 4, 1431 (1962).
26. Bolton, J.R.; Frankel, G.K. J. Chem. Phys., 40, 3307 (1964).
27. Anderson, P.W.; Weiss, P.R. Rev. Modern Phys., 25, 269
28. Richards, P.M. Phys. Rev., 142, 196 (1966).
29. Masnda, K.; Mamba, S. In "Energy and Charge Transfer in Organic Semiconductors," K. Masuda and M. Silver, Eds., Plenum, New York, 1974, p. 53.
30. Yang, N.-L.; Snow, A.W.; Haubenstock, H. Polym. Prep. Am. Chem. Soc., Div. Polym. Chem., 19, 476 (1978).
31. Kybett, B.D.; Johnson, G.K.; Barker, C.K.; Margrave, J. J. Phys. Chem. 69, 3603 (1965).
32. Woztczak, J.; Weiman, L.; Konarski, J.M. Monatsh. Chem., 99, 501 (1968).
33. Williams, E.A.; Cargiola, J.D.; Ewo, A. Chem. Comm. 366 (1975).
34. Smith, W.H.; Leroi, G.E. J. Chem. Phys. 45, 1767 (1966).
35. Olsen, J.F.; Burnelle, L. J. Phys. Chem. 73, 2298 (1967).
36. Geline, U.; Allan, C.J.; Allison, D.A.; Siegbahn, H.; Siegbahan, K. Chemical Phys. Lett., 11, 224 (1971).
37. Sabin, J.R.; Kim, H. J. Chem. Phys. 56, 2195 (1972).
38. Dashkerich, L.B. Zh. Obshch. Khim. 32, 2346 (1962).
39. Ulrich, H. "Cycloaddition Reactions of Heterocumulenes," Academic Press, New York, 1967, Chapter 3.
40. Ried, W.; Mengler, H. Ann., 678, 113 (1964).
41. Baron, W.J.; DeCamp, M.R.; Hendrick, M.E.; Jones, M., Jr.; Levin, R.H.; Sohn, M.B. In "Carbenes," Ed. M. Jones, Jr. and R.A. Moss, Vol. 1, J. Wiley and Sons, New York, 1973, p. 107.
42. Gaspar, P.P.; Hammond, G.S. In "Carbenes," Ed. M. Jones, Jr. and R.A. Moss, Vol. II, J. Wiley and Sons, New York, 1975, p. 319ff.
43. Tanaka, M.; Nagai, T.; Tokura, N. J. Org. Chem., 38, 1602 (1973).
44. Bukowski, A.; Porejko, S., J. Polym. Sci., A-1, 8, 2491 (1970).
45. Bukowski, A.; Porejko, S., J. Polym. Sci., A-1, 8, 2501 (1970).

RECEIVED February 4, 1982.

INDEX

INDEX

A

B

C

I

K

L

M

T

U

V

W

X

Y

Z

Jacket design by Martha Sewall.
Production by Deborah Corson and Cynthia Hale.

Elements typeset by Service Composition Co., Baltimore, MD.
Printed and bound by Maple Press Co., York, PA.